Heat and Mass Transfer

Series Editors: D. Mewes and F. Mayinger

Carsten Baumgarten

Mixture Formation in Internal Combustion Engines

With 180 Figures and 9 Tables

 Springer

Dr.-Ing. Carsten Baumgarten,
MTU Friedrichshafen GmbH
Maybachplatz 1
88045 Friedrichshafen
Germany

Series Editors

Prof. Dr.-Ing. Dieter Mewes
Universität Hannover
Institut für Verfahrenstechnik
Callinstr. 36
30167 Hannover, Germany

Prof. em. Dr.-Ing. E.h. Franz Mayinger
Technische Universität München
Lehrstuhl für Thermodynamik
Boltzmannstr. 15
85748 Garching, Germany

Library of Congress Control Number: 2005937086

ISSN - 1860-4846
ISBN-10 3-540-30835-0 Springer-Verlag Berlin Heidelberg New York
ISBN-13 978-3-540-30835-5

Springer-Verlag Berlin Heidelberg New York
a member of BertelsmannSpringer Science+Business Media GmbH
springer.com

Typesetting: Digital data supplied by editor
Cover design: deblik Berlin
Printed on acid free paper   62/3020/SPI Publisher Services - 5 4 3 2 1 0

Preface

A systematic control of mixture formation with modern high-pressure injection systems enables us to achieve considerable improvements of the combustion process in terms of reduced fuel consumption and engine-out raw emissions. However, because of the growing number of free parameters due to more flexible injection systems, variable valve trains, the application of different combustion concepts within different regions of the engine map, etc., the prediction of spray and mixture formation becomes increasingly complex. For this reason, the optimization of the in-cylinder processes using 3D computational fluid dynamics (CFD) becomes increasingly important.

In these CFD codes, the detailed modeling of spray and mixture formation is a prerequisite for the correct calculation of the subsequent processes like ignition, combustion and formation of emissions. Although such simulation tools can be viewed as standard tools today, the predictive quality of the sub-models is constantly enhanced by a more accurate and detailed modeling of the relevant processes, and by the inclusion of new important mechanisms and effects that come along with the development of new injection systems and have not been considered so far.

In this book the most widely used mathematical models for the simulation of spray and mixture formation in 3D CFD calculations are described and discussed. In order to give the reader an introduction into the complex processes, the book starts with a description of the fundamental mechanisms and categories of fuel injection, spray break-up, and mixture formation in internal combustion engines. They are presented in a comprehensive way using data from experimental investigations. Next, the basic equations needed for the simulation of mixture formation processes are derived and discussed in order to give the reader the basic knowledge needed to understand the theory and to follow the description of the detailed sub-models presented in the following chapters. These chapters include the modeling of primary and secondary spray break-up, droplet drag, droplet collision, wall impingement, and wall film formation, evaporation, ignition, etc. Different modeling approaches are compared and discussed with respect to the theory and underlying assumptions, and examples are given in order to demonstrate the capabilities of today's simulation models as well as their shortcomings. Further on, the influence of the computational grid on the numerical computation of spray processes is discussed. The last chapter is about modern and future mixture formation and combustion processes. It includes a discussion of the potentials and future developments of high-pressure direct injection diesel, gasoline, and homogeneous charge compression ignition engines.

This book may serve both as a graduate level textbook for combustion engineering students and as a reference for professionals employed in the field of combustion engine modeling.

The research necessary to write this book was carried out during my employment as a postdoctoral scientist at the Institute of Technical Combustion (ITV) at the University of Hannover, Germany. The text was accepted in partial fulfillment of the requirements for the postdoctoral Habilitation-degree by the Department of Mechanical Engineering at the University of Hannover.

There are many people who helped me in various ways while I was working on this book. First, I would like to thank Prof. Dr.-Ing. habil. Günter P. Merker, the director of the Institute of Technical Combustion, for supporting my work in every possible respect. Prof. Dr.-Ing. Ulrich Spicher, the director of the Institute of Reciprocating Engines, University of Karlsruhe, and Prof. Dr.-Ing. habil. Dieter Mewes, the director of the Institute of Process Engineering, University of Hannover, contributed to this work by their critical reviews and constructive comments.

I would also like to thank my colleagues and friends at the University of Hannover who gave me both, information and helpful criticism, and who provided an inspiring environment in which to carry out my work. Special thanks go to Mrs. Christina Brauer for carrying out all the schematic illustrations and technical drawings contained in this book.

Hannover, October 2005   Carsten Baumgarten

Contents

Preface.. V

Contents.. VII

Nomenclature.. XI

1 Introduction... 1
  1.1 Modeling of Spray and Mixture Formation Processes........................... 1
  1.2 Future Demands... 3

2 Fundamentals of Mixture Formation in Engines................................... 5
  2.1 Basics... 5
    2.1.1 Break-Up Regimes of Liquid Jets... 5
    2.1.2 Break-Up Regimes of Liquid Drops.. 8
    2.1.3 Structure of Engine Sprays.. 10
    2.1.4 Spray-Wall Interaction... 29
  2.2 Injection Systems and Nozzle Types... 32
    2.2.1 Direct Injection Diesel Engines... 32
    2.2.2 Gasoline Engines... 38
  References... 43

3 Basic Equations.. 47
  3.1 Description of the Continuous Phase... 47
    3.1.1 Eulerian Description and Material Derivate............................... 47
    3.1.2 Conservation Equations for One-Dimensional Flows................ 49
    3.1.3 Conservation Equations for Multi-Dimensional Flows.............. 54
    3.1.4 Turbulent Flows... 66
    3.1.5 Application to In-Cylinder Processes... 79
  3.2 Description of the Disperse Phase... 81
    3.2.1 Spray Equation... 81
    3.2.2 Monte-Carlo Method.. 82
    3.2.3 Stochastic-Parcel Method.. 82
    3.2.4 Eulerian-Lagrangian Description... 83
  References... 83

4 Modeling Spray and Mixture Formation.. 85
  4.1 Primary Break-Up.. 85

4.1.1 Blob-Method.. 86
4.1.2 Distribution Functions.. 90
4.1.3 Turbulence-Induced Break-Up.. 94
4.1.4 Cavitation-Induced Break-Up... 98
4.1.5 Cavitation and Turbulence-Induced Break-Up....................... 100
4.1.6 Sheet Atomization Model for Hollow-Cone Sprays.................. 109
4.2 Secondary Break-Up... 114
4.2.1 Phenomenological Models... 115
4.2.2 Taylor Analogy Break-Up Model... 116
4.2.3 Droplet Deformation and Break-Up Model............................ 122
4.2.4 Kelvin-Helmholtz Break-Up Model...................................... 125
4.2.5 Rayleigh-Taylor Break-Up Model.. 128
4.3 Combined Models... 130
4.3.1 Blob-KH/RT Model... 130
4.3.2 Blob-KH/DDB Model... 131
4.3.3 Further Combined Models... 132
4.3.4 LISA-TAB Model.. 133
4.3.5 LISA-DDB Model... 135
4.4 Droplet Drag Modeling.. 136
4.4.1 Spherical Drops... 136
4.4.2 Dynamic Drag Modeling... 136
4.5 Evaporation... 139
4.5.1 Evaporation of Single-Component Droplets........................... 140
4.5.2 Evaporation of Multi-Component Droplets............................. 144
4.5.3 Flash-Boiling.. 158
4.5.4 Wall Film Evaporation... 162
4.6 Turbulent Dispersion.. 166
4.7 Collision and Coalescence.. 169
4.7.1 Droplet Collision Regimes.. 169
4.7.2 Collision Modeling... 172
4.7.3 Implementation in CFD Codes.. 178
4.8 Wall Impingement.. 180
4.8.1 Impingement Regimes.. 181
4.8.2 Impingement Modeling... 183
4.8.3 Wall Film Modeling... 191
4.9 Ignition.. 197
4.9.1 Auto-Ignition... 197
4.9.2 Spark-Ignition.. 200
References.. 203

5 Grid Dependencies... 211
5.1 General Problem.. 211
5.2 Improved Inter-Phase Coupling.. 216
5.3 Improved Collision Modeling... 220
5.4 Eulerian-Eulerian Approaches.. 221
References.. 223

6 Modern Concepts.. **225**
  6.1 Introduction... 225
  6.2 DI Diesel Engines... 226
    6.2.1 Conventional Diesel Combustion..................................... 226
    6.2.2 Multiple Injection and Injection Rate Shaping..................... 230
    6.2.3 Piezo Injectors.. 234
    6.2.4 Variable Nozzle Concept... 236
    6.2.5 Increase of Injection Pressure... 237
    6.2.6 Pressure Modulation.. 239
    6.2.7 Future Demands... 241
  6.3 DI Gasoline Engines.. 242
    6.3.1 Introduction.. 242
    6.3.2 Operating Modes.. 244
    6.3.3 Stratified-Charge Combustion Concepts............................. 246
    6.3.4 Future Demands... 251
  6.4 Homogeneous Charge Compression Ignition (HCCI)..................... 253
    6.4.1 Introduction.. 253
    6.4.2 HCCI Chemistry.. 256
    6.4.3 Emission Behavior.. 261
    6.4.4 Basic Challenges.. 264
    6.4.5 Influence Parameters and Control of HCCI Combustion........... 270
    6.4.6 Transient Behavior – Control Strategies............................. 279
    6.4.7 Future HCCI Engine Applications.................................... 279
  References.. 280

7 Conclusions.. **287**

Index..**291**

Nomenclature

Abbreviations

ATDC	after top dead center
B	Spalding transfer number
BMEP	break mean effective pressure
BTDC	before top dead center
CAI	controlled auto-ignition
CAN	controlled auto-ignition number
CFD	computational fluid dynamics
CI	compression ignition
CN	cetane number,
	cavitation number
CR	compression ratio,
	common rail
DDB	droplet deformation and break-up model
DDM	discrete droplet model
DI	direct injection
DISI	direct injection spark ignition
DNS	direct numerical simulation
EGR	exhaust gas recirculation
GDI	gasoline direct injection
HCCI	homogeneous charge compression ignition
HTO	high temperature oxidation
ICAS	interactive cross-sectionally averaged spray
IMEP	indicated mean effective pressure
K	cavitation number
KH	Kelvin-Helmholtz model
La	Laplace number
LES	large eddy simulation
LHF	lower heating value
LISA	linearized instability sheet atomization model
LTO	low temperature oxidation
M	third body species in chemical reactions
MEF	maximum entropy formalism
MW	molecular weight
NTC	negative temperature coefficient
Nu	Nusselt number

ON	octane number
PDF	probability density function
PFI	port fuel injection
PM	particulate matter (soot)
Pr	Prandtl number
RANS	Reynolds averaged Navier-Stokes equations
Re	Reynolds number
RT	Rayleigh-Taylor model
Sc	Schmidt number
Sh	Sherwood number
SI	spark ignition
SMD	Sauter mean diameter
SOC	start of combustion
SR	swirl ratio
St	Stokes number
T	Taylor number
TAB	Taylor-analogy break-up model
TDC	top dead center
UIS	unit injector system
UPS	unit pump system
VCO	valve covered orifice
VVT	variable valve train
We	Weber number
Z	Ohnesorge number

Symbols

a	sound speed [m/s],
	acceleration [m^2/s^2],
	thermal diffusivity [m^2/s],
	major semi axis of ellipsoid [m]
A	area [m^2],
	constant [/]
b	minor semi axis of ellipsoid [m],
	spray width [m]
B	non-dimensional impact parameter [/]
c	molar density, concentration [mol/m^3]
C	constant [/]
C_c	contraction coefficient [/]
C_d	discharge coefficient [/]
C_D	drag coefficient [/]
c_f	wall friction coefficient [/]

c_p	specific heat capacity at constant pressure [J/(kg K)]
c_v	specific heat capacity at constant volume [J/(kg K)]
$\bar{c}_v$	molar specific heat at constant volume [J/mol K]
$\bar{c}_p$	molar specific heat at constant pressure [J/mol K]
d	diameter [m],
	damping constant [kg/s]
D	nozzle hole diameter [m],
	blob diameter [m],
	binary diffusivity [m²/s]
$\bar{D}, \tilde{D}, \hat{D}$	binary diffusion coefficients (cont. thermodynamics) [m²/s]
e	specific internal energy [J/kg]
E	energy [J]
f	function,
	body force [N/m³]
F	force [N]
h	enthalpy [J/kg],
	liquid film thickness [m]
h_{f0}	latent heat of vaporization [J/kg]
$\bar{h}_{fg}$	molar heat of vaporization [J/mol]
I	mod. Bessel function of first kind,
	distribution variable, usually molecular weight [kg/kmol]
J	moment of inertia [kg m²]
k	wave number [m⁻¹],
	specific turbulent kinetic energy [J/kg],
	loss coefficient [/],
	spring constant [N/m],
	constant [/],
	k-factor [μm]
K	wave number of fastest growing wave [m⁻¹],
	modified Bessel function of second kind,
	constant [/]
K_C	form loss coefficient [/]
l	length [m]
L	length of nozzle hole [m],
	angular momentum [(kg m²)/s]
L_A	atomization length scale [m]
L_t	turbulence length scale [m]
m	mass [kg]
M	momentum [N·m]
n	engine speed [min⁻¹],
	number, quantity [/]

$\dot{n}$	molar flux [mol/(m^2 s)]
$\vec{n}$	unit vector normal to a surface
N	number, quantity [/]
p	pressure [Pa]
P	probability [/]
$\dot{q}$	heat flux per unit area, [W/m^2],
	distribution parameter (Rosin-Rammler dist.) [/]
Q	heat, [J]
$\dot{Q}$	heat flux [W]
r	radius [m]
R	radius of bubble or drop [m],
	gas constant [J/(kg K)]
$\bar{R}$	(universal) molar gas constant [J/(mol K)], $\bar{R} = 8.314151$ J/(mol K)
s	entropy [J/(kg K)]
S	spray penetration length [m]
$\bar{s}_{fg}$	molar entropy of evaporation [J/(mol K)]
S	Shannon entropy [/]
t	time, [s]
T	temperature [K]
T^+	dimensionless temperature [/]
T_b	boiling temperature [K]
u	velocity component, usually in x-direction [m/s]
u_1, u_2, u_3	velocity components in a Cartesian coordinate system [m/s]
U	velocity [m/s]
u^+	non-dimensional velocity [/]
v	velocity component, usually in y-direction [m/s]
V	volume [m^3]
w	velocity component, usually in z-direction [m/s]
W	work [J]
x	coordinate [m],
	mole fraction in liquid phase [/]
x_1, x_2, x_3	coordinates in a Cartesian system [m]
X	impact parameter [m]
y	coordinate [m],
	mole fraction in gas phase [/],
	non-dimensional droplet deformation [/]
y^+	non-dimensional distance from wall [/]
Y	deformation [m],
	mass fraction in gas phase [/]
z	coordinate [m]

Greek Letters

α void fraction [/],
convection heat transfer coefficient [W/(m^2 K)],
spray angle [deg],
shape parameter of gamma function [/]

β shape parameter of gamma function [/],

spray angle [deg]

γ shape parameter of gamma function [/]

Γ gamma function [/]

δ thickness [m]

Δ difference [/],
diameter ratio [/]

ε compression ratio [/],
dissipation rate of turbulent kinetic energy [m^2/s^3]

η efficiency [/],

disturbance on gas/liquid interface [m]

θ spray cone angle [rad], [deg],
first moment (mean value) of a distribution

κ energy ratio [/],
adiabatic exponent [/],
constant [/]

λ air-fuel equivalence ratio ($= 1 / \phi$) [/],

wave length [m],
thermal conductivity [W/(m K)]
Lagrange multiplier (MEF)

Λ wave length of fastest growing wave [m]

μ dynamic viscosity [(N s)/m^2]

ν kinematic viscosity [m^2/s],
collision frequency [s^{-1}]

ξ random number [/]

ρ density [kg/m^3]

σ surface tension [N/m],
variance of a distribution

τ characteristic time scale [s],
shear stress [N/m^2]

τ_A atomization time scale [s]

τ_t turbulence time scale [s]

φ angle [rad], [deg]

ϕ fuel-air equivalence ratio [/],

spray cone angle [rad], [deg]

Φ angle [rad], [deg],

	turbulence energy spectrum [/],
	viscous dissipation [W],
	dissipation function [W/m^3]
ψ	angle [rad],
	second moment of a distribution
ω	growth rate [s^{-1}],
	angular frequency [s^{-1}]
Ω	growth rate of most unstable wave [s^{-1}]

Subscripts and Superscripts

0	reference value, initial condition
∞	condition at infinity or ambient
a	atomization
ad	adiabatic
$aero$	aerodynamic
amb	ambient
av	average
ax	axial
b	break-up
bu	break-up
cav	cavitation
$coll$	collapse, collision
$cond$	conduction
$conv$	convection
$crit$	critical value, at critical point
cs	control surface
cv	control volume
cyl	cylinder
eff	effective
EGR	recycled gas
eq	equilibrium
$evap$	evaporation
f	fuel
g	gas
i	variable index, imaginary part of imaginary number
ign	ignition
imp	impingement
in	incoming
inj	injection

k	variable index,
	kernel
kin	kinetic
l	liquid,
	laminar
lam	laminar
m	variable index,
	model,
	mass
max	maximum
min	minimum
mix	mixture
n	normal,
	variable index
osc	oscillation
out	outgoing
pl	plasma
r	radial,
	real part of imaginary number
R	at radius R
rel	relative
ref	reference
res	resulting,
	residence
rot	rotation
s	surface,
	source term,
	splash,
	sac hole
sat	saturation
sp	spark
sto	stoichiometric
surf	surface
t	turbulent,
	tangential,
	total
turb	turbulent
u	unburned
vap	vapor
w	wall
σ	surface tension
$\cdot$	$d(\ )/dt$
$\cdot\cdot$	$d^2(\ )/dt^2$
$-$	averaged value
$\rightarrow$	vector

1 Introduction

1.1 Modeling of Spray and Mixture Formation Processes

Due to the growing importance of future emission restrictions, manufacturers of internal combustion engines are forced continuously to improve the mixture formation and combustion processes in order to reduce engine raw emissions. In many applications, even an additional reduction of the remaining emissions with after-treatment systems like catalysts and filters will be necessary in order to achieve the required exhaust gas quality in the future.

In this context, the numerical simulation and optimization of mixture formation and combustion processes is today becoming more and more important. One advantage of using simulation models is that in contrast to experiments, results can often be achieved faster and cheaper. Much more important is the fact that despite the higher uncertainty compared to experiments, the numerical simulation of mixture formation and combustion processes can give much more extensive information about complex in-cylinder processes than experiments could ever provide. Using numerical simulations, it is possible to calculate the temporal behavior of every variable of interest at any place inside the computational domain. This allows the obtainment of a detailed knowledge of the relevant processes and is a prerequisite for their improvement.

Furthermore, numerical simulation can be used to investigate processes that take place at time and length scales or in places that are not accessible and thus cannot be investigated using experimental techniques. In the case of high-pressure diesel injection for example, the spray break-up near the nozzle is mainly influenced by the flow conditions inside the injection holes. However, because of the small hole diameters (less than 200 μm for passenger cars) and the high flow velocities (about 600 m/s and more), the three-dimensional turbulent and cavitating two-phase flow is not accessible by measurement techniques. One very costly and time-consuming possibility of getting some insight into these processes is to manufacture a glass nozzle in real-size geometry and to use laser-optical measurement techniques. Outside the nozzle in the very dense spray measurements of the three-dimensional spray structure (droplet sizes, velocities etc.) become even more complicated, because the dense spray does not allow any sufficient optical access of the inner spray core. In these and other similar cases numerical simulations can give valuable information and can help to improve and optimize the processes of interest.

Finally, the enormous research work which is necessary to develop and continuously improve the numerical models must be mentioned. This research work continuously increases our knowledge about the relevant processes, reveals new and unknown mechanisms, and is also a source of new, unconventional ideas and improvements.

There are three classes of models that can be used in numerical simulations of in-cylinder processes. If very short calculation times are necessary, so-called thermodynamic models are used. These zero-dimensional models, which do not include any spatial resolution, only describe the most relevant processes without providing insight into local sub-processes. Very simple sub-models are used, and a prediction of pollutant formation is not possible. The second class of models are the phenomenological models, which consider some kind of quasi-spatial resolution of the combustion chamber and which use more detailed sub-models for the description of the relevant processes like mixture formation, ignition and combustion. These phenomenological models may be used to predict integral quantities like heat release rate and formation of nitric oxides (NO_x). The third class of models are the computational fluid dynamics (CFD) models. In CFD codes, the most detailed sub-models are used, and every sub-process of interest is considered. For example, in case of mixture formation, the sub-processes injection, break-up and evaporation of single liquid droplets, collisions of droplets, impingement of droplets on the wall etc. are modeled and calculated for every individual droplet, dependent on its position inside the three-dimensional combustion chamber. Thus, this class of models is the most expensive regarding the consumption of computational power and time. The turbulent three-dimensional flow field is solved using the conservation equations for mass, momentum and energy in combination with an appropriate turbulence model. The CFD codes are especially suited for the investigation of three-dimensional effects on the in-cylinder processes, like the effect of tumble and swirl, the influence of combustion chamber geometry, position of injection nozzle, spray angle, number of injection holes, etc.

Although all of the three model categories mentioned above are needed and are being used today, the anticipated further increase of computer power will especially support the use of the more detailed CFD models in the future. As far as modeling of in-cylinder processes is concerned, most of the research work today concentrates on the development of CFD sub-models.

Summarizing the situation today, it must be pointed out that the predictive quality of the models currently used in CFD codes has already reached a very high level, and that the use of CFD simulations for the research and development activities of engine manufacturers with respect to the design of new and enhanced mixture formation and combustion concepts is not only practical but already necessary. Today, the complex task of developing advanced mixture formation and combustion concepts can only be achieved with a combination of experimental and numerical studies.

1.2 Future Demands

Fulfilling emission restrictions will be of growing importance in the future and is even expected to become the most challenging task of future engine development. However, the development of the fuel cell, which is often proposed as a possible future alternative to the internal combustion engine, will last at least for the next two or three decades. Thus, the internal combustion engine will keep its leading position and will continuously be improved in order to fulfill future requirements.

Because a systematic control of mixture formation with modern high-pressure injection systems enables considerable improvements in the combustion process in terms of reduced fuel consumption and raw emissions, the optimization of injection system and mixture formation is becoming more and more important today. In this respect, the development and improvement of highly flexible direct injection (DI) systems for gasoline as well as diesel injection currently has a key position.

While DI technology has already become the standard concept for passenger car diesel engines, most of today's spark ignition engines still rely on port fuel injection, where the fuel is injected into the intake manifold and most of the mixture formation process is already completed when the charge has entered the combustion chamber. Only very recently have direct injection spark ignition (DISI) engines become of interest, because the direct injection of gasoline offers the opportunity to run the engine in the stratified-charge mode under part load conditions and to reduce significantly the well-known throttling losses of homogeneously operated SI engines. Furthermore, the evaporation of fuel inside the combustion chamber cools the charge down and allows an increase of the compression ratio, which improves the efficiency at full load. However, the stable ignition of the charge in the stratified-charge mode is one of the most challenging tasks that still has to be solved. The motion of the in-cylinder charge must be controlled in such a way that, at the moment of ignition, the fuel-rich and ignitable zones of the cylinder charge stay at the spark plug. Various techniques like the wall-guided, the air-guided and the spray-guided techniques are the focus of current research. According to the different demands, different sprays have to be produced, and new injection systems and injection nozzles have to be designed.

A considerable amount of research work is already spent on developing appropriate CFD models for the description of spray and mixture formation in the case of direct injection of both gasoline and diesel fuel. Important effects that have to be described by these models are the high-pressure injection of gasoline using multi-hole injectors, flexible injection rate shaping (e.g. multi-pulse injection), the modulation of injection pressure during the injection event, etc.

Models describing the relevant processes as well as their interactions and interdependencies are needed. Usually, the output data of a sub-model is used as input data for the subsequent process. For this reason, a detailed and accurate description of the relevant mechanisms and processes is absolutely necessary in order to guarantee a high level of predictive quality in the final result. For example, the detailed and accurate description of the disintegration of the coherent liquid inside the injection nozzle into millions of small droplets in the combustion chamber is a prerequisite for the correct calculation of subsequent processes like evaporation, ignition, combustion, and formation of emissions. Because, in the case of high-pressure injection, the flow conditions inside the injectors (e.g. turbulence, cavitation, flash-boiling) are of growing importance for the spray break-up, enhanced spray models must also include the effect of the injection system.

Considering the fact that the importance of synthetic and so-called tailored fuels as well as that of new combustion concepts with auto-ignition of homogeneous fuel-air mixtures will significantly increase in the near future, simulation models describing the spray and mixture formation of multi-component fuels must also be developed.

Altogether, the internal combustion engine has currently reached a high level of sophistication. However, important improvements especially with regard to the spray and mixture formation process have to be realized in the near future in order to fulfill emission restrictions. The development of highly flexible injection systems for diesel as well as gasoline direct injection and the use of new combustion concepts like the auto-ignition of homogeneous fuel-air mixtures and synthetic fuels increases the need to improve and develop appropriate CFD models, which are able to describe the relevant processes during spray break-up and mixture formation, and which can be used in order to design and optimize future injection strategies. This book shall contribute to this future development.

2 Fundamentals of Mixture Formation in Engines

2.1 Basics

2.1.1 Break-Up Regimes of Liquid Jets

Dependent on the relative velocity and the properties of the liquid and surrounding gas, the break-up of a liquid jet is governed by different break-up mechanisms. These different mechanisms are usually characterized by the distance between the nozzle and the point of first droplet formation, the so-called break-up length, and the size of the droplets that are produced. According to Reitz and Bracco [44], four regimes, the Rayleigh regime, the first and second wind-induced regime, and the atomization regime, can be distinguished.

In order to give a quantitative description of the jet break-up process, Ohnesorge [37] performed measurements of the intact jet length and showed that the disintegration process can be described by the liquid Weber number

$$We_l = \frac{u^2 D \rho_l}{\sigma} \tag{2.1}$$

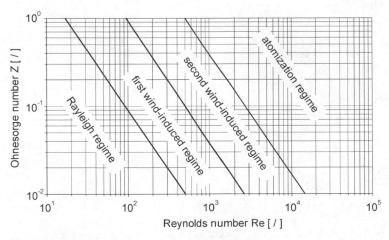

Fig. 2.1. Ohnesorge diagram: jet break-up regimes

and the Reynolds number

$$Re = \frac{uD\rho_l}{\mu_l}.$$ (2.2)

Eliminating the jet velocity u, Ohnesorge derived the dimensionless Ohnesorge number,

$$Z = \frac{\sqrt{We_l}}{Re} = \frac{\mu_l}{\sqrt{\sigma \rho_l D}},$$ (2.3)

which includes all relevant fluid properties (σ: surface tension at the liquid-gas interface, ρ_l: density of liquid μ_l: dynamic viscosity of liquid) as well as the nozzle hole diameter D. Figure 2.1 shows the Ohnesorge diagram, where Z is given as a function of Re. For stationary conditions, the boundaries between the four different jet break-up regimes can be drawn in. However, it has turned out that including only the liquid phase properties in the description of the regimes is not sufficient, because atomization can be enhanced by increasing the gas density (e.g. Torda [56], Hiroyasu and Arai [20]). Thus, Reitz [43] suggested to include the gas-to-liquid density ratio and to extend the two-dimensional Ohnesorge diagram into a three-dimensional one as shown in Fig. 2.2.

A schematic description of the different jet break-up regimes is given in Fig. 2.3. If the nozzle geometry is fixed and the liquid properties are not varied, the only variable is the liquid velocity u. Figure 2.4 shows the corresponding break-up curve, which describes the length of the unbroken jet as a function of jet velocity u.

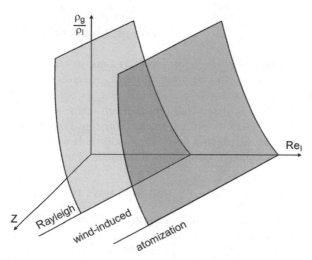

Fig. 2.2. Schematic diagram including the effect of gas density on jet break-up

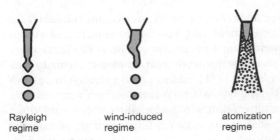

Rayleigh   wind-induced   atomization
regime   regime   regime

Fig. 2.3. Schematic description of jet break-up regimes

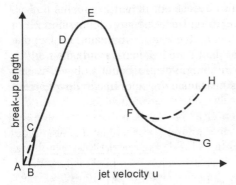

Fig. 2.4. Jet surface break-up length as function of jet velocity u [44]. ABC: Drip flow, CD: Rayleigh break-up, EF: first wind-induced break-up, FG (FH): second wind-induced break-up, beyond G (H): atomization regime

At very low velocities, drip flow occurs and no jet is formed. An increase of u results in the formation of an unbroken jet length, which increases with increasing velocity. This regime is called Rayleigh break-up. Break-up occurs due to the growth of axis-symmetric oscillations of the complete jet volume, initiated by liquid inertia and surface tension forces. The droplets are pinched off the jet, and their size is greater than the nozzle hole diameter D. This flow has already been described theoretically by Rayleigh [42]. Further advanced analyses have been published by Yuen [62], Nayfey [36], and Rutland and Jameson [48] for example.

A further increase in jet velocity results in a decrease of the break-up length, but it is still a multiple of the nozzle diameter. The average droplet size decreases and is now in the range of the nozzle diameter. In this first wind-induced regime, the relevant forces of the Rayleigh regime are amplified by aerodynamic forces. The relevant parameter is the gas phase Weber number $We_g = u^2{}_{rel}D\rho_g/\sigma$, which describes the influence of the surrounding gas phase. A detailed theoretical analysis is given in Reitz and Bracco [44].

In the second wind-induced break-up regime, the flow inside the nozzle becomes turbulent. Jet break-up now occurs due to the instable growth of short wavelength surface waves that are initiated by jet turbulence and amplified by aerodynamic forces due to the relative velocity between gas and jet. The diameter of the resulting droplets is smaller than the nozzle diameter, and the break-up length decreases with an increasing Reynolds number, line FG in Fig. 2.4. A detailed theoretical analysis is again given in Reitz and Bracco [44]. The jet now no longer breaks up as a whole. Due to the separation of small droplets from the jet surface, the disintegration process begins at the surface and gradually erodes the jet until it is completely broken up. Now two break-up lengths, the length describing the beginning of surface break-up (intact surface length) and the length describing the end of jet break-up (core length) should be accounted for. While the intact surface length decreases with increasing jet velocity, the core length may increase. However, it must be pointed out that measurements of both lengths become extremely difficult at increased Reynolds numbers, and, for this reason, experimental results from different authors may differ in this regime.

The atomization regime is reached if the intact surface length approaches zero. A conical spray develops, and the spray divergence begins immediately after the jet leaves the nozzle, i.e. the vertex of the spray cone is located inside the nozzle. An intact core or at least a dense core consisting of large liquid fragments may still be present several nozzle diameters downstream the nozzle. This is the relevant regime for engine sprays. The resulting droplets are much smaller than the nozzle diameter. The theoretical description of jet break-up in the atomization regime is much more complex than in any other regime, because the disintegration process strongly depends on the flow conditions inside the nozzle hole, which are usually unknown and of a chaotic nature. The validation of models is also difficult, because experiments are extremely complicated due to the high velocities, the small dimensions, and the very dense spray.

2.1.2 Break-Up Regimes of Liquid Drops

The break-up of drops in a spray is caused by aerodynamic forces (friction and pressure) induced by the relative velocity u_{rel} between droplet and surrounding gas. The aerodynamic forces result in an instable growing of waves on the gas/liquid interface or of the whole droplet itself, which finally leads to disintegration and to the formation of smaller droplets. These droplets are again subject to further aerodynamically induced break-up. The surface tension force on the other hand tries to keep the droplet spherical and counteracts the deformation force. The surface tension force depends on the curvature of the surface: the smaller the droplet, the bigger the surface tension force and the bigger the critical relative velocity, which leads to an instable droplet deformation and to disintegration. This behavior is expressed by the gas phase Weber number,

$$We_g = \left(\rho_g \cdot u_{rel}^2 \cdot d\right)/\sigma \, , \qquad (2.4)$$

where d is the droplet diameter before break-up, σ is the surface tension between liquid and gas, u_{rel} is the relative velocity between droplet and gas, and ρ_g is the gas density. The Weber number represents the ratio of aerodynamic (dynamic pressure) and surface tension forces.

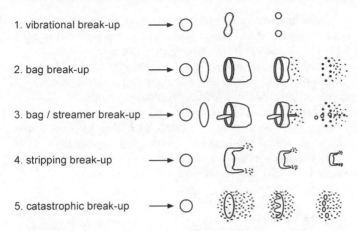

1. vibrational break-up

2. bag break-up

3. bag / streamer break-up

4. stripping break-up

5. catastrophic break-up

Fig. 2.5. Drop break-up regimes according to Wierzba [59]

From experimental investigations it is known that, depending on the Weber number, different droplet break-up modes exist. A detailed description is given in Hwang et al. [27] and Krzeczkowski [30] for example. Figure 2.5 summarizes the relevant mechanisms of drop break-up. It must be pointed out that the transition Weber numbers that are published in the literature are not consistent. This holds especially true for break-up mechanisms at high Weber numbers, where some authors also distinguish between additional sub-regimes. While the transition Weber numbers of Wierzba [59] are in the same range as the ones of Krzeczkowski [30], Arcoumanis et al. [4] distinguish between two different kinds of stripping break-up, Table 2.1, that cover the Weber number range from 100 to 1000, and the chaotic break-up is beyond $We_g = 1000$, see also Chap. 4, Sect. 4.2.1.

The vibrational mode occurs at very low Weber numbers near the critical value of $We_g \approx 12$, below which droplet deformation does not result in break-up. Bag break-up results in a disintegration of the drop due to a bag-like deformation. The rim disintegrates into larger droplets, while the rest of the bag breaks up into smaller droplets, resulting in a bimodal size distribution. An additional jet appears in the bag-streamer regime. In the stripping regime, the drop diameter gradually

Table 2.1. Transition Weber numbers of the different drop break-up regimes

Wierzba [59]	Weber number	Arcoumanis et al. [4]	Weber number
1. Vibrational	≈ 12	1. Vibrational	≈ 12
2. Bag	< 20	2. Bag	< 18
3. Bag-jet (Bag-streamer)	< 50	3. Bag-jet (bag-streamer)	< 45
		4. Chaotic break-up	< 100
4. Stripping	< 100	5. Sheet stripping	< 350
		6. Wave crest stripping	< 1000
5. Catastrophic	> 100	7. Catastrophic	> 1000

reduces because very small droplets are continuously shed from the boundary layer due to shear forces. This break-up mode also results in a bimodal droplet size distribution. Catastrophic break-up shows two stages: Because of a strong deceleration, droplet oscillations with large amplitude and wavelength lead to a disintegration in a few large product droplets, while at the same time surface waves with short wavelengths are stripped off and form small product droplets.

In engine sprays, all of these break-up mechanisms occur. However, most of the disintegration processes take place near the nozzle at high Weber numbers, while further downstream the Weber numbers are significantly smaller because of reduced droplet diameters due to evaporation and previous break up, and because of a reduction of the relative velocity due to drag forces.

2.1.3 Structure of Engine Sprays

2.1.3.1 Full-Cone Sprays

A schematic description of a full-cone high-pressure spray is given in Fig. 2.6. The graphic shows the lower part of an injection nozzle with needle, sac hole, and injection hole. Modern injectors for passenger cars have hole diameters of about 180 μm and less, while the length of the injection holes is about 1 mm.

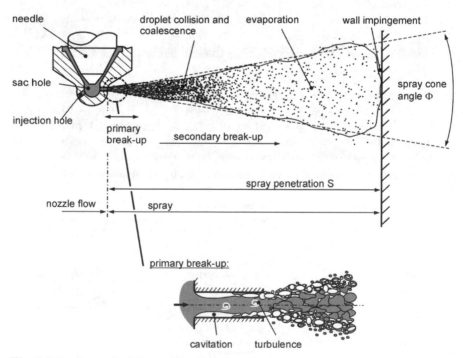

Fig. 2.6. Break-up of a full-cone diesel spray

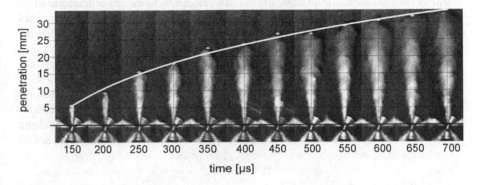

Fig. 2.7. Spray development during injection [53], p_{rail} = 70 MPa, p_{back} = 5 MPa, T_{air} = 890 K

Today, injection pressures of up to 200 MPa are used. The liquid enters the combustion chamber with velocities of 500 m/s and more, and the jet breaks up according to the mechanisms of the atomization regime.

Immediately after leaving the nozzle hole, the jet starts to break up into a conical spray. This first break-up of the liquid is called primary break-up and results in large ligaments and droplets that form the dense spray near the nozzle. In case of high-pressure injection, cavitation and turbulence, which are generated inside the injection holes, are the main break-up mechanisms. The subsequent break-up processes of already existing droplets into smaller ones are called secondary break-up and are due to aerodynamic forces caused by the relative velocity between droplets and surrounding gas, as described in the previous section.

The aerodynamic forces decelerate the droplets. The drops at the spray tip experience the strongest drag force and are much more decelerated than droplets that follow in their wake. For this reason the droplets at the spray tip are continuously replaced by new ones, and the spray penetration S increases, see Fig. 2.7. The droplets with low kinetic energy are pushed aside and form the outer spray region.

Altogether, a conical full-cone spray (spray cone angle Φ) is formed that is more and more diluted downstream the nozzle by the entrainment of air. Most of the liquid mass is concentrated near the spray axis, while the outer spray regions contain less liquid mass and more fuel vapor, see Fig. 2.8. Droplet velocities are maximal at the spray axis and decrease in the radial direction due to interaction with the entrained gas. In the dense spray, the probability of droplet collisions is high. These collisions can result in a change of droplet velocity and size. Droplets can break up into smaller ones, but they can also combine to form larger drops, which is called droplet coalescence.

In the dilute spray further downstream the main factors of influence on further spray disintegration and evaporation are the boundary conditions imposed by the combustion chamber such as gas temperature and density as well as gas flow (tumble, swirl). The penetration length is limited by the distance between the nozzle and the piston bowl. In the case of high injection pressure and long injection duration (full load) or low gas densities (early injection) the spray may impinge on the wall, and the formation of a liquid wall film is possible. Liquid wall films usually have a negative influence on emissions, because the wall film evaporates slower and may only be partially burnt.

Numerous fundamental experiments and semi-empirical relations about the general behavior of the relevant spray parameters of full-cone diesel sprays such as spray cone angle, spray penetration, break-up length, and average droplet diameter as a function of the boundary conditions have been performed and published by many different authors. Because these experiments have usually been performed with quasi-stationary sprays, most of the results can only be used to describe the main injection phase (full needle lift) of full-cone sprays. In the following, the most relevant relations will be presented. The most detailed investigations are from Hiroyasu and Arai [20].

According to Hiroyasu and Arai [20], the time-dependent development of the spray penetration length S can be divided into two phases. The first phase starts at the beginning of injection ($t = 0$, needle begins to open) and ends at the moment the liquid jet emerging from the nozzle hole begins to disintegrate ($t = t_{break}$). Because of the small needle lift and the low mass flow at the beginning of injection, the injection velocity is small, and the first jet break-up needs not always occur immediately after the liquid leaves the nozzle. During this time, a linear growth of S over t is observed, Eq. 2.5a. During the second phase ($t > t_{break}$), the spray tip consists of droplets, and the tip velocity is smaller than during the first phase. The spray tip continues to penetrate into the gas due to new droplets with high kinetic energy that follow in the wake of the slower droplets at the tip (high exchange of momentum with the gas) and replace them. The longer the penetration length, the

Fig. 2.8. Distribution of liquid (black) and vapor (gray) in an evaporating high-pressure diesel spray from a multi-hole nozzle under engine like conditions. Measurement technique: superposition of Schlieren technique (vapor and liquid) and Mie scattering (liquid)

smaller the energy of the new droplets at the tip and the slower the tip velocity. Altogether, the authors give the following relations:

$$t < t_{break} : S = 0.39 \cdot \left(\frac{2 \Delta p}{\rho_l} \right)^{0.5} \cdot t , \tag{2.5a}$$

$$t > t_{break} : S = 2.95 \cdot \left(\frac{\Delta p}{\rho_g} \right)^{0.25} \cdot (D \cdot t)^{0.5} , \tag{2.5b}$$

where

$$t_{break} = \frac{28.65 \cdot \rho_l D}{\left(\rho_g \Delta p \right)^{0.5}} . \tag{2.5c}$$

In Eq. 2.5, Δp in [Pa] is the difference of injection pressure and chamber pressure, ρ_l and ρ_g are the liquid and gas densities in [kg/m^3], t is the time in [s], and D is the nozzle hole diameter in [m]. A higher injection pressure results in increased penetration, while an increase in gas density reduces penetration. An increase in the nozzle diameter increases the momentum of the jet and increases penetration. Up to gas temperatures of 590 K, no effect of the gas temperature on spray penetration could be detected. Further empirical equations are published by Dent [10] and Fujimoto et al. [13]. Dent [10] also includes the effect of gas temperature T_g, which shortens the penetration if the spray is injected in hot combustion chambers (all quantities in SI units):

$$S = 3.07 \cdot \left(\frac{\Delta p}{\rho_g} \right)^{0.25} \cdot (t \cdot D)^{0.5} \cdot \left(\frac{294}{T_g} \right)^{0.25} . \tag{2.6}$$

The spray cone angle is another characteristic parameter of a full-cone spray that has been investigated by Hiroyasu and Arai [20]. For sac hole nozzles, the authors give the following relation for the stationary spray cone angle (full needle lift):

$$\Phi = 83.5 \left(\frac{L}{D} \right)^{-0.22} \left(\frac{D}{D_s} \right)^{0.15} \left(\frac{\rho_g}{\rho_l} \right)^{0.26} . \tag{2.7}$$

In Eq. 2.7, Φ is the spray cone angle in [deg], D_s is the sac hole diameter in [m], and L is the length of the nozzle hole in [m]. In case of small L/D ratios cavitation structures do not collapse inside the injection holes but enter the combustion chamber, collapse outside the nozzle and increase the spray cone angle. A large value of D/D_s promotes the reduction of effective cross-sectional area at the entrance of the nozzle hole (vena contracta), reduces the static pressure at this point and facilitates the inception of cavitation. The most important influence parameter is the density ratio. The higher the gas density, the smaller the penetration and the more the fuel mass inside the combustion chamber is pushed aside by the new droplets.

Another relation for the spray cone angle is given by Heywood [19],

$$\tan\left(\frac{\Phi}{2}\right) = \frac{4\pi}{A}\left(\frac{\rho_g}{\rho_l}\right)^{0.5} f(\Upsilon),\tag{2.8}$$

where A is a constant depending on the nozzle design and may be extracted from experiments or approximated by $A = 3.0 + 0.28(L/D)$. The last term on the right hand side of Eq. 2.8 is a weak function of the physical properties of the liquid and the injection velocity [9]:

$$f(\Upsilon) = \frac{\sqrt{3}}{6}\left(1 - \exp(-10\Upsilon)\right), \quad \Upsilon = \left(\frac{Re_l}{We_l}\right)^2 \frac{\rho_l}{\rho_g}.\tag{2.9}$$

For high-pressure sprays and thus increasing values of Υ, $f(\Upsilon)$ becomes asymptotically equal to $3^{1/2}/6$ [44]. However, Kuensberg et al. [31] have shown that in case of high-pressure injection the predicted spray cone angles are underpredicted compared to experimental results.

One quantity characterizing the average droplet size of a spray, and thus the success of spray break-up, is the Sauter mean diameter (SMD). The SMD is the diameter of a model drop (index: m) whose volume-to-surface-area ratio is equal to the ratio of the sum of all droplet volumes (V) in the spray to the sum of all droplet surface areas (A):

$$\left(\frac{V}{A}\right)_m = \frac{(\pi/6)SMD^3}{\pi SMD^2} = \frac{SMD}{6},\tag{2.10a}$$

$$\left(\frac{V}{A}\right)_{spray} = \left(\sum_{i=1}^{n} d_i^3\right) \Big/ \left(6\sum_{i=1}^{n} d_i^2\right).\tag{2.10b}$$

Equating Eq. 2.10a and Eq. 2.10b yields

$$SMD = \frac{\displaystyle\sum_{i=1}^{n} d_i^3}{\displaystyle\sum_{i=1}^{n} d_i^2}.\tag{2.11}$$

The smaller the SMD, the more surface per unit volume. The more surface, the more effective evaporation and mixture formation. Although the SMD is a well-known quantity in characterizing the spray formation process, it is important to remember that it does not provide any information about the droplet size distribution of the spray. In other words, two sprays with equal SMD can have significantly different droplet size distributions.

Based on their experimental work, Hiroyasu and Arai [20, 21] give the following relation for the SMD:

$$\frac{SMD}{D} = 0.38 \, Re^{0.25} \, We_l^{-0.32} \left(\frac{\mu_l}{\mu_g}\right)^{0.37} \left(\frac{\rho_l}{\rho_g}\right)^{-0.47} .$$

(2.12)

In Eq. 2.12, SMD is in [m], and μ is the dynamic viscosity in [N·s/m^2]. The units of the other quantities are already given in the equations above. The Sauter mean diameter increases with increasing gas density due to the higher number of collisions (coalescence) and with increased nozzle hole diameter (larger initial drops). An increase in injection pressure results in improved atomization and thus in a decrease of the SMD.

However, it must be pointed out that the measurement of droplet sizes is only possible in the dilute spray regions at the edge of the spray or at greater distances from the nozzle. Correlations describing the SMD of a complete spray always include a high degree of uncertainty and can only be used to get a qualitative estimation.

At the beginning of experimental investigations of the inner structure of high-pressure full-cone diesel sprays, it was unclear whether the spray core directly at the nozzle is an intact liquid core whose diameter is reduced downstream due to the separation of droplets, or whether it already consists of large ligaments and droplets. The idea of an intact liquid core was based on electrical conductivity measurements that have been performed by Hiroyasu et al. [22] in order to draw conclusions about the inner structure of the spray. The authors measured the electrical resistance between the nozzle and a fine wire detector located in the spray jet. Chehroudi et al. [8] performed similar experiments, but they showed that the conductivity of dense spray regions (droplets) is comparable to those consisting of pure liquid, and that the measurement technique is not suitable to prove the existence of an intact liquid core. The authors measured core lengths

$$L_C = C \cdot D \cdot \sqrt{\frac{\rho_l}{\rho_g}}$$

(2.13)

that were only half as large as the ones detected by Hiroyasu et al. [22]. Equation 2.13 expresses the fact that the core length is dependent on the ratio of liquid and gas density, and that it is directly proportional to the nozzle hole diameter D. The empirical constant C expresses the influence of the nozzle flow conditions and other effects that cannot be described in detail and is in the range of $C = 3.3$ to $C = 11$. Youle and Saltes [61] have shown that the radial extent of the core region increases with increasing distance from the nozzle, and that it cannot consist of pure liquid. The authors use the expression break-up zone instead of intact core and draw the conclusion that this region consists of a very dense cluster of ligaments and drops. Gülder et al. [15, 16] have performed laser-optical investigations of the inner spray structure of high-pressure sprays directly at the nozzle hole and have proven that this break-up zone consists of areas with a very high content of liquid, and that these areas are clearly separated from each other by gaseous zones. Finally, optical measurements in combination with transparent nozzles in real size geometry [5, 7, 52] could prove the fact that due to the turbulent and often cavitating flow the disintegration of high-pressure diesel sprays begins already inside the nozzle holes, and that the jet leaving the nozzle hole consists of a very dense spray of ligaments and droplets. Nevertheless, Eq. 2.13 can be used to describe the length of this break-up zone. Another more detailed expression is given by Hiroyasu and Arai [20],

$$L_b = 7 \cdot D \cdot \left(1 + 0.4 \frac{r}{D}\right) \cdot \left(\frac{p_g}{\rho_l u^2}\right)^{0.05} \cdot \left(\frac{L}{D}\right)^{0.13} \cdot \left(\frac{\rho_l}{\rho_g}\right)^{0.5}. \qquad (2.14)$$

Both relations result in similar break-up lengths. In Eq. 2.14, u is the initial jet velocity in [m/s], and r in [m] is the radius of the inlet edge of the hole. The units of the remaining quantities are already given in the equations above. In addition to Eq. 2.13, Eq. 2.14 includes the effect of the inlet edge rounding. A rounded inlet edge shifts the inception of cavitation to higher injection pressures and increases L_b. The influence of cavitation and turbulence is also included via the cavitation number $p_g/(\rho_l u^2)$.

In the following, the mechanisms of primary break-up of high-pressure full-cone sprays shall be described in detail. Primary break-up is the first disintegration of the coherent liquid into ligaments and large drops. Figure 2.9 summarizes possible break-up mechanisms.

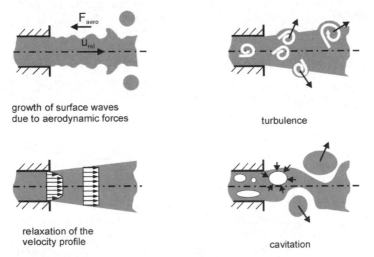

growth of surface waves
due to aerodynamic forces

turbulence

relaxation of the
velocity profile

cavitation

Fig. 2.9. Mechanisms of primary break-up

The very high relative velocities between jet and gas phase induce aerodynamic shear forces at the gas-liquid interface. Due to the liquid turbulence that is created inside the nozzle, the jet surface is covered with a spectrum of infinitesimally small surface waves. Some of these waves are amplified by the aerodynamic shear forces, become instable, are separated from the jet, and form primary droplets. However, the instable growth of waves due to aerodynamic forces is a time-dependent process and cannot explain the immediate break-up of the jet at the nozzle exit. Furthermore, aerodynamic forces can only affect the edge of the jet, but not its inner structure, which has been shown to be also in train of disintegration. Hence, aerodynamic break-up, which is the relevant mechanism of secondary droplet disintegration, is of secondary importance.

A second possible break-up mechanism is turbulence-induced disintegration. If the radial turbulent velocity fluctuations inside the jet, which are generated inside the nozzle, are strong enough, turbulent eddies can overcome the surface tension and leave the jet to form primary drops as discussed by Wu et al. [60]. Turbulence-induced primary break-up is regarded as one of the most important break-up mechanisms of high-pressure sprays.

A further potential primary break-up mechanism is the relaxation of the velocity profile. In the case of fully developed turbulent pipe flow (large L/D ratios, no cavitation), the velocity profile may change at the moment the jet enters the combustion chamber. Because there is no longer a wall boundary condition, the viscous forces inside the jet cause an acceleration of the outer jet region, and the velocity profile turns into a block profile. This acceleration may result in instabilities and in break-up of the outer jet region. However, in the case of high-pressure injection, cavitation occurs, L/D ratios are small, and the development of the velocity profile described above is very unlikely.

Another very important primary break-up mechanism is the cavitation-induced disintegration of the jet. Cavitation structures develop inside the nozzle holes because of the decrease of static pressure due to the strong acceleration of the liquid (axial pressure gradient) combined with the strong curvature of the streamlines (additional radial pressure gradient) at the inlet edge. Hence, a two-phase flow exists inside the nozzle holes. The intensity and spatial structure of the cavitation zones depends on nozzle geometry and pressure boundary conditions. The cavitation bubbles implode when leaving the nozzle because of the high ambient pressure inside the cylinder. Different opinions exist regarding whether the energy that is released during these bubble collapses contributes to the primary break-up either by increasing the turbulent kinetic energy of the jet or by causing a direct local jet break-up. However, experimental investigations have shown that the transition from a pure turbulent to a cavitating nozzle hole flow results in an increase of spray cone angle and in a decrease of penetration length (Arai et al. [2], Hiroyasu et al. [22], Bode [6], Soteriou et al. [51], Tamaki et al. [55]). Implosions of cavitation bubbles inside the nozzle holes increase the turbulence level and thus also intensify the spray disintegration. Hence, the two main break-up mechanisms in the case of high-pressure full-cone jets are turbulence and cavitation. Usually, both mechanisms occur simultaneously and cannot be clearly separated from each other.

Because of the importance of so-called hydrodynamic cavitation in injection nozzles, its development shall be described in detail. Hydrodynamic cavitation is the formation of bubbles and cavities in a liquid due to the decrease of static pressure below the vapor pressure, caused by the geometry through which the liquid flows. Usually, liquids cannot stand negative pressures, and if the vapor pressure is reached, the liquid evaporates. The growth of cavitation bubbles and films starts from small nuclei, which are either already present in the liquid (micro-bubbles filled with gas or gas that adheres to the surface of solid particles) or at the wall (surface roughness and imperfections, small gaps filled with gas).

Figure 2.10 shows the difference between boiling and hydrodynamic cavitation. In the case of boiling, the temperature is increased at constant pressure, while in the case of hydrodynamic cavitation the temperature is not altered, and the pressure decreases. Because fuels usually consist of many different components with different vapor pressure curves, the components with the highest vapor pressures evaporate first and fill the cavitation zones.

Up to now, only a small number of authors, e.g. Bode [6], Badock [5], Busch [7] and Arcoumanis et al. [3], have investigated the phenomenon of cavitation in transparent nozzles in real-size geometry (optical access, shadowgraphy, laser-optical techniques). According to these authors, the inception of cavitation can be explained as follows. The liquid entering the injection hole is strongly accelerated due to the reduction of cross-sectional area. Assuming a simplified one-dimensional, stationary, frictionless, incompressible, and isothermal flow, the Bernoulli equation,

$$p_1 + \frac{\rho}{2}u_1^2 = p_2 + \frac{\rho}{2}u_2^2 , \qquad (2.15)$$

can be used to explain the fact that an increase in flow velocity u from a point 1 to

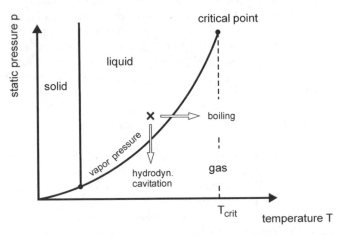

Fig. 2.10. Hydrodynamic cavitation, example: a single-component liquid

a point 2 further downstream results in a decrease in static pressure p (axial pressure gradient). At the inlet of the injection hole, the inertial forces, caused by the curvature of the streamlines, result in an additional radial pressure gradient, which is superimposed on the axial one. The lowest static pressures are reached at the inlet edges in the recirculation zones of the so-called vena contracta, see Fig. 2.12. If the pressure at the vena contracta reaches the vapor pressure of the liquid, the recirculation zones fill with vapor. An additional effect enhancing the onset of cavitation in this low-pressure zone is the strong shear flow that is caused by the large velocity gradients in the region between recirculation zone and main flow. This shear flow produces small turbulent vortices. Due to centrifugal forces, the static pressure in the centers of these eddies is lower than in the surrounding liquid, and cavitation bubbles may be generated. The cavitation zones develop along the walls, can separate from the walls, disintegrate finally into bubble clusters, and may already begin to collapse inside the nozzle hole. A good description of the possible cases is given in Kuensberg et al. [31]. In the case of high-pressure diesel injection, the cavitation structures usually leave the hole and collapse in the primary spray.

The nozzle geometry at the inlet of the injection holes is of great importance concerning the development of cavitation. The more the inlet edges are rounded, the smaller the flow contraction and the smoother the decrease of static pressure. Schugger et al. [50, 49] have performed experimental investigations using nozzles with different inlet edge roundings and have shown that during full needle lift sharp-edged inlets produce stronger cavitation, smaller ligaments near the nozzle, and larger spray cone angles than nozzles with rounded inlet edges. Further investigations are published in Su et al. [54] and König et al. [29]. Another geometrical influence parameter is the angle between the needle axis and the hole axis. The bigger this angle, the more the main flow direction is changed at the entrance of the hole, and the more the centrifugal forces push the liquid to the bottom of the injection hole. The lowest static pressures are then reached at the upper part of the inlet edge, where cavitation structures start to grow. This can result in asymmetric three-dimensional flow structures, where the upper part of the hole is occupied by cavitation and the lower part is filled with liquid.

An asymmetric nozzle hole flow results in an asymmetric primary spray [11, 6, 28, 7]. As an example, Fig. 2.11 shows such a flow in a single-hole transparent

Fig. 2.11. Effect of an asymmetric distribution of cavitation inside the nozzle holes on the primary spray [36], p_{rail} : 60 MPa, $p_{chamber}$: 5 MPa, shadowgraphy

test nozzle in real-size geometry with cavitation on the upper side (due to secondary flow, some cavitation structures are also transported into the lower half). The white areas inside the hole are pure liquid, while the black areas are cavitation structures. It is obvious that the concentration of cavitation in the upper half of the nozzle hole directly affects the primary spray structure. Due to the increased break-up energy per unit mass, the upper part of the spray diverges stronger than the lower one. Altogether, this example very clearly shows the fact that in case of high-pressure injection the flow inside the injection holes directly influences the primary jet break-up. This must be taken into account when developing primary break-up models.

A second source of cavitation is the needle seat. During opening and closing the smallest cross-sectional flow area is no longer located at the inlet of the holes, but at the needle seat. The cavitation structures that are produced in this region either collapse before entering the holes and increase the turbulence of the flow, or enter the holes and alter the flow conditions there. Optical investigations of this effect are published in Busch [7] for example. However, with the exception of small needle lifts, the smallest cross-sectional flow area is always located at the inlet of the injection holes.

Different opinions exist concerning whether the presence of cavitation has a positive or negative effect on engine performance and emissions. On the one hand, cavitation reduces the effective cross-sectional flow area and complicates the injection of large fuel masses (full load) through small nozzle holes. On the other hand, cavitation enhances mixture formation and cleans the exit of the nozzle hole from deposits that are caused by carbonization (injector fouling). In order to reduce the extent of cavitation, low local pressures due to a sudden reduction of cross-sectional area have to be avoided. The effective cross-sectional area must be smoothly decreased until it reaches its minimal value at the hole exit. Then, the static pressure cannot fall below the combustion chamber pressure, and the formation of cavitation bubbles is significantly reduced, see Sect. 2.2.1. In order to suppress cavitation completely, any imperfections of the wall, especially at the hole entrance, have to be avoided. Furthermore, the formation of strong vortices, which can also produce cavitation, as well as low pressures at the needle seat, must be suppressed. Altogether, it is possible to reduce the extent of cavitation significantly, but it is hardly possible to produce completely cavitation-free injectors for engine applications.

Because of the very small dimensions, the high flow velocities, and the very dense spray, which does not allow optical access to the inner spray directly at the nozzle tip, no detailed experimental investigations about the structure and size of the cavitation bubbles in the primary spray have been published up to now. First investigations at low and thus not engine-like injection pressures are published by Fath [12] and Heimgärtner et al. [18]. Hence, statements about the behavior and size of cavitation bubbles in the primary spray under engine-like conditions are solely based on mathematical models. A good summary of mathematical models describing the dynamics of bubble growth and collapse is given in Prosperetti and Lezzi [41].

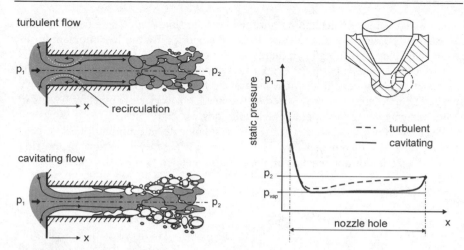

Fig. 2.12. Cavitating and non-cavitating nozzle hole flow

Whether cavitation occurs in a nozzle or not can be estimated using a dimensionless characteristic number, the so-called cavitation number K. Different definitions of K exist in the literature, so that the cavitation number either increases or decreases with increasing cavitation. The most-used form is

$$K_1 = \frac{p_1 - p_2}{p_2 - p_{vap}} \approx \frac{p_1 - p_2}{p_2} , \tag{2.16}$$

where p_{vap} is the vapor pressure. The other pressures are explained in Fig. 2.12. This cavitation number represents the ratio of pressure decrease inside the hole (increase of flow velocity) to the backpressure. A strong decrease of pressure inside the hole enhances cavitation, while a high level of backpressure suppresses cavitation. The larger the value of K_1, the more intensive the cavitation. The inception of cavitation is strongly dependent on the nozzle geometry.

A second well-known definition of the cavitation number is

$$K_2 = \frac{p_1 - p_{vap}}{p_1 - p_2} \approx \frac{p_1}{p_1 - p_2} , \tag{2.17}$$

where the intensity of cavitation increases with decreasing values of K_2. Further definitions of cavitation numbers as well as relations showing the interrelation among each other are given in He and Ruiz [17].

Finally, the temporal development of a typical full-cone diesel spray shall be discussed. The injection can be divided into three phases. During the first phase, the needle opens. During this early phase, the small cross-sectional flow area at the needle seat is the main throttle reducing the mass flow through the injector. Cavitation at the needle seat usually produces a highly turbulent nozzle hole flow. This holds especially true for common rail systems, where high injection pressures are already present at the start of injection. Due to the low axial velocity and the strong radial velocity fluctuations (turbulence), the first spray angle near the nozzle is usually large, Fig. 2.13. This effect is supported by the low momentum of the injected mass, resulting in an increasing amount of mass near the nozzle that is pushed aside by the subsequent droplets. As soon as the axial velocity increases, the resulting spray cone angle near the nozzle becomes smaller. Hence, the early spray structure depends on the speed of the needle: a very slow opening results in larger spray angles, a fast opening in smaller angles.

A second class of injection systems are systems with intermittent pressure generation, Sect. 2.2.1. Whether cavitation occurs in these systems during the first phase of injection or not depends on the pressure needed to open the spring-loaded needle.

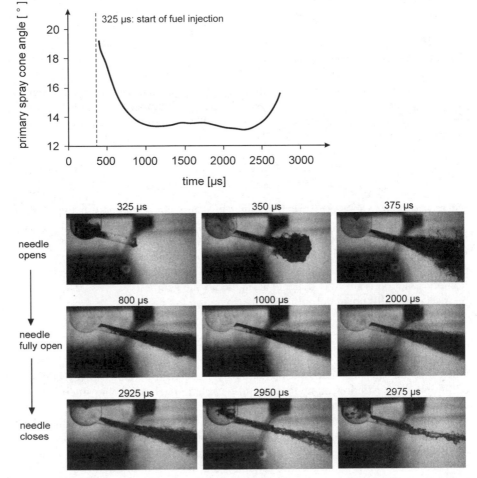

Fig. 2.13. Spray formation during injection. Data from [7], CR injector, $p_{rail} = 60$ MPa, $p_{back} = 0.1$ MPa, $T_{air} = 293$ K

As soon as the cross-sectional flow area at the needle seat is larger than the sum of the nozzle hole areas, the nozzle hole inlets become the main throttle of the system. The extent of cavitation now depends on the hole geometry. Strongly cavitating nozzle flows produce larger overall spray cone angles and smaller penetration lengths than non-cavitating ones. The spray penetration increases with time due to the effect that new droplets with high kinetic energy continuously replace the slow droplets at the spray tip.

At the end of injection, the needle closes and the injection velocity decreases to zero, resulting in a disruption of the spray in the axial direction. Due to the decreasing injection velocity, droplet and ligament sizes increase and atomization deteriorates. It is obvious that a rapid closing of the needle is advantageous in order to minimize the negative influence of these large liquid drops on hydrocarbon and soot emissions.

2.1.3.2 Hollow-Cone Sprays

In order to achieve maximum dispersion of the liquid at moderate injection pressures and low ambient pressures, hollow-cone sprays are usually used. Hollow-cone sprays are typically characterized by small droplet diameters, effective fuel-air mixing, reduced penetration, and consequently high atomization efficiencies. These sprays are used in conventional gasoline engines, where the fuel is injected into the manifold, and in direct injection spark ignited (DISI) engines as well, see Sect. 2.2.2.

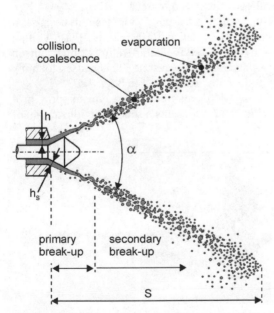

Fig. 2.14. Hollow-cone spray. Example: outwardly opening nozzle

Fig. 2.14 shows the typical structure of such a spray. The liquid emerging from the nozzle forms a free cone-shaped liquid sheet inside the combustion chamber, which thins out because of the conservation of mass as it departs from the nozzle and subsequently disintegrates into droplets. Two nozzle concepts exist: the inwardly opening pressure-swirl atomizer and the outwardly opening nozzle. In the case of a swirl-atomizer, a cylindrical and strongly rotating liquid film leaves the nozzle. The radial velocity component, which is caused by the rotational motion, results in the formation of the free cone-shaped liquid sheet. In the case of an outwardly opening nozzle, the geometry of the needle causes the liquid to form the cone-shaped liquid sheet. More details are discussed in Sect. 2.2.2.

The primary break-up of the liquid sheet is induced by turbulence and aerodynamic forces. First, the liquid film with initial thickness h_s and spray angle α becomes thinner because of the conservation of mass as it departs from the nozzle. The turbulence, which is produced inside the nozzle, causes the formation of initial perturbations on the liquid surface, the frequency and amplitude spectrum of which is dependent on the internal nozzle hole flow. These waves grow instable due to the aerodynamic interaction with the surrounding gas. At critical amplitudes, the liquid film breaks up into ligaments, which, under the influence of surface tension and gas forces, rapidly disintegrate into drops. Besides these oscillation phenomena, a spontaneous separation of small satellite droplets directly at the nozzle exit is possible, if sufficient turbulent kinetic energy is present. This effect mainly occurs at high injection pressures.

The secondary break-up of the droplets is aerodynamically induced, and is governed by the break-up mechanisms already described in Sect. 2.1.2.

A special case of spray break-up, which is usually not desired because it changes the spray structure completely, is so-called flash-boiling. Flash-boiling may occur at low ambient pressures and increased fuel temperatures, and results in a boiling of the fuel at the exit cross section of the nozzle. This causes foaming and results in a complete alteration of the spray structure, see Sect. 4.5.3.

Figures 2.15 and 2.16 show the temporal development of a typical spray produced by a pressure-swirl atomizer. Fig. 2.17 represents a schematic illustration of the fully developed spray.

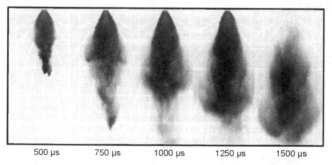

500 μs   750 μs   1000 μs   1250 μs   1500 μs

Fig. 2.15. Temporal development of a spray from a pressure-swirl atomizer [25], $p_{rail} = 10$ MPa, $p_{chamber} = 0.24$ MPa, $T_{chamber} = 393$ K

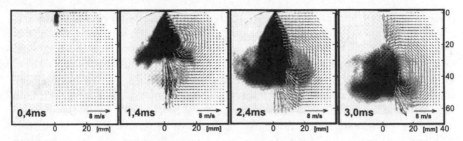

Fig. 2.16. Temporal development of the secondary gas flow around a spray from a pressure-swirl atomizer [14], $p_{rail} = 7$ MPa, $p_{chamber} = 0.102$ MPa, $T_{chamber} = 298$ K

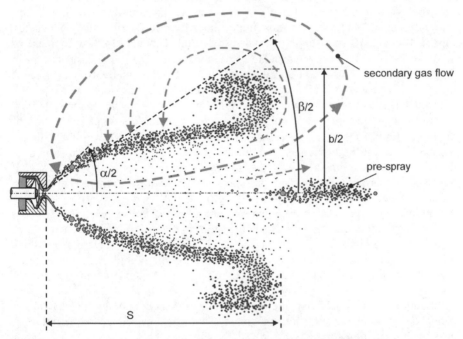

Fig. 2.17. Typical spray from a pressure-swirl atomizer (schematic illustration)

Due to the small flow velocity inside the nozzle at the beginning of injection, the first amount of fuel entering the swirl chamber inside the nozzle does not receive sufficient rotational motion, leaves the nozzle with nearly zero swirl, and forms a kind of solid-cone spray with narrow spray cone angle and large drops, the so-called pre-spray. As the fuel velocity increases, the liquid inside the swirl chamber forms a hollow-cylinder structure. This structure is then transformed into a hollow-cone spray as it leaves the nozzle. The hollow-cone spray starts to penetrate into the gas atmosphere with an initial cone angle α, see Fig. 2.17. Due to momentum transfer between liquid droplets and gas, the air inside the hollow cone is accelerated in the axial direction. A strong initial acceleration is already achieved by the compact full-cone pre-spray. This acceleration results in the formation of a region of low static pressure inside the hollow cone. As a consequence, a secondary airflow develops, transporting gas from outside the spray into its center. This secondary flow reduces the radial velocity components of the droplets and results in the typical decrease of the spray cone angle β with increasing distance from the nozzle. Hence, the spray width b stagnates while the penetration length S continues to increase, forming the typical bell-shaped spray. Small droplets follow the secondary gas flow, fill the center of the spray, and improve fuel-air mixing. At the leading edge of the spray, the aerodynamic drag forces result in a strong deceleration of the drops and cause the formation of a three-dimensional torus-like vortex at the spray tip, which also enhances mixture formation.

The temporal development and structure of hollow-cone sprays produced by pressure-swirl atomizers is strongly dependent on the boundary conditions imposed by the ambient gas. The most important influence factor is the gas density. Figs. 2.18 and 2.19 show experimental results from Gindele [14]. An increase of ambient pressure strongly reduces the spray cone angle β and increases the mean droplet size (SMD), while the spray penetration S remains relatively unchanged if the injection pressure is high enough and is reduced in case of lower injection pressures. It is assumed that due to the higher gas density the secondary gas flow is more effective at increased ambient pressures, causing a stronger reduction of the radial droplet velocities (decrease of β) and supporting the flow in the axial direction inside the hollow cone. For this reason, the spray penetration is not reduced at moderate ambient pressures. However, experiments from Homberg [25] have shown that at higher ambient pressures (1 MPa to 3 MPa) there is also a significant reduction of spray penetration length. At sufficiently high gas densities, the hollow-cone spray structure collapses into a kind of full-cone spray. This behavior is extremely disadvantageous with regard to the use of pressure-swirl atomizers in DISI engines. Depending on the start of injection, the chamber pressures differ strongly enough in these engines to produce the complete range of spray structures and to cause significant changes regarding the quality of mixture formation.

In contrast to this behavior, the sprays from outwardly opening nozzles do not collapse if the ambient pressure is increased, but do show reduced penetration. A possible explanation for such differing behavior may be a reduced overall influence of the secondary flow effects. The outwardly opening nozzle does not need any rotational motion of the liquid in order to form a hollow-cone spray, and therefore there is no pre-spray at the beginning of injection and no strong initial acceleration of the gas inside the hollow cone. Hence, the development of a low-pressure gas region inside the cone, inducing the strong secondary flow, is much slower and weaker compared to the spray from a pressure-swirl injector.

In contrast to full-cone sprays, only few semi-empirical relations describing the behavior of the main spray parameters have been published in the literature. The most-described parameter is the droplet size. An increase of injection decreases the Sauter mean diameter (SMD), while the spray penetration is slightly increased [38, 23, 24], Fig. 2.19. An increase of ambient pressure increases the SMD [32, 40]. However, the exact values do strongly depend on the design and type of injec-

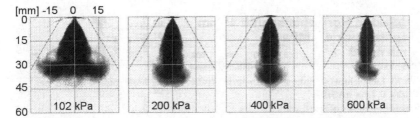

Fig. 2.18. Influence of ambient pressure on the spray structure (pressure-swirl atomizer) [14], pictures at 2 ms after energizing the injector, $T_{chamber}$ = 298 K, p_{rail} = 7 MPa

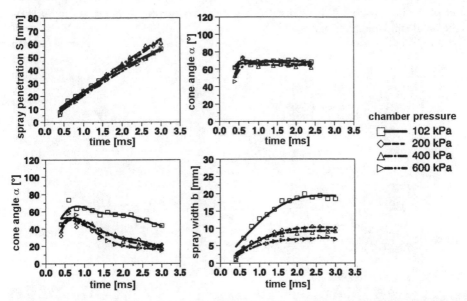

Fig. 2.19. Influence of ambient pressure and injection pressure on the spray structure (pressure-swirl atomizer) [14], $T_{chamber}$ = 298 K, p_{rail} = 7 MPa, t = 0.0 ms: energizing of injector

tor. The most-used design is the pressure-swirl atomizer. The SMD of such an injector can be expressed by the semi-empirical relation [33]:

$$SMD = 2.25 \cdot \sigma^{0.25} \cdot \mu_l^{0.25} \cdot \dot{m}_l^{0.25} \cdot \Delta p^{-0.5} \cdot \rho_g^{-0.25} .   (2.18)$$

The exponents are dependent on the special nozzle geometry. A more detailed relation, also including the effect of the spray cone angle, is given in Lefebvre [33]:

$$SMD = 4.52 \left[\frac{\sigma \mu_l^2}{\rho_g \Delta p^2} \right]^{0.25} \left(h \cos \frac{\alpha}{2} \right)^{0.25} + 0.39 \left[\frac{\sigma \mu_l}{\rho_g \Delta p} \right]^{0.25} \left(h \cos \frac{\alpha}{2} \right)^{0.75} .   (2.19)$$

In Eq. 2.19

$$h = 2.7 \left[\frac{D \dot{m}_l \mu_l}{\rho_l \Delta p_l} \right]^{0.25} \qquad (2.20)$$

is the thickness of the liquid sheet at the nozzle exit (Fig. 2.14), α is the initial spray cone angle, and D is the nozzle hole diameter. More details about droplet sizes from pressure-swirl atomizers and other nozzle types are given in [33].

An increase in gas temperature causes a reduction of spray width b and an increase in spray penetration S, Fig. 2.20, and thus results in a similar effect as the increase of ambient pressure. One possible explanation is as follows. First, a faster evaporation reduces the overall spray width. Second, due to the increased gas viscosity, the secondary flow is more effective and entrains more drops into the center of the hollow cone. This causes an additional reduction of b and an increase in the axial spray penetration.

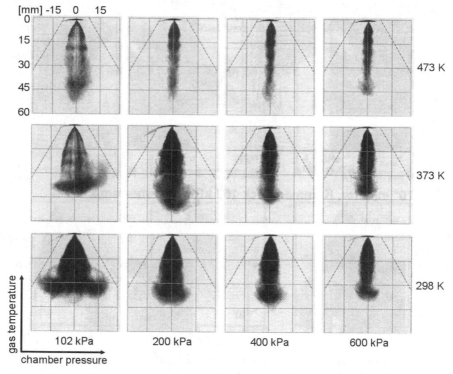

Fig. 2.20. Influence of ambient temperature on the spray structure (pressure-swirl atomizer) [14], pictures at 2 ms after energizing the injector, $p_{rail} = 7$ MPa

2.1.4 Spray-Wall Interaction

Spray-wall interactions occur if a spray penetrating into a gaseous atmosphere impacts a wall, which can be the backside of the intake valve or the wall of the induction system in case of port fuel injection, or the combustion chamber wall in case of a direct injection engine. Two main physical processes can be involved: wall-spray development and wall film evolution. Both processes may strongly influence combustion efficiency and the formation of pollutants. Whether wall impingement occurs or not depends on the penetration length of the spray and on the distance between injection nozzle and wall. High injection pressures as well as low gas densities and temperatures increase penetration and the possibility of wall impact.

Depending on wall temperature and on the amount of liquid deposited on the wall, spray-wall impingement can have both negative and positive effects. In the case of low wall temperature, under cold starting conditions for instance, the formation of a liquid wall film will significantly increase the unburned hydrocarbon and soot emissions because of partial burning due to the very slow evaporation of the wall film. On the other hand, shattering of droplets at the wall may intensify droplet disintegration and increase the total spray surface. The formation of a large-scale gas vortex near the wall may also promote air entrainment and enhance mixture formation. Furthermore, contact with a hot wall intensifies evaporation. In some DI engines, the fuel spray has been intentionally impinged on a surface as a method to control combustion. One early system which used such a technique was the M-combustion developed by MAN [57], in which the fuel is injected on the surface of the piston bowl, forming a liquid film. Vaporization of this film controls the rate of combustion allowing the use of low cetane number fuels.

In the case of port-injected engines, the low gas densities promote spray penetration, and a liquid film may form on the cold walls of the induction system. This causes difficulties in the transient control of the engine, because only a part of the injected fuel enters the combustion chamber during the corresponding cycle, and

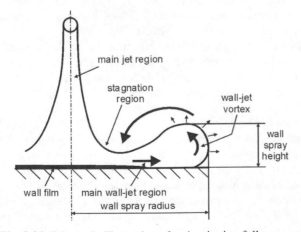

Fig. 2.21. Schematic illustration of an impinging full-cone spray

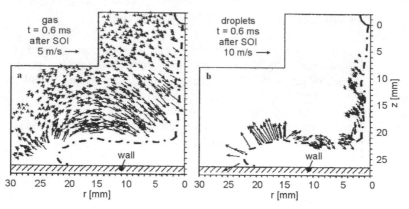

Fig. 2.22. a Typical flow field around an impinging full-cone spray, **b** droplet velocities [35], $p_{inj} = 100$ MPa, $p_{chamber} = 1.0$ MPa (N_2), $d_{nozzle} = 0.25$ mm, $T_{wall} = 298$ K

the rest is added to the wall film and slowly transported to the valve. Here it enters the combustion chamber many cycles later and adds an unknown and often not well-dispersed amount of fuel to the actual injection. This can cause decreased engine response, increased fuel consumption and increased hydrocarbon emissions.

Figures 2.21 and 2.22 show a schematic illustration of a full-cone spray impinging on a wall and a typical gas flow field around such a spray [35]. In Fig. 2.23 the temporal development of the wall spray is shown. As is known from non-impinging sprays, the main jet region forms a dense full-cone spray surrounded by a region of increased air entrainment containing finely atomized droplets. When the spray impinges on the wall, a large number of droplets is formed at the periphery of the spray tip, which develops along the wall. Whether the drops stick to the wall and continue to evaporate, spread out to form a liquid film, reflect, or break up into smaller droplets, depends on the kinetic energy of the incident drops and on the wall temperature, see also Sect. 4.8. A large-scale vortex is formed around the wall spray entraining a considerable amount of hot gas into the spray. In the case of small distances between nozzle and wall as well as high injection rates and cold walls, a liquid wall film can be generated, Fig. 2.21.

Detailed investigations of wall-spray development are reported in Mohammadi et al. [35] and Allocca et al. [1] for example. Increasing the injection pressure increases the wall spray penetration tangential and normal to the surface and promotes air entrainment into the spray, see Figs. 2.24a and 2.24b. An increase of chamber pressure results in a decrease of wall-spray penetration, similar to the effect of gas density on the penetration length of a free spray, see Figs. 2.24c and 2.24d. Even though the entrained air contains more mass per unit volume, the overall entrained gas mass can be reduced by this effect. An increase of wall temperature, Figs. 2.24e and 2.24f, reduces wall wetting and promotes droplet rebound and reflection because of the vapor layer formed between droplet and wall. Especially at high wall temperature, this effect may result in the existence of larger droplets with higher momentum in the wall-spray region and in increase in penetration.

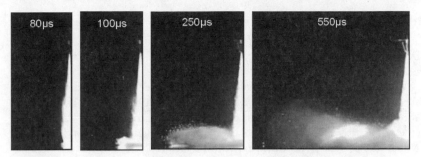

Fig. 2.23. Time evolution of an impinging spray, [1], p_{inj} = 120 MPa, $p_{chamber}$ = 0.1 MPa (N$_2$), d_{nozzle} = 0.18 mm, T_{wall} = 298 K

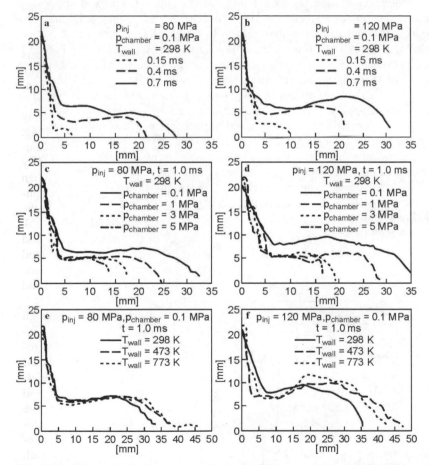

Fig. 2.24. Effect of injection pressure, chamber pressure and wall temperature on wall-spray impingement, data from [1], p_{inj} : 80 MPa and 120 MPa, chamber gas: N$_2$, d_{nozzle} = 0.18 mm

2.2 Injection Systems and Nozzle Types

2.2.1 Direct Injection Diesel Engines

The task of the injection system is to achieve a high degree of atomization in order to enable sufficient evaporation in a very short time and to achieve sufficient spray penetration in order to utilize the full air charge. The fuel injection system must be able to meter the desired amount of fuel, depending on engine speed and load, and to inject that fuel at the correct time and with the desired rate. Further on, depending on the particular combustion chamber, the appropriate spray shape and structure must be produced.

Usually a supply pump draws the fuel from the fuel tank and carries it through a filter to the high-pressure injection pump. Dependent on the area of application and engine size, pressures between 100 and 200 MPa are generated. The high-pressure injection pump carries the fuel through high-pressure pipes to the injection nozzles in the cylinder head. Excess fuel is transported back into the fuel tank.

Today, two main groups of high-pressure injection systems exist. Concepts belonging to the first one are the so-called common rail injection systems, Fig. 2.25. Here, pressure generation and the injection event are not coupled, and the injection pressure is not dependent on engine speed. Compared to the injection systems that are driven by a camshaft, this enables a significantly greater flexibility of injection and mixture formation. Fuel under high pressure is stored inside the rail, which usually consists of a thick-walled closed pipe. A high-pressure fuel pump continuously feeds the rail. A pressure sensor adjusts the desired rail pressure via an additional valve that controls the mass flow of excess fuel back to the fuel tank. Hence, the rail pressure is not dependent on engine speed, and an optimal adjustment to the actual operating point of the engine can be achieved. Short pipes connect the rail with the injectors.

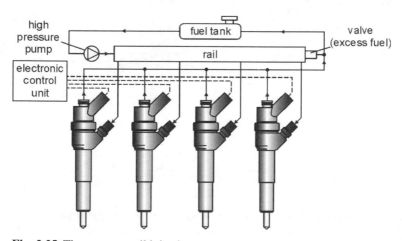

Fig. 2.25. The common rail injection system

The volume of the rail is large enough to suppress pressure fluctuations due to injection. Injection timing and duration are controlled by solenoid valves and are independent of the pressure generation. Hence, the common rail injection system is capable of keeping the injection pressure at the desired level and of performing pre-injections (reduction of noise and nitric oxides), main injections, and post-injections (reduction of soot raw emissions, heating of catalysts) with variable duration and timing according to the demands of the actual operating point.

Fig. 2.26 shows a typical common rail injector. The needle movement is controlled by a solenoid valve. In Fig. 2.26, the needle is closed. The spring force F_{spring} and the hydraulic force F_1 of the high-pressure fuel on top of the control rod are larger than the hydraulic force F_2 on the circular ring area, and the needle is kept closed. As soon as the solenoid valve opens, the pressure in the control chamber above the control rod decreases and the needle begins to open, because the inlet throttle connecting the control chamber with the high-pressure fuel supply is smaller than the outlet throttle. Excess fuel passes off through the outlet throttle and then flows back to the fuel tank. Hence, the motion of the needle is a hydraulically controlled process. The opening speed of the needle is determined by the size ratio of both throttles. The closing process is initiated by closing the solenoid valve. The pressure inside the control chamber increases and the control rod closes the needle. The closing speed is again influenced by the size of the throttle.

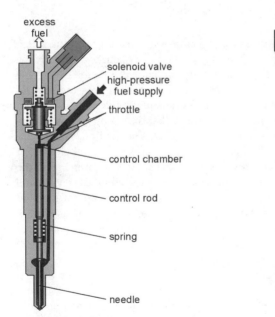
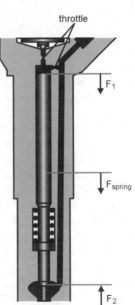

Fig. 2.26. A typical common rail injector [47]

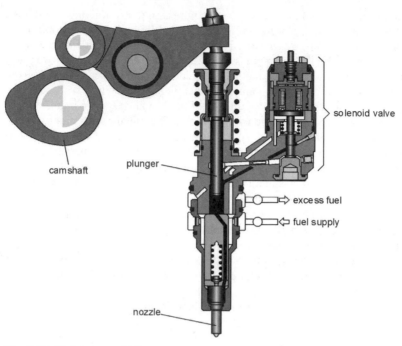

Fig. 2.27. Unit Injector [46]

The second main group of high-pressure injection systems are those in which the generation of injection pressure and the injection itself occur synchronously. These systems are driven by a camshaft, which is mechanically coupled with the engine. A basic characteristic of these systems is the intermittent pressure generation: high pressure is only available during a small crank angle interval.

In the case of the so-called unit injector system (UIS), the pump and the injection nozzle are combined into a single unit, see Fig. 2.27. Each cylinder of the engine is equipped with such a unit, which is driven by the cam via a rocker arm for example. The omission of the high-pressure pipes between the pump and the injector allows significantly higher peak injection pressures (about 200 MPa and more) than in the case of the unit pump system, see below. The shape of the cam determines the motion of the plunger and thus the generation of pressure as a function of crank angle. The spring at the upper part of the injector presses the plunger against the rocker arm and the rocker arm against the cam, and guaranteeing force closure during operation.

Fig. 2.28a shows a simplified picture of the unit injector with a closed needle. The solenoid valve is opened, and as soon as the plunger is pushed down, excess fuel is carried into the scavenge line. According to the desired start of injection, the control unit of the engine closes the solenoid valve, and the plunger compresses the fuel. As soon as the pressure is large enough to lift the spring-loaded needle, injection begins, Fig. 2.28b. Due to the large feed rate of the plunger, the

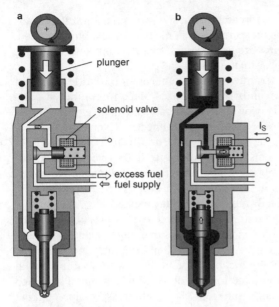

Fig. 2.28. Unit injector: schematic of injection control [46]

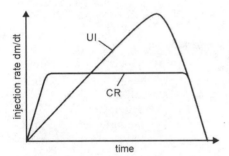

Fig. 2.29. Typical injection rate profiles of common rail and unit injector system

injection pressure strongly increases during injection, resulting in a triangle-shaped injection rate profile, Fig. 2.29. In contrast to the common rail system, in which the maximum injection pressure is already present at the start of injection (resulting in a rectangular-shaped injection rate profile and in maximum spray atomization during the complete injection event), the atomization of the early spray is considerably slower and enhances during injection. At the end of injection, the solenoid valve opens again, excess fuel is carried into the scavenge line, and the injection pressure decreases. As soon as the spring is able to close the needle, the injection is stopped. Finally, the plunger moves up and the chamber below the plunger fills again with fuel from the low-pressure supply.

Because the motion of the plunger is directly coupled with the engine, the maximum injection pressure increases with engine speed. At high engine speed, the fuel volume below the plunger is forced much faster through the nozzle, resulting in significantly higher injection pressures than in the case of low engine speed. Furthermore, high injection pressures are not available at any crank angle, complicating the generation of pre- and post-injections.

It is obvious that, due to the intermittent pressure generation, the flexibility of high-pressure injection systems driven by a camshaft is much more limited compared to the common rail system. However, a considerable advantage of these systems are the increase of injection pressure during the injection event and the very high maximum injection pressures at the end of injection, which cannot be provided by today's common rail systems and are known to have a very positive influence on mixture formation, combustion and reduction of pollutants. However, an insufficient atomization due to the low injection pressures causes problems in the case of pre-injections. More details are discussed in Chap. 6.

The functionality of the so-called unit pump system (UPS) is practically identical to that of the unit injector system and offers the same advantages and disadvantages. However, the pump and nozzle are not combined into one unit. The high-pressure pump is again driven by a camshaft and thus directly coupled with the engine speed, see Fig. 2.30. The injection nozzle is located inside a so-called nozzle holder in the cylinder head and connected via a high-pressure pipe with the pump. An advantage of this system is that the pump and nozzle must not be installed at the same place. This reduces the size of the components that have to be

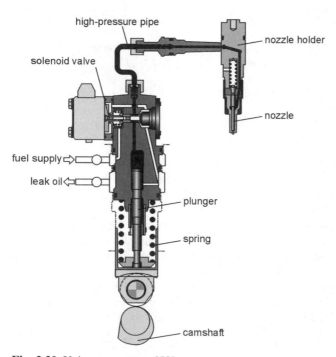

Fig. 2.30. Unit pump system [58]

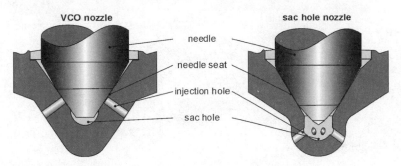

Fig. 2.31. Nozzle types

integrated into the cylinder head and simplifies the assembly of the injection system. The maximum attainable injection pressures of the UPS are smaller than those of the UIS.

The most important part of the injection system is the nozzle. The fuel is injected through the nozzle holes into the combustion chamber. The number and size of the holes depends on the amount of fuel that has to be injected, the combustion chamber geometry, and the air motion (swirl) inside the cylinder. In direct injection diesel engines, two main nozzle types, the sac hole nozzle and the valve covered orifice nozzle (VCO), are used, see Fig. 2.31. Compared to the VCO nozzle, the sac hole nozzle has an additional volume below the needle seat. Due to the increased distance between needle seat and injection hole, an eccentricity or radial motion of the needle tip does not influence the mass flow through the different holes, and a very symmetric overall spray is produced. However, the large liquid volume between the needle seat and the combustion chamber causes problems in terms of increased hydrocarbon emissions. It is important to keep this volume as small as possible, because otherwise some of this fuel can enter the cylinder after the end of injection. This fuel is usually not well dispersed and increases soot emissions. A late evaporation of the fuel inside the sac hole increases the hydrocarbon emissions of the engine. From this point of view, the VCO nozzle is superior to the sac hole nozzle. However, special constructive actions must be taken in order to suppress any radial eccentricity of the needle tip, because an eccentricity directly results in an uncontrollable variation of the discharge through the different nozzle holes and thus strongly deteriorates the overall spray quality.

The inlets of the nozzle holes are usually rounded (hydro-grinding) in order to enhance the inflow conditions and to produce abrasion in advance, which would otherwise occur during operation and change the spray characteristics. Depending on the particular application, different nozzle hole geometries are used today, Fig. 2.32. The cylindrical hole produces the strongest cavitation and results in an increased spray break up with a large spray divergence near the nozzle. The axissymmetric conical geometry suppresses cavitation by gradually reducing the effective cross-sectional area along the hole.

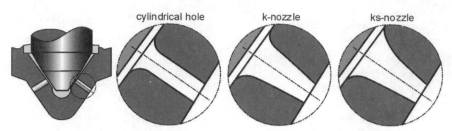

Fig. 2.32. Nozzle hole geometries

The sprays from conical holes usually show an increased penetration. The degree of conicity, the so-called k-factor [39], is defined by Bosch as

$$k = \frac{D_{inlet}\left[\mu m\right] - D_{exit}\left[\mu m\right]}{10} \ . \tag{2.21}$$

The last category are the conical and flow optimized geometries (ks nozzles, German: konisch strömungsoptimiert), where the reduction of cross-sectional area is dependent on the distribution of mass flow, and which are designed to completely suppress cavitation. For example, if most of the mass flow enters the hole from the upper side, the largest cross-sectional areas and the strongest rounding of the inlet are produced at the upper wall of the hole. The sprays generated by these nozzles are characterized by small spray cone angles, especially near the nozzle, and large penetration lengths. Detailed optical investigations of the effect of nozzle hole geometry on the primary spray structure are published by Schugger et al. [49] and König et al. [29], for example.

2.2.2 Gasoline Engines

In contrast to modern diesel engines, which are exclusively equipped with high-pressure direct injection systems, two main categories of injection techniques exist for gasoline engines. Injection systems belonging to the first category (port fuel injection) inject the fuel into the manifold, injection systems of the second category (direct injection) inject the fuel directly into the cylinder. Depending on the position of the injector, the first category is divided into the two sub-categories single-point injection and multi-point injection, Fig. 2.33.

2.2.2.1 Port Fuel Injection

In the case of single-point injection, a central injection unit replaces the well-known carburetor. The more or less well-prepared mixture of fuel vapor, fuel droplets, and air is delivered to the cylinders via the intake manifold. Depending on the operating point, a part of the mixture is deposited on the walls, leading to the formation of a liquid wall film, which flows at substantial lower velocities than the remaining mixture stream to the individual cylinders. Any change in load point results in a change of the wall film thickness. Consequently, the mixture varies constantly both in quality and quantity under transient operating conditions. In the case of single-point injection, the manifold has to distribute the centrally prepared mixture to the individual cylinders. Due to the complex geometry of the manifold and the many parameters determining the multiphase flow mixture along the relatively long way from the point of central mixture preparation to the individual cylinders, it is obvious that it is impossible to achieve a completely uniform distribution of the mixture.

A much more sophisticated system is the multi-point injection, in which one injector per cylinder is placed near the intake valve and is used in order to inject the desired amount of fuel at the desired time into the part of the manifold belonging to the individual cylinder, see Fig. 2.33. This avoids the problems related to the unequal fuel distribution of single-point injection systems. Hence, the major part of the induction system is only filled with air, the numerous problems of multiphase flow do not occur upstream of the injector, and the intake system can be optimized with regard to gas-dynamic supercharging effects. Downstream of the in-

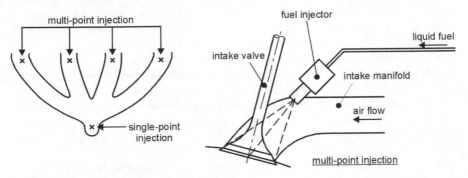

Fig. 2.33. Single-point and multi-point injection system

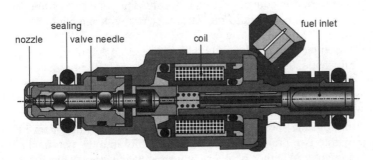

Fig. 2.34. Solenoid injector [34]

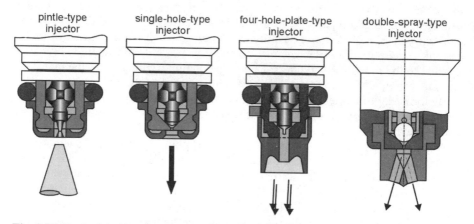

Fig. 2.35. Typical design of injectors, classified by the spray shape [34]

jector, the mixture formation process is similar to that of the single-point injection system. Immediately after the start of injection, droplet vaporization and wall film formation begin with varying intensity, depending on temperature and pressure boundary conditions. However, because of the small wettable surface area and the higher temperatures near the inlet valve, the amount of fuel stored in the liquid film is much smaller than in the case of the single-point injection, resulting in a much more accurate metering of the desired fuel quantity to the actual cycle, and thus in advantages regarding fuel consumption and emissions. Most port injection today occurs with the intake valve closed. This improves evaporation and mixture formation outside the cylinder and prevents large droplets from directly entering the combustion chamber. If large drops enter the cylinder, they may be deposited on the walls, due to the low pressure and density of the gas, and increase HC emissions.

The classical gasoline injector today is the solenoid injector. Despite the various nozzle types, the basic function is the same for all injectors, Fig. 2.34. While the injector is closed, a spring presses the needle on the seal seat. In order to start the injection, a current, sent by the control unit, is applied to the coil in the valve body and lifts an armature that is connected to the needle. The majority of injectors used these days are low- and medium-pressure injectors with injection pressures in the range of 0.5 to 1.5 MPa.

Injectors for port fuel injection differ mainly in the type of nozzle used to generate the desired spray shape, see Fig. 2.35. The pintle-type injector produces a cone-shaped spray, the single-hole injector a pencil-shaped spray, and the multi-hole plate injector produces multiple sprays according to the number of holes. An additional type of injector is the swirl injector, which has already been discussed in Sect. 2.1.3.2. Further on, the spray can be influenced by a front attachment in order to split the spray into two branches and to inject the fuel into the two intake ports of a four-valve engine. Whether a special injector design is suitable for a given port fuel application depends on many influence parameters, such as port geometry, injector position, injection timing (injection during closed or opened inlet valve), maximum gas temperatures, fuel quality (deposit formation at the nozzle tip), minimum gas pressure inside the manifold (prevention of flash-boiling), etc. More information about port fuel injection systems and nozzle types are included, for example, in the detailed review of Zhao et al. [63] and in Lenz [34].

2.2.2.2 Direct Injection

The direct injection spark ignition (DISI) engine promises significant advantages over the port fuel injection engine, especially in improving fuel consumption and reducing CO_2 emissions, see also Sect. 6.3. In order to provide highly dispersed sprays in the very short amount of time available for spray disintegration and mixture formation inside the cylinder, common-rail injection systems are used, and the injection pressures are significantly increased. Compared to diesel fuel, gasoline however does not serve well as lubricant. In order to avoid excessive pump wear, the maximum injection pressures are limited to approx. 15 MPa today.

Compared to the conventional diesel injection near top dead center, the direct injection of gasoline may occur already during the induction stroke in the case of homogeneous mixture formation (full load), or very late during the compression stroke in the case of stratified charge operation (part load), see also Sect. 6.3. Hence, the gas pressure inside the cylinder varies significantly, and fuel injectors are needed that are able to produce the desired spray quality. Fig. 2.36 shows three possible injector categories that are suitable for use in DISI engines.

The first kind of nozzle is the multi-hole nozzle. The functional principle as well as the shape and structure of the individual sprays produced by each hole are well-known from the high-pressure diesel injection. This nozzle produces a number of compact sprays with relatively large penetration (low gas density) and large droplet sizes (low injection pressure compared to diesel applications). Hence, the overall spray is strongly inhomogeneous, consisting of fuel-rich zones that are separated by very lean regions, Fig. 2.37. During combustion, such inhomogeneity of the mixture may result in different burning velocities as well as increased emissions of soot and nitric oxides. An advantage of the multi-hole injector is that the spray structure and the spray cone angle do not change with increasing backpressure, which is an important criterion for the realization of the so-called spray-guided DISI engine, see Sect. 6.3. Further on, the overall spray shape can be easily influenced by the position and number of the individual holes.

The other two nozzle categories, the outwardly opening nozzle and the inwardly opening pressure-swirl injector, produce a hollow-cone spray. The key advantage of hollow-cone sprays over solid-cone sprays is the high area-to-volume ratio, which leads to the required level of atomization without large penetration. The average droplet size is usually smaller than that of the multi-hole injector. Details of the spray structure have already been discussed in Sect. 2.1.3.2.

In a high-pressure swirl atomizer, the fuel passes through tangentially arranged swirl ports and gets a rotational motion inside the swirl chamber, Fig. 2.38. The centrifugal motion of the liquid forms a hollow air core. Because the area of the swirl chamber reduces to a nozzle, the rotational motion is further increased. The liquid passes through the nozzle and forms a free cone-shaped liquid sheet inside the combustion chamber, which thins because of the conservation of mass as it departs from the nozzle and subsequently disintegrates into droplets.

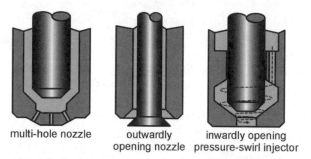

Fig. 2.36. Direct injection of gasoline: injector geometries [14]

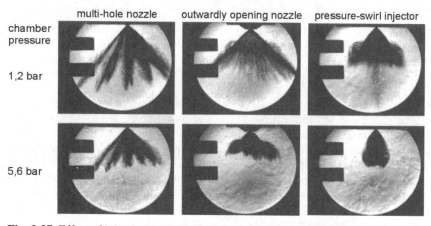

Fig. 2.37. Effect of injection pressure on the spray structure [26], p_{rail} = 10 MPa

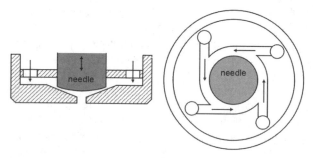

Fig .2.38. Example of a high-pressure swirl atomizer with tangentially arranged swirl ports

In contrast to the swirl atomizer, the hollow-cone spray of an outwardly opening nozzle does not collapse at higher backpressures and does not have any pre-spray, Fig. 2.37. This is an important advantage with regard to the mixture formation process in DISI engines, Sect. 6.3. However, the formation of deposits due to high temperatures at the needle tip and at the seat may be critical [14].

Today, the pressure-swirl injector is the most-used injector for the first generation of series production DISI engines. However, in order to realize the full benefit of direct injection and to minimize fuel consumption, the next generation of DISI engines needs sprays whose general structure and spray cone angle is independent of backpressure.

References

[1] Allocca L, De Vita A, Di Angelo L (2002) Wall-Impingement Analysis of a Spray from a Common Rail Injection System for Diesel Engines. THIESEL 2002 Conference on Thermo- and Fluid Dynamic Processes in Diesel Engines, pp. 67–76

[2] Arai M, Shimizu M, Hiroyasu H (1991) Similarity between the Breakup Lengths of a High Speed Liquid Jet in Atmospheric and Pressurized Conditions. Los Alamos National Laboratory

[3] Arcoumanis C, Badami M, Flora H, Gavaises M (2000) Cavitation in Real-Size Multi-Hole Diesel Injector Nozzles. SAE paper 2000-01-1249

[4] Arcoumanis C, Gavaises M, French B (1997) Effect of Fuel Injection Process on the Structure of Diesel Sprays. SAE paper 970799

[5] Badock C (1999) Untersuchungen zum Einfluß der Kavitation auf den primären Strahlzerfall bei der dieselmotorischen Einspritzung. Ph.D. Thesis, University of Darmstadt, Germany

[6] Bode J (1991) Zum Kavitationseinfluß auf den Zerfall von Flüssigkeitsstrahlen. Ph.D. Thesis, Max-Planck-Institut für Strömungsforschung, Göttingen, Germany

[7] Busch R (2001) Untersuchung von Kavitationsphänomenen in Dieseleinspritzdüsen. Ph.D. Thesis, University of Hannover, Germany

[8] Chehroudi B, Chen SH, Bracco FV, Onuma Y (1985) On the Intact Core of Full-Cone Sprays. SAE paper 850126

[9] Dan T, Takagishi S, Senda J, Fujimoto H (1997) Organized Structure and Motion on a Diesel Spray. SAE paper 970641

[10] Dent JC (1971) A Basis for the Comparison of Various Experimental Methods for Studying Spray Penetration. SAE paper 710571

[11] Eifler W (1990) Untersuchungen zur Struktur des instationären Dieselöleinspritz-strahles im Düsennahbereich mit der Methode der Hochfrequenz-Kinematographie. Ph.D. Thesis, University of Kaiserslautern, Germany

[12] Fath (1997) Charakterisierung des Strahlaufbruch-Prozesses bei der instationären Druckzerstäubung. Berichte zur Energie- und Verfahrenstechnik (BEV), Schriftenreihe Heft 97.3, Ph.D. Thesis, University of Erlangen-Nürnberg, Germany

[13] Fujimoto H, Sugihara H, Tanabe H, Sato GT (1981) Investigation on Combustion in Medium-Speed Marine Diesel Engines Using Model Chambers. CIMAC-Congress 1981, Helsinki

[14] Gindele J (2001) Untersuchung zur Ladungsbewegung und Gemischbildung im Otto-motor mit Direkteinspritzung. Ph.D. Thesis, University of Karlsruhe, Germany, Logos-Verlag, Berlin, ISBN 3-89722-727-4

[15] Gülder ÖL, Smallwood GJ, Snelling DR (1992) Diesel Spray Structure Investigation by Laser Diffraction and Sheet Illumination. SAE paper 920577

[16] Gülder ÖL, Smallwood GJ, Snelling DR (1994) Internal Structure of Transient Full-Cone Dense Diesel Sprays. Commodia 94, pp 355–360

[17] He L, Ruiz F (1995) Effect of Cavitation on Flow and Turbulence in Plain Orifices for High-Speed Atomization. Atomization and Sprays, vol 5 pp 569–584

[18] Heimgärtner C, Leipertz A (2000) Investigation of Primary Spray Breakup Close to the Nozzle of a Common Rail High Pressure Injection System. Eight International Confer-ence on Liquid Atomization and Spray Systems, Pasadena, CA, USA

[19] Heywood JB (1988) Internal Combustion Engine Fundamentals. McGraw-Hill Book Company

[20] Hiroyasu H, Arai M (1990) Structures of Fuel Sprays in Diesel Engines. SAE-paper 900475

[21] Hiroyasu H, Arai M, Tabata M (1989) Empirical Equations for the Sauter Mean Di-ameter of a Diesel Spray. SAE paper 890464

[22] Hiroyasu H, Shimizu M, Arai M (1991) Breakup Length of a Liquid Jet and Internal Flow in a Nozzle. ICLASS-91

[23] Hochgreb S, VanDer Wege B (1998) The Effect of Fuel Volatility on Early Spray De-velopment from High Pressure Swirl Injectors. In Direkteinspritzung im Ottomotor, Editor: Spicher U, Expert-Verlag, Rennigen-Malmsheim

[24] Hochgreb S, VanDer Wege B (1999) Investigation of the Effect of Fuel Volatility on Operating Conditions on DISI Sprays. In Direkteinspritzung im Ottomotor II, Editor: Spicher U, Expert-Verlag, Rennigen-Malmsheim

[25] Homburg A (2002) Optische Untersuchungen zur Strahlausbreitung und Gemis-chbildung beim DI-Benzin-Brennverfahren. Ph.D. Thesis, University of Braun-schweig, Germany

[26] Hübel M, Günther H, Ortmann R, Stein J, Yildirim F (2001) Einspritzventile für die Benzin-Direkteinspritzung – ein systematischer Vergleich verschiedener Aktor-konzepte. Wiener Motorensymposium 2001

[27] Hwang SS, Liu Z, Reitz RD (1996) Breakup Mechanisms and Drag Coefficients of High-Speed Vaporizing Liquid Drops. Atomization and Sprays, vol 6, pp 353–376

[28] Iida H, Matsumura E, Tanaka K, Senda J, Fujimoto H, Maly RR (2000) Effect of In-ternal Flow in a Simulated Diesel Injection Nozzle on Spray Atomization. ICLASS 2000, Pasadena, CA, USA

[29] König G, Blessing M, Krüger C, Michels U, Schwarz V (2002) Analysis of Flow and Cavitation Phenomena in Diesel Injection Nozzles and its Effects on Spray and Mix-ture Formation. 5. Int Symp für Verbrennungsdiagnistik der AVL Deutschland, Baden-Baden

[30] Krzeczkowski SA (1980) Measurement of Liquid Droplet Disintegration Mechanisms. Int J Multiphase Flow, vol 6, pp 227–239

[31] Kuensberg Sarre C, Kong SC, Reitz RD (1999) Modelling the Effects of Injector Noz-zle Geometry on Diesel Sprays. SAE paper 1999-01-0912

[32] Lai MC, Yoo JH, Kim SK (1997) High Pressure Gasoline Spray Structure and ist Im-plications to In-Cylinder Mixture Formation. Tagung Direkteinspritzung im Ottomo-tor, Haus der Technik, Essen, Germany

[33] Lefebvre AH (1989) Atomization and Sprays. Hemisphere Publishing Corporation, New York, Washington, Philadelphia, London

[34] Lenz HP (1992) Mixture Formation in Spark-Ignition Engines. Springer-Verlag Wien, New York, ISBN 3-211-82331-X (Springer) and ISBN 1-56091-188-3 (SAE)

[35] Mohammadi A, Kidoguchi Y, Miwa K (2002) Effect of Injection Parameters and Wall-Impingement on Atomization and Gas Entrainment Processes in Diesel Sprays. SAE paper 2002-01-0497

[36] Nayfey AH (1968) Nonlinear Stability of a Liquid Jet. Physics of Fluids, vol 13, no 4, pp 841–847

[37] Ohnesorge W (1931) Die Bildung von Tropfen an Düsen und die Auflösung flüssiger Strahlen. Zeitschrift für angewandte Mathematik und Mechanik, Bd.16, Heft 6, pp 355–358

[38] Ohyama Y, Ohsuga M, Nogi T, Fujieda M Shirashi T (1998) Mixture Formation in Gasoline Direct Injection Engines. In Direkteinspritzung im Ottomotor, Editor: Spicher U, Expert-Verlag, Rennigen-Malmsheim

[39] Potz D, Christ W, Dittus B (2000) Diesel Nozzle – the Determining Interface Between Injection System and Combustion Chamber. THIESEL 2000, pp 249–258

[40] Preussner C, Döring C, Fehler S, Kampmann S (1998) GDI: Interaction Between Mixture Preparation, Combustion System and Injector Performance. SAE paper 980498

[41] Prosperetti A, Lezzi A (1986) Bubble Dynamics in a Compressible Liquid. Part 1. First-Order Theory. J Fluid Mech, vol 168, pp 457–478

[42] Rayleigh Lord FRS (1878) On the Stability of Liquid Jets. Proc. of the Royal Society London

[43] Reitz RD (1978) Atomization and other Breakup Regimes of a Liquid Jet. Ph.D. Thesis, Princeton University

[44] Reitz RD, Bracco FV (1986) Mechanisms of Breakup of Round Liquid Jets. Enzyclopedia of Fluid Mechanics, Gulf Pub, NJ, 3, pp 233–249

[45] Reitz RD, Diwakar R (1987) Structure of High-Pressure Fuel Sprays. SAE paper 870598

[46] Robert Bosch GmbH (1999) Diesel Einspritzsysteme – Unit Injector System / Unit Pump System. Technische Unterrichtung

[47] Robert Bosch GmbH (1999) Diesel-Speichereinspritzsystem Common Rail. ISBN 3-934584-13-6

[48] Rutland DF Jameson GJ (1970) Theoretical Prediction of the Size of Drops Formed in the Breakup of Capillary Jets. Chem Eng Science, vol 25, p 1689

[49] Schugger C, Renz (2003) Experimental Investigations on the Primary Breakup Zone of High Pressure Diesel Sprays from Multi-Orifice Nozzles. ICLASS 2003

[50] Schugger C, Renz U (2001) Experimental Investigations on the Primary Breakup of High Pressure Diesel Sprays. ILASS-Europe 2001

[51] Soteriou C, Andrews R, Smith M (1995) Direct Injection Diesel Sprays and the Effect of Cavitation and Hydraulic Flip on Atomization. SAER paper 950080

[52] Soteriou C, Andrews R, Torres N, Smith M, Kunkulagunta R (2001) Through the Diesel Nozzle – a Journey of Discovery II. ILASS-Europe 2001, Zürich

[53] Stegemann J, Seebode J, Baltes J, Baumgarten C, Merker GP (2002) Influence of Throttle Effects at the Needle Seat on the Spray Characteristics of a Multihole Injection Nozzle. ILASS-Europe 2002, Zaragoza, Spain

[54] Su TF, Farrell PV, Nagarajan RT (1995) Nozzle Effect on High Pressure Diesel Injection. SAE paper 950083

[55] Tamaki N, Shimizu M, Hiroyasu H (2000) Enhanced Atomization of a Liquid Jet by Cavitation in a Nozzle Hole. 8th Int Conf on Liquid Atomization and Spray Systems, Pasadena, CA, USA

[56] Torda TP (1973) Evaporation of Drops and Breakup of Sprays. Astronautica Acta, vol 18, p 383

[57] Urlaub AG, Chmela FG (1974) High-Speed Multifuel Engine: L9204 FMV, SAE paper 740122

[58] van Basshuysen R, Schäfer F (2002) Handbuch Verbrennungsmotor. ISBN 3-528-03933-7, Vieweg-Verlag, Braunschweig, Wiesbaden

[59] Wierzba A (1993) Deformation and Breakup of Liquid Drops in a Gas Stream at Nearly Critical Weber Numbers. Experiments in Fluids, vol 9, pp 59–64

[60] Wu PK, Faeth GM (1995) Onset and End of Drop Formation along the Surface of Turbulent Liquid Jets in Still Gases. Phys. of Fluids 7 (11)

[61] Youle AJ, Salters DG (1994) A Conductivity Probe Technique for Investigating the Breakup of Diesel Sprays. Atomization and Sprays, vol 4, pp 253–262

[62] Yuen MC (1968) Non-linear Capillary Instability of a Liquid Jet. J Fluid Mech, vol 33, p 151

[63] Zhao FQ, Lai MC, Harrington D (1995) The Spray Characteristics of Automotive Port Fuel Injection – A Critical Review. SAE paper 950506

3 Basic Equations

3.1 Description of the Continuous Phase

3.1.1 Eulerian Description and Material Derivate

In this section, the basic equations for the description of multi-dimensional flow fields will be derived. In internal combustion engines, such flows are the airflow inside the induction system, the gas flow inside the cylinder, and the flow of burnt gases through the exhaust system. The flow of fuel through the three-dimensional injection nozzle geometry is an example of a liquid flow field.

In solid rigid body mechanics, the position in space of a special particle as function of time t is usually the quantity of interest, and from this information all the other questions, such as the amount of velocity and acceleration, may be answered. Hence, all flow quantities F are given as function of particle and time,

$$F = F(particle, t), \qquad (3.1)$$

which is called a Lagrangian description. If the vector $\bar{x} = \bar{x}\,(particle, t)$ denotes the position of the particular particle, velocity and acceleration are simply given by $\bar{u} = d\bar{x}\,(particle,\ t)/dt$ and $\bar{a} = d^2\bar{x}\,(particle,\ t)/dt^2$. In order to describe the spatial and temporal development of pressure, density, temperature, magnitude, and direction of velocity etc. in a complete flow field using the Lagrangian approach, the position, pressure, density, temperature and velocity of every liquid element inside the flow field have to be calculated. Then, if for example the temporal development of pressure at a fixed point inside the flow field is required, the pressures of the liquid elements that passed this point during the time span of interest have to be listed in the correct order. It is obvious that the Lagrangian approach is well suited for the description of disperse phases (e.g. sprays consisting of liquid droplets), but not for that of continuous fluids.

In a continuum, the flow quantities change continuously in space. In contrast to a dispersed phase, in which the density of the fluid of interest is zero in the volume between two fluid elements, and for which the exact position and size of the elements must be known in order to describe the spatial and temporal distribution of the relevant flow quantities, it is not necessary to distinguish between different fluid elements in the case of a continuum. It is more convenient to describe the flow quantities as a function of a point in space (related to a fixed coordinate system) and time, regardless of what element of liquid happens to be there at any particular time:

$$F = F\left(\vec{x}, t\right). \tag{3.2}$$

In order to describe a complete flow field, this approach is applied to all points $\vec{x}$ in space. Such a description is called a Eulerian description.

In differential equations, often the time derivate is used. Using the Lagrangian description, the time derivate of F is

$$\frac{dF\left(particle,t\right)}{dt} = \frac{DF}{Dt}. \tag{3.3}$$

This so-called substantial or material derivate describes the complete derivate of the property F of a liquid element. Remember that all flow quantities are only a function of particle and time. In contrast to this, the Eulerian approach describes the flow quantities F as function of point $\vec{x} = (x_1, x_2, x_3)$ in space and time t. Using the chain rule, the quantity dF is

$$dF\left(t, x_1, x_2, x_3\right) = \frac{\partial F}{\partial t} dt + \frac{\partial F}{\partial x_1} dx_1 + \frac{\partial F}{\partial x_2} dx_2 + \frac{\partial F}{\partial x_3} dx_3. \tag{3.4}$$

The time derivate dF/dt reads

$$\frac{dF\left(t, x_1, x_2, x_3\right)}{dt} = \frac{\partial F}{\partial t} \frac{dt}{dt} + \frac{\partial F}{\partial x_1} \frac{dx_1}{dt} + \frac{\partial F}{\partial x_2} \frac{dx_2}{dt} + \frac{\partial F}{\partial x_3} \frac{dx_3}{dt}. \tag{3.5}$$

In order to emphasize that the material derivate is meant, the expression DF/dt is usually used instead of dF/dt. Because $dx_i/dt = u_i$, the material derivate now reads

$$\frac{DF}{Dt} = \frac{\partial F}{\partial t} + u_1 \frac{\partial F}{\partial x_1} + u_2 \frac{\partial F}{\partial x_2} + u_3 \frac{\partial F}{\partial x_3}, \tag{3.6}$$

where $\vec{u} = (u_1, u_2, u_3)$ is the velocity vector. The Lagrangian and the Eulerian approaches, if applied to a special problem, must have the same overall result. For example, the resulting acceleration of a liquid element passing any point in space (Lagrangian description) must be identical to the resulting acceleration at this point predicted by the Eulerian approach. In order to explain the different terms in Eq. 3.6, the flow through a converging nozzle shall be regarded. Due to the reduction of cross-sectional area, the flow velocity is larger at an upstream position (point A) than at a downstream position (point B). The first partial derivate on the right hand side of Eq. 3.6 is the local derivate. It describes the temporal change of F at a fixed point in space and is equal to zero at points A and B if the flow is steady. However, the flow velocity at point B is always larger than at point A, and a liquid element passing point A will experience an additional acceleration which is not included in the local derivate. This additional acceleration is expressed by the remaining three terms on the right hand side of Eq. 3.6, the so-called convective derivate. Hence, the Eulerian description enables us to distinguish between a local and a convective part of the material derivate, while the Lagrangian description does not.

Eq. 3.6 can be expressed in a much shorter form,

$$\frac{DF}{Dt} = \frac{\partial F}{\partial t} + u_i \frac{\partial F}{\partial x_i},$$

(3.7)

using the so-called index notation (Einstein notation). This convention states that whenever the same index appears twice in a term (here: i), summation over the range of that index (here: $i = 1, 2, 3$) is implied.

If $\vec{F}$ is a vector field, e.g. a velocity field, the equation for every component (coordinate direction $j = 1, 2, 3$, Cartesian coordinate system) reads

$$\frac{DF_j}{Dt} = \frac{\partial F_j}{\partial t} + u_i \frac{\partial F_j}{\partial x_i},$$

(3.8)

while the more generalized vector form valid in any coordinate system is

$$\frac{D\vec{F}}{Dt} = \frac{\partial \vec{F}}{\partial t} + (\vec{u} \cdot \nabla) \vec{F}.$$

(3.9)

3.1.2 Conservation Equations for One-Dimensional Flows

Multi-dimensional flows can often be treated in a simplified manner as one-dimensional. Then, the flow quantities only change in the main flow direction. In this case, it is convenient to apply the conservation equation in integral form, which means that the basic fluid equations are applied to a control volume including the complete area of interest. In contrast to this, the conservation equations in differential form are obtained if the basic fluid equations are applied to an infinitesimal elemental volume. The differential equations, which are derived in Sect. 3.1.3, are usually appropriate when the distributive conditions (three-dimensional flow fields) are desired, e.g. the velocity, temperature, and pressure fields inside a combustion chamber. In the following, the integral conservation equations are derived and simplified for one-dimensional flows.

3.1.2.1 Conservation of Mass

Consider a fixed collection of fluid particles as denoted by the dotted lines in Fig. 3.1. Such a collection of fluid particles (Lagrangian approach) is called a system. The system is shown at times t and $t+\Delta t$. According to the law stating that mass must be conserved, the mass of the system remains constant:

$$\frac{Dm_{sys}}{Dt} = 0.$$

(3.10)

However, the Eulerian description is usually used in fluid dynamics, and an expression for mass conservation using a control volume that is fixed in space and

passed by the flow without resistance must be found. Such a fixed control volume
(cv), also shown in Fig. 3.1, equals the volume of the system at time t. Using the
masses contained in the volumes 1, 2, and 3 at times t and $t + \Delta t$, the left hand side
of Eq. 3.10 becomes

$$\frac{Dm_{sys}}{Dt} = \lim_{\Delta t \to 0} \frac{m_3\left(t + \Delta t\right) + m_2\left(t + \Delta t\right) - m_2\left(t\right) - m_1\left(t\right)}{\Delta t}$$

$$= \lim_{\Delta t \to 0} \frac{m_2\left(t + \Delta t\right) + m_1\left(t + \Delta t\right) - m_2\left(t\right) - m_1\left(t\right)}{\Delta t}$$

$$+ \lim_{\Delta t \to 0} \frac{m_3\left(t + \Delta t\right) - m_1\left(t + \Delta t\right)}{\Delta t}$$

$$= \frac{\partial m_{cv}}{\partial t} + \lim_{\Delta t \to 0} \frac{m_3\left(t + \Delta t\right) - m_1\left(t + \Delta t\right)}{\Delta t} . \tag{3.11}$$

In order to find expressions for the masses m_3 $(t+\Delta t)$ and m_1 $(t+\Delta t)$ contained in the
volumes 1 and 3, the volumes V_1 and V_3 are needed. These volumes can be calculated from

$$dV_1 = \vec{n} \cdot \vec{u} \Delta t dA_1 \quad \text{and} \quad dV_3 = \vec{n} \cdot \vec{u} \Delta t dA_3 , \tag{3.12}$$

Fig. 3.1, and the desired expressions can be written as

$$m_3\left(t + \Delta t\right) = \int_{A_3} \rho|\vec{u}|\cos\alpha \Delta t dA_3 = \int_{A_3} \rho\vec{n} \cdot \vec{u} \Delta t dA_3 ,$$

$$m_1\left(t + \Delta t\right) = \int_{A_1} \rho|\vec{u}|\cos\alpha \Delta t dA_1 = -\int_{A_1} \rho\vec{n} \cdot \vec{u} \Delta t dA_1 . \tag{3.13}$$

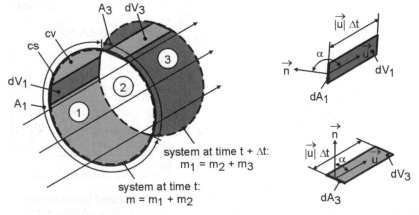

Fig. 3.1. System-to-control-volume transformation: system moving through a control volume, two-dimensional flow and control volume

Remember that the unit vector $\bar{n}$ is normal to the surface element dA and always points out of the control volume (cv). Because A_3 plus A_1 completely surrounds the control volume ($A_3 + A_1 = cv$), Eq. 3.11 is equivalent to

$$\frac{Dm_{sys}}{Dt} = \frac{\partial m_{cv}}{\partial t} + \int_{cs} \rho\bar{n}\cdot\bar{u}dA = \frac{\partial}{\partial t}\int_{cv}\rho dV + \int_{cs}\rho\bar{n}\cdot\bar{u}dA. \tag{3.14}$$

The left hand side of this equation describes the rate of change of mass in a Lagrangian frame of reference, while the right hand side represents the respective Eulerian description for a fixed control volume. Hence, Eq. 3.14 is the desired system-to-control-volume transformation.

A more general expression for such a transformation is given by the Reynolds transport theorem,

$$\frac{D}{Dt}\left(N_{sys}\right) = \frac{\partial}{\partial t}\int_{cv}\eta\rho dV + \int_{cs}\eta\rho\bar{n}\cdot\bar{u}dA, \tag{3.15}$$

where N_{sys} is the extensive quantity (e.g. mass, energy, momentum) and η is the intensive property associated with N_{sys}, the property per unit mass. The time derivate of the first term on the right hand side can be moved inside the integral, since for a fixed control volume the limits on the volume integral are independent of time.

Combining Eqs. 3.10 and 3.14 results in the continuity equation in integral form:

$$\frac{\partial}{\partial t}\int_{cv}\rho dV = -\int_{cs}\rho\bar{n}\cdot\bar{u}dA. \tag{3.16}$$

For steady flow, the left hand side is equal to zero. Assuming that the velocity is also normal to all surfaces where fluid crosses (e.g. pipe flow, Fig. 3.2) results in

$$\int_{A_2}\rho_2 u_2 dA - \int_{A_1}\rho_1 u_1 dA = 0, \tag{3.17}$$

where $\left|\bar{u}_i\right| = u_i$.

If the densities and velocities are uniform over their respective areas, the flow can be treated as one-dimensional, and the continuity equation finally reads

$$\rho_2 u_2 A_2 = \rho_1 u_1 A_1. \tag{3.18}$$

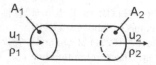

Fig. 3.2. Conservation of mass: one-dimensional steady pipe flow

3.1.2.2 Conservation of Momentum

Newton's second law, the momentum equation, states that the sum of all external forces acting on a system equals the rate of change of momentum of the system:

$$\frac{D\vec{M}}{Dt} = \sum \vec{F} . \tag{3.19}$$

In terms of a control volume, Eq. 3.19 becomes

$$\frac{\partial}{\partial t} \int_{cv} \rho \vec{u} dV + \int_{cs} \rho \vec{u} (\vec{n} \cdot \vec{u}) dA = \sum \vec{F} \tag{3.20}$$

by applying Eq. 3.15, where $N_{sys} = \vec{M} = m \cdot \vec{u}$, $\eta = \vec{u}$. The external forces are all forces acting on the surfaces of the control volume (pressure, shear, additional surface forces from a solid wall) and body forces (e.g. gravity) acting on the mass inside the control volume.

Eq. 3.20 can be simplified significantly if the control volume has entrances and exits across which the flow quantities may be assumed to be uniform, and if the flow is steady:

$$\sum_{i=1}^{N} \left(\rho \vec{u} (\vec{n} \cdot \vec{u}) A \right)_i = \sum \vec{F} , \tag{3.21}$$

where N is the number of exit and entrance areas. If there is only one entrance and one exit, and if the velocity vector is normal to the entrance and exit areas, the momentum equation reduces to

$$\rho_2 A_2 u_2 \vec{u}_2 - \rho_1 A_1 u_1 \vec{u}_1 = \sum \vec{F} . \tag{3.22}$$

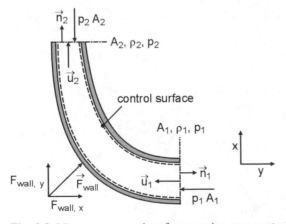

Fig. 3.3. Momentum equation: forces acting on a control volume

At the entrance $\vec{u} \cdot \vec{n} = -u$, since $\vec{n}$ and $\vec{u}$ point in opposite directions. Note that the momentum equation is a vector equation and consists of one equation for each coordinate direction. If only external forces caused by static pressure and the wall are considered, Fig. 3.3, Eq. 3.22 reads

$$x - direction: \quad \rho_1 A_1 u_1^2 = F_{wall,x} - p_1 A_1$$

$$y - direction: \quad \rho_2 A_2 u_2^2 = F_{wall,y} - p_2 A_2 .$$

$$(3.23)$$

3.1.2.3 Conservation of Energy

Many problems involving fluid motion such as compressible and non-isothermal flows require the use of the energy equation. The energy equation for a system reads

$$\frac{DE_t}{Dt} = \dot{Q} + \dot{W} ,$$

$$(3.24)$$

where dQ/dt is the amount of heat per unit time and dW/dt is the work per unit time transferred to the system. E_t is the total energy consisting of internal energy $\rho e V$ (V: volume of the system, e: internal energy per unit mass), kinetic energy $V\rho u^2/2$, and potential energy $-V\rho \vec{g} \cdot \vec{x}$ ($\vec{x}$: position of system at time t, $\vec{g}$: vector of gravitational acceleration). In terms of a control volume, Eq. 3.24 becomes

$$\frac{\partial}{\partial t} \int_{cv} e_t \rho dV + \int_{cs} \rho e_t \vec{u} \cdot \vec{n} dA = \dot{Q} + \dot{W}$$

$$(3.25)$$

by applying Eq. 3.15, where $N_{sys} = E_t$, $\eta = e_t$, and

$$e_t = \frac{E_t}{m} = e + \frac{|\vec{u}|^2}{2} - \vec{g} \cdot \vec{x} = e + \frac{u^2}{2} - \vec{g} \cdot \vec{x} .$$

$$(3.26)$$

The work-rate term includes the work on the boundaries due to pressure and tangential stresses plus the work added by source terms (e.g. shaft work) minus the reduction due to energy dissipation. The heat-rate term includes the rate-of-energy transfer across the control face due to a temperature difference and the energy production or reduction inside the control volume due to energy sources or sinks such as chemical reactions and radiation.

An important special case of the energy equation occurs in one-dimensional steady flow, see Fig. 3.4. The shear work done at the boundary is zero because the velocity is either zero or normal to the shear force. The work done by pressure,

$$\dot{W}_p = -\int_{cs} p \, \vec{u} \cdot \vec{n} dA ,$$

$$(3.27)$$

is only performed at the inflow and outflow boundaries where the flow velocity is not zero. If dissipation and chemical reactions are neglected, Eq. 3.25 reads

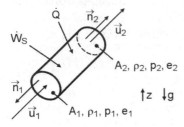

Fig. 3.4. One-dimensional pipe flow with heat and work

$$\dot{Q} + \dot{W}_s = \int\limits_{A_{in}+A_{out}} \rho\left(\frac{p}{\rho} + e + \frac{u^2}{2} + gz\right)\vec{u}\cdot\vec{n}dA .$$ (3.28)

Since the flow is one-dimensional, p, ρ, u, and e are uniform over A_1 and A_2. Further on, the velocity is also normal to all surfaces where fluid crosses, resulting in

$$\dot{Q} + \dot{W}_s = \left(\frac{p_2}{\rho_2} + e_2 + \frac{u_2^2}{2} + gz_2\right)\rho_2 u_2 A_2 - \left(\frac{p_1}{\rho_1} + e_1 + \frac{u_1^2}{2} + gz_1\right)\rho_1 u_1 A_1 .$$ (3.29)

Using the continuity equation $\rho_1 u_1 A_1 = \rho_2 u_2 A_2 = \dot{m}$ for one-dimensional steady flow, we obtain

$$\frac{\dot{Q}}{\dot{m}} + \frac{\dot{W}_s}{\dot{m}} = \left(\frac{p_2}{\rho_2} + e_2 + \frac{u_2^2}{2} + gz_2\right) - \left(\frac{p_1}{\rho_1} + e_1 + \frac{u_1^2}{2} + gz_1\right) .$$ (3.30)

Introducing the enthalpy $h = e + p/\rho$ yields

$$\frac{Q}{m} + \frac{W_s}{m} = \left(h_2 + \frac{u_2^2}{2} + gz_2\right) - \left(h_1 + \frac{u_1^2}{2} + gz_1\right) .$$ (3.31)

If the energy transfer caused by the left hand side of Eq. 3.30 is zero and the internal energy of the fluid does not change (isothermal flow), the energy equation becomes equal to the Bernoulli equation (Sect. 3.1.3.2):

$$\frac{p_2}{\rho_2} + \frac{u_2^2}{2} + gz_2 = \frac{p_1}{\rho_1} + \frac{u_1^2}{2} + gz_1 .$$ (3.32)

3.1.3 Conservation Equations for Multi-Dimensional Flows

In order to describe three-dimensional compressible and non-isothermal flows, the differential conservation equations of mass (one equation), momentum (three equations), and energy (one equation) are used to calculate the three unknown velocity components u_1, u_2, u_3 as well as the two thermodynamic variables pressure p and temperature T. The remaining thermodynamic variables and transport properties density ρ, enthalpy h (or internal energy e), dynamic viscosity μ, and heat conductivity λ, which appear in the final forms of the conservation equations, are given by the state relations $\rho = \rho\,(p,\,T)$, $h = h\,(p,\,T)$, $\mu = \mu\,(p,\,T)$ and $\lambda = \lambda\,(p,\,T)$, which must be known in order to complete the system of equations. In the following, the differential conservation equations will be derived.

3.1.3.1 Conservation of Mass

The continuity equation in differential form can be derived from a mass balance at an infinitesimal volume element $dV = dx_1 dx_2 dx_3$, as shown in Fig. 3.5. Due to the Eulerian approach, the coordinate system as well as the control volume are fixed in space. Thus, the volume element is solely a control volume that is passed by the flow without resistance, and that is used only in order to quantify the amount of mass inside the volume at a given time t. The mass will increase if the inflow exceeds the outflow and will decrease in the opposite case. In both cases, the mass density inside the control volume will change. The total mass balance yields

$$\frac{\partial}{\partial t}\left(dx_1 dx_2 dx_3 \rho\right) = d\dot{m}_{x_1} + d\dot{m}_{x_2} + d\dot{m}_{x_3}\,. \tag{3.33}$$

The mass balance in the x_1-direction gives

$$d\dot{m}_{x_1} = \left(\dot{m}_{x_1}\right)_{x_1} - \left(\dot{m}_{x_1}\right)_{x_1 + dx_1} = dx_2 dx_3 \left(\rho u_1\right)_{x_1} - dx_2 dx_3 \left(\rho u_1\right)_{x_1 + dx_1}\,. \tag{3.34}$$

Using a Taylor series for the second term on the right hand side,

$$\left(\rho u_1\right)_{x_1 + dx_1} = \left(\rho u_1\right)_{x_1} + \frac{\partial}{\partial x_1}\left(\rho u_1\right)_{x_1} dx_1\,, \tag{3.35}$$

the mass balance in the x_1-direction finally reads

$$d\dot{m}_{x_1} = -dx_2 dx_3 \frac{\partial}{\partial x_1}\left(\rho u_1\right)_{x_1} dx_1\,. \tag{3.36}$$

The other coordinate directions are treated accordingly. Because the mass balances hold true for any position (x_1, x_2, x_3) of the control volume inside the three-dimensional flow field, the indices are omitted in the following. Using the mass balances in Eq. 3.33 and eliminating the constant volume $dx_1 dx_2 dx_3$ finally yields the general form of the continuity equation

$$\frac{\partial \rho}{\partial t} + \frac{\partial}{\partial x_1}\left(\rho u_1\right) + \frac{\partial}{\partial x_2}\left(\rho u_2\right) + \frac{\partial}{\partial x_3}\left(\rho u_3\right) = 0\,, \tag{3.37}$$

which can also be written as

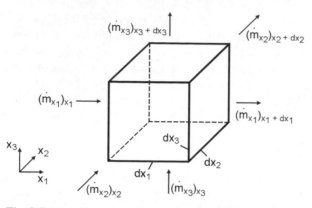

Fig. 3.5. Mass fluxes entering and exiting the control volume

$$\frac{\partial \rho}{\partial t} + \text{div}\left(\rho \vec{u}\right) = 0 ,\tag{3.38}$$

or as

$$\frac{\partial \rho}{\partial t} + \frac{\partial}{\partial x_i}\left(\rho u_i\right) = 0 ,\tag{3.39}$$

if the Einstein notation is used. If the product rule is applied to the second term in Eq. 3.39, the continuity equation can be transformed into

$$\frac{D\rho}{Dt} + \rho \frac{\partial u_i}{\partial x_i} = 0 ,\tag{3.40}$$

where $D\rho/Dt$ is the substantial derivate of ρ.

If the fluid density is constant (incompressible flow), the continuity equation reduces to

$$\nabla \cdot \vec{u} = \text{div}\left(\vec{u}\right) = \frac{\partial u_i}{\partial x_i} = 0 .\tag{3.41}$$

3.1.3.2 Conservation of Momentum

The momentum conservation equations can be derived in a manner similar to the mass conservation equation. Fig. 3.6 shows the corresponding infinitesimal control volume, which is again fixed in space and passed by the flow without resistance. The momentum equation is based on the principle that the temporal change

$$\frac{\partial \vec{M}}{\partial t} = \frac{\partial}{\partial t}\left(\rho \vec{u}\right) dx_1 dx_2 dx_3 \tag{3.42}$$

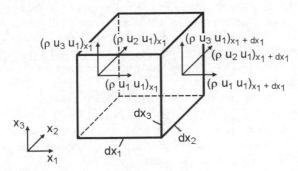

Fig. 3.6. Momentum fluxes entering and exiting the control volume

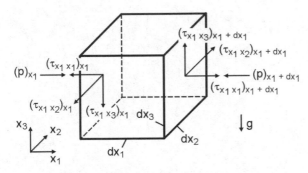

Fig. 3.7. Normal and shear stresses

of the momentum of the fluid inside the control volume at time t equals the sum of all momentum fluxes entering the volume (due to mass entering the volume) minus the sum of all momentum fluxes exiting the control volume, plus the sum of all external forces, which are forces due to static pressure, friction inside the fluid, and gravitation (body forces).

First of all, the momentum fluxes due to mass transport shall be derived. As an example, Fig 3.6 shows the momentum fluxes entering and leaving the control volume through the two faces that are perpendicular to the x_1-axis. Because the mass flows through both of these faces have velocity components in all three coordinate directions, momentum also having components in all three directions is transported. This becomes obvious regarding the mass flow entering the control volume through the left face: due to the velocity components u_1, u_2, and u_3, the mass flow transports momentum ρu_1, ρu_2, and ρu_3, which enters the control volume with an effective velocity (perpendicular to the surface) u_1. For this reason, the momentum fluxes at the left face in Figure 3.6 are $dx_2 dx_3 \rho u_1 u_1$, $dx_2 dx_3 \rho u_2 u_1$, and $dx_2 dx_3 \rho u_3 u_1$. The momentum fluxes through the remaining five faces are treated accordingly.

Next, the normal and shear stresses, as shown in Fig. 3.7, have to be considered. The meaning of the indices of the tensions is as follows: the first index characterizes the face of the control volume. For example, if the unit vector, which is normal to the surface, is in x_1-direction, the first index is x_1. The second index indicates the direction of the force caused by the tension. The force due to static pressure p is always directed into the control volume.

Finally, the body force $\vec{f} = (f_1, f_2, f_3)$ must be considered. This is usually the gravitational force, whose components are $f_1 = 0$, $f_2 = 0$, $f_3 = -\rho g$, see Fig. 3.7.

In contrast to the continuity equation, the momentum is not a scalar but a vector. For this reason, one resulting equation for each of the Cartesian dimensions x_1, x_2, and x_3 has to be written. For the x_1-dimension, the equation reads

$$
\begin{aligned}
\frac{\partial}{\partial t}(\rho u_1) dx_1 dx_2 dx_3 =& \left(\rho u_1 u_1 - \left(\rho u_1 u_1 + \frac{\partial(\rho u_1 u_1)}{\partial x_1} dx_1 \right) \right) dx_2 dx_3 \\
&+ \left(\rho u_1 u_2 - \left(\rho u_1 u_2 + \frac{\partial(\rho u_1 u_2)}{\partial x_2} dx_2 \right) \right) dx_1 dx_3 \\
&+ \left(\rho u_1 u_3 - \left(\rho u_1 u_3 + \frac{\partial(\rho u_1 u_3)}{\partial x_3} dx_3 \right) \right) dx_1 dx_2 \\
&+ \left(-\tau_{x_1 x_1} + \left(\tau_{x_1 x_1} + \frac{\partial \tau_{x_1 x_1}}{\partial x_1} dx_1 \right) \right) dx_2 dx_3 \\
&+ \left(-\tau_{x_2 x_1} + \left(\tau_{x_2 x_1} + \frac{\partial \tau_{x_2 x_1}}{\partial x_2} dx_2 \right) \right) dx_1 dx_3 \\
&+ \left(-\tau_{x_3 x_1} + \left(\tau_{x_3 x_1} + \frac{\partial \tau_{x_3 x_1}}{\partial x_3} dx_3 \right) \right) dx_1 dx_2 \\
&+ \left(p_{x_1} - \left(p_{x_1} + \frac{\partial p_{x_1}}{\partial x_1} dx_1 \right) \right) dx_2 dx_3 + dx_1 dx_2 dx_3 f_1.
\end{aligned}
$$

(3.43)

The static pressure p has the same value in all three coordinate directions, and the indices can be omitted. Eq. 3.43 can be simplified:

$$
\frac{\partial(\rho u_1)}{\partial t} + \frac{\partial(\rho u_1 u_1)}{\partial x_1} + \frac{\partial(\rho u_1 u_2)}{\partial x_2} + \frac{\partial(\rho u_1 u_3)}{\partial x_3}
$$

$$
= -\frac{\partial p}{\partial x_1} + \frac{\partial \tau_{x_1 x_1}}{\partial x_1} + \frac{\partial \tau_{x_2 x_1}}{\partial x_2} + \frac{\partial \tau_{x_3 x_1}}{\partial x_3} + f_1.
$$

(3.44)

Applying the product rule to the left hand side of Eq. 3.44 yields:

$$\frac{\partial(\rho u_1)}{\partial t} + \frac{\partial(\rho u_1 u_1)}{\partial x_1} + \frac{\partial(\rho u_1 u_2)}{\partial x_2} + \frac{\partial(\rho u_1 u_3)}{\partial x_3} = \rho\frac{\partial u_1}{\partial t} + u_1\frac{\partial \rho}{\partial t}$$

$$+u_1\frac{\partial(\rho u_1)}{\partial x_1} + \rho u_1\frac{\partial u_1}{\partial x_1} + u_1\frac{\partial(\rho u_2)}{\partial x_2} + \rho u_2\frac{\partial u_1}{\partial x_2} + u_1\frac{\partial(\rho u_3)}{\partial x_3} + \rho u_3\frac{\partial u_3}{\partial x_3}$$

$$= \rho\left(\frac{\partial u_1}{\partial t} + u_1\frac{\partial u_1}{\partial x_1} + u_2\frac{\partial u_1}{\partial x_2} + u_3\frac{\partial u_1}{\partial x_3}\right)$$

$$+u_1\left(\frac{\partial \rho}{\partial t} + \frac{\partial(\rho u_1)}{\partial x_1} + \frac{\partial(\rho u_2)}{\partial x_2} + \frac{\partial(\rho u_3)}{\partial x_3}\right).$$

$$(3.45)$$

The last term of Eq. 3.45 contains the left hand side of the continuity equation and is equal to zero. Hence, Eq. 3.44 can be rewritten:

$$\rho\frac{Du_1}{Dt} = \rho\left(\frac{\partial u_1}{\partial t} + u_1\frac{\partial u_1}{\partial x_1} + u_2\frac{\partial u_1}{\partial x_2} + u_3\frac{\partial u_1}{\partial x_3}\right)$$

$$= -\frac{\partial p}{\partial x_1} + \frac{\partial \tau_{x_1 x_1}}{\partial x_1} + \frac{\partial \tau_{x_2 x_1}}{\partial x_2} + \frac{\partial \tau_{x_3 x_1}}{\partial x_3} + f_1.$$

$$(3.46)$$

The corresponding equations for the x_2- and the x_3-directions are

$$\rho\frac{Du_2}{Dt} = -\frac{\partial p}{\partial x_2} + \frac{\partial \tau_{x_1 x_2}}{\partial x_1} + \frac{\partial \tau_{x_2 x_2}}{\partial x_2} + \frac{\partial \tau_{x_3 x_2}}{\partial x_3} + f_2,$$

$$(3.47)$$

$$\rho\frac{Du_3}{Dt} = -\frac{\partial p}{\partial x_3} + \frac{\partial \tau_{x_1 x_3}}{\partial x_1} + \frac{\partial \tau_{x_2 x_3}}{\partial x_2} + \frac{\partial \tau_{x_3 x_3}}{\partial x_3} + f_3.$$

$$(3.48)$$

A much shorter way of writing these three ($j = 1, 2, 3$) momentum equations is achieved using the Einstein notation:

$$\rho\frac{Du_j}{Dt} = \rho\left(\frac{\partial u_j}{\partial t} + u_i\frac{\partial u_j}{\partial x_i}\right) = -\frac{\partial p}{\partial x_j} + \frac{\partial \tau_{ij}}{\partial x_i} + f_j.$$

$$(3.49)$$

The second term on the right hand side includes the components of the stress tensor

$$T_\tau = \begin{pmatrix} \tau_{x_1 x_1} & \tau_{x_2 x_1} & \tau_{x_3 x_1} \\ \tau_{x_1 x_2} & \tau_{x_2 x_2} & \tau_{x_3 x_2} \\ \tau_{x_1 x_3} & \tau_{x_2 x_3} & \tau_{x_3 x_3} \end{pmatrix},$$

$$(3.50)$$

which accounts for momentum transfer due to friction. For Newtonian fluids ($\mu \neq f(\tau)$), the well-known linear relation between shear stress and velocity gradient for one-dimensional flow has been extended for use in three-dimensional flows (Stokes' hypothesis (1845), e.g. [13, 12, 20]):

$$\tau_{ij} = \mu \left(\frac{\partial u_i}{\partial x_j} + \frac{\partial u_j}{\partial x_i} \right) - \delta_{ij} \frac{2}{3} \mu \frac{\partial u_i}{\partial x_i} . \tag{3.51}$$

The stress tensor is symmetric, which means that $\tau_{ij} = \tau_{ji}$. In Eq. 3.51, δ_{ij} is the Kronecker symbol ($\delta_{ij} = 1$ if $i = j$ and $\delta_{ij} = 0$ otherwise). For non-Newtonian fluids, the relation between stress tensor and velocity is usually much more complex and has to be defined by a set of partial differential equations. This subject is just at the beginning of being explored. However, most of the fluids of interest can be treated as Newtonian fluids.

The so-called Navier-Stokes equations are the complete equations of motion for a viscous Newtonian fluid. Using Eq. 3.51 in Eq. 3.49 finally yields

$$\rho \frac{Du_j}{Dt} = -\frac{\partial p}{\partial x_j} + \frac{\partial}{\partial x_i} \left(\mu \left(\frac{\partial u_i}{\partial x_j} + \frac{\partial u_j}{\partial x_i} - \delta_{ij} \frac{2}{3} \frac{\partial u_i}{\partial x_i} \right) \right) + f_j .$$

$$\tag{3.52}$$

As an example, the first of the three Navier-Stokes equations (x_1-direction) reads:

$$\rho \frac{Du_1}{Dt} = \rho \left(\frac{\partial u_1}{\partial t} + u_1 \frac{\partial u_1}{\partial x_1} + u_2 \frac{\partial u_1}{\partial x_2} + u_3 \frac{\partial u_1}{\partial x_3} \right)$$

$$= -\frac{\partial p}{\partial x_1} + \frac{\partial}{\partial x_1} \left(\mu \left(\frac{\partial u_1}{\partial x_1} + \frac{\partial u_1}{\partial x_1} - \frac{2}{3} \left(\frac{\partial u_1}{\partial x_1} + \frac{\partial u_2}{\partial x_2} + \frac{\partial u_3}{\partial x_3} \right) \right) \right)$$

$$+ \frac{\partial}{\partial x_2} \left(\mu \left(\frac{\partial u_2}{\partial x_1} + \frac{\partial u_1}{\partial x_2} \right) \right) + \frac{\partial}{\partial x_3} \left(\mu \left(\frac{\partial u_3}{\partial x_1} + \frac{\partial u_1}{\partial x_3} \right) \right) + f_1 .$$

$$\tag{3.53}$$

For incompressible flows, the first and the last term inside the brackets on the right hand side of Eq. 3.52 cancel out (Eq. 3.41), such that Eq. 3.52 becomes

$$\rho \frac{Du_j}{Dt} = -\frac{\partial p}{\partial x_j} + \mu \frac{\partial^2 u_j}{\partial x_i^2} + f_j . \tag{3.54}$$

In a coordinate-free vector form, Eq. 3.54 reads

$$\rho \frac{D\vec{u}}{Dt} = \rho \left(\frac{\partial \vec{u}}{\partial t} + (\vec{u} \cdot \nabla) \vec{u} \right) = -\nabla p + \mu \nabla^2 \vec{u} + \vec{f} . \tag{3.55}$$

In frictionless flows, there are no shear stresses, and the second term on the right hand side of Eq. 3.52 vanishes completely. This form of the momentum equation is the so-called Euler equation. If the flow is also steady, a special form of the momentum equation, the Bernoulli equation, can be derived by integrating between two points along a streamline (coordinate: s)

$$\int_{1}^{2} u \frac{\partial u}{\partial s} ds = \int_{1}^{2} \frac{\partial \left(u^2 / 2 \right)}{\partial s} ds = \frac{u_2^2 - u_1^2}{2} = -\int_{1}^{2} \frac{1}{\rho} \frac{\partial p}{\partial s} ds - g \left(z_2 - z_1 \right), \qquad (3.56)$$

where z is the elevation above an arbitrary datum. For incompressible flow, the Bernoulli equation reads:

$$\frac{u_2^2 - u_1^2}{2} + \frac{p_2 - p_1}{\rho} + g \left(z_2 - z_1 \right) = 0. \qquad (3.57)$$

3.1.3.3 Conservation of Energy

In order to derive the energy equation, the temporal change of the energy contained in an infinitesimal volume element is regarded, Fig. 3.8. The total energy,

$$E_t = \left(\rho e + \rho \frac{|\vec{u}|^2}{2} \right) dx_1 dx_2 dx_3, \qquad (3.58)$$

consists of the internal energy e [J/kg] and the kinetic energy. The velocity term in Eq. 3.58 can also be expressed as $|\vec{u}|^2 = \vec{u} \cdot \vec{u} = u_1^2 + u_2^2 + u_3^2$. The change of energy inside the control volume is equal to the sum of energy fluxes dE_m/dt entering and exiting the control volume due to mass entering or exiting, plus the sum of the energy fluxes dQ/dt due to heat conduction, plus the sum of the work per unit time done by pressure (dW_p /dt), viscous forces (dW_τ /dt), and body forces (dW_g /dt), plus energy supply dW_s /dt due to shaft work, radiation, or chemical processes inside the volume (e.g. combustion). Hence, the energy equation reads

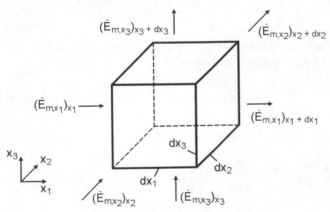

Fig. 3.8. Convective energy fluxes through a control volume

$$\frac{\partial}{\partial t}\left(\rho e + \rho \frac{|\vec{u}|^2}{2}\right) dx_1 dx_2 dx_3 = \dot{E}_m + \dot{Q} + \dot{W}_p + \dot{W}_\tau + \dot{W}_g + \dot{W}_s \,. \tag{3.59}$$

Remember that energy is not a vector, but a scalar. Hence, there is only one equation, as in the case of the continuity equation.

First, the energy flux due to mass flow will be derived. Fig. 3.8 shows the energy fluxes entering and exiting the control volume. In the x_1-direction, the resulting energy flux entering the volume is

$$\begin{aligned}
\dot{E}_{m,x_1} &= \left(\dot{E}_{m,x_1}\right)_{x_1} - \left(\dot{E}_{m,x_1}\right)_{x_1+dx_1} = \left(\rho e + \rho \frac{|\vec{u}|^2}{2}\right) u_1 dx_2 dx_3 \\
&- \left(\left(\rho e + \rho \frac{|\vec{u}|^2}{2}\right) u_1 + \frac{\partial}{\partial x_1}\left(\left(\rho e + \rho \frac{|\vec{u}|^2}{2}\right) u_1\right) dx_1\right) dx_2 dx_3 \\
&= -\frac{\partial}{\partial x_1}\left(\left(\rho e + \rho \frac{|\vec{u}|^2}{2}\right) u_1\right) dx_1 dx_2 dx_3 \,.
\end{aligned} \tag{3.60}$$

The energy fluxes in the two other coordinate directions are derived accordingly. The resulting energy flux entering the volume due to mass flow is

$$\begin{aligned}
\dot{E}_m &= -\left[\frac{\partial}{\partial x_1}\left(\left(\rho e + \rho \frac{|\vec{u}|^2}{2}\right) u_1\right) + \frac{\partial}{\partial x_2}\left(\left(\rho e + \rho \frac{|\vec{u}|^2}{2}\right) u_2\right)\right. \\
&\left. + \frac{\partial}{\partial x_3}\left(\left(\rho e + \rho \frac{|\vec{u}|^2}{2}\right) u_3\right)\right] dx_1 dx_2 dx_3 = -\nabla \cdot \left[\left(\rho e + \rho \frac{|\vec{u}|^2}{2}\right) \vec{u}\right] dx_1 dx_2 dx_3 \,.
\end{aligned} \tag{3.61}$$

Next, the energy fluxes due to heat conduction must be derived. Assuming that the heat transfer to the element is given by Fourier's law,

$$\vec{\dot{q}} = -\lambda \nabla T, \quad or \quad \dot{q}_j = -\lambda \frac{\partial T}{\partial x_j}, \tag{3.62}$$

where $\dot{q}$ is in [W/m^2], the heat flux in the x_1-direction is

$$\dot{Q}_{x_1} = \left[-\lambda \frac{\partial T}{\partial x_1} - \left(-\lambda \frac{\partial T}{\partial x_1} + \frac{\partial}{\partial x_1}\left(-\lambda \frac{\partial T}{\partial x_1}\right) dx_1\right)\right] dx_2 dx_3 \,, \tag{3.63}$$

and the resulting heat flux due to heat conduction can be expressed as

$$\dot{Q} = \left[\frac{\partial}{\partial x_1}\left(\lambda \frac{\partial T}{\partial x_1} \right) + \frac{\partial}{\partial x_2}\left(\lambda \frac{\partial T}{\partial x_2} \right) + \frac{\partial}{\partial x_3}\left(\lambda \frac{\partial T}{\partial x_3} \right) \right] dx_1 dx_2 dx_3$$

$$= -\nabla \cdot \vec{q}\, dx_1 dx_2 dx_3 = \nabla \cdot (\lambda \nabla T)\, dx_1 dx_2 dx_3 \; .$$

(3.64)

Next, the relations for the work done by pressure as well as normal and shear stresses are discussed. At each face of the control volume, two shear stresses, one normal stress and the static pressure, have to be accounted for, see Fig. 3.7. At first, only the two opposing faces normal to the x_1-axis shall be regarded. The rate of work done to the element per unit time due to static pressure is

$$\dot{W}_{p,x_1} = \left[pu_1 - \left(pu_1 + \frac{\partial (pu_1)}{\partial x_1} dx_1 \right) \right] dx_2 dx_3 = -\frac{\partial (pu_1)}{\partial x_1} dx_1 dx_2 dx_3 \; ,$$

(3.65)

and the work done due to viscous forces is

$$\dot{W}_{\tau,x_1} = \left[-\tau_{x_1 x_1} u_1 + \left(\tau_{x_1 x_1} u_1 + \frac{\partial \left(\tau_{x_1 x_1} u_1 \right)}{\partial x_1} dx_1 \right) \right] dx_2 dx_3$$

$$+ \left[-\tau_{x_1 x_2} u_2 + \left(\tau_{x_1 x_2} u_2 + \frac{\partial \left(\tau_{x_1 x_2} u_2 \right)}{\partial x_1} dx_1 \right) \right] dx_2 dx_3$$

$$+ \left[-\tau_{x_1 x_3} u_3 + \left(\tau_{x_1 x_3} u_3 + \frac{\partial \left(\tau_{x_1 x_3} u_3 \right)}{\partial x_1} dx_1 \right) \right] dx_2 dx_3$$

$$= \left(\frac{\partial \left(\tau_{x_1 x_1} u_1 \right)}{\partial x_1} + \frac{\partial \left(\tau_{x_1 x_2} u_2 \right)}{\partial x_1} + \frac{\partial \left(\tau_{x_1 x_3} u_3 \right)}{\partial x_1} \right) dx_1 dx_2 dx_3 \; .$$

(3.66)

In Eq. 3.66, the work per unit time, which is the product of velocity and force, is given a positive sign if velocity (directed in positive coordinate direction) and force (directed by the corresponding normal or shear stress, Fig. 3.7) have the same direction, and a negative sign in the case of opposite directions.

The expressions for the other two coordinate directions are derived accordingly. The resulting expressions for the work done per unit time by pressure and viscous forces are

$$\dot{W}_p = -\left(\frac{\partial (pu_1)}{\partial x_1} + \frac{\partial (pu_2)}{\partial x_2} + \frac{\partial (pu_3)}{\partial x_3} \right) dx_1 dx_2 dx_3 = -\nabla \cdot (p\vec{u})\, dx_1 dx_2 dx_3 \; ,$$

(3.67)

and

$$\dot{W}_\tau = \left[\frac{\partial\left(\tau_{x_1 x_1} u_1 + \tau_{x_1 x_2} u_2 + \tau_{x_1 x_3} u_3\right)}{\partial x_1} + \frac{\partial\left(\tau_{x_2 x_1} u_1 + \tau_{x_2 x_2} u_2 + \tau_{x_2 x_3} u_3\right)}{\partial x_2} \right.$$

$$\left. + \frac{\partial\left(\tau_{x_3 x_1} u_1 + \tau_{x_3 x_2} u_2 + \tau_{x_3 x_3} u_3\right)}{\partial x_3} \right] dx_1 dx_2 dx_3 = \nabla \cdot \left(\vec{u} \cdot T_\tau\right) dx_1 dx_2 dx_3 \quad .$$

(3.68)

The change of energy due to the effect of body forces is

$$\dot{W}_g = \left(\vec{f} \cdot \vec{u}\right) dx_1 dx_2 dx_3 \, ,$$

(3.69)

where $\vec{f} = \rho \vec{g}$ in the case of gravitation.

Using Eqs. 3.61, 3.64, 3.67, 3.68, and 3.69 in Eq. 3.59 and applying the product rule results in

$$\left(e + \frac{|\vec{u}|^2}{2}\right)\left(\frac{\partial \rho}{\partial t} + \frac{\partial(\rho u_1)}{\partial x_1} + \frac{\partial(\rho u_2)}{\partial x_2} + \frac{\partial(\rho u_3)}{\partial x_3}\right) + \rho \frac{De}{Dt} + \rho \frac{D}{Dt}\left(\frac{|\vec{u}|^2}{2}\right)$$

$$= \nabla \cdot (\lambda \nabla T) - \nabla \cdot (p\vec{u}) + \nabla \cdot (\vec{u} \cdot T_\tau) + \vec{f} \cdot \vec{u} + \frac{\dot{Q}_s}{dx_1 dx_2 dx_3} \quad .$$

(3.70)

Because of the continuity equation, the first term in Eq. 3.70 is zero. The use of

$$\nabla \cdot (\vec{u} \cdot T_\tau) = \vec{u} \cdot (\nabla \cdot T_\tau) + \tau_{ij} \frac{\partial u_j}{\partial x_i}$$

(3.71)

and

$$\nabla \cdot (p\vec{u}) = p(\nabla \cdot \vec{u}) + \vec{u} \cdot (\nabla p)$$

(3.72)

results in:

$$\rho\left(\frac{De}{Dt} + \frac{D}{Dt}\left(\frac{|\vec{u}|^2}{2}\right)\right) = \nabla \cdot (\lambda \nabla T) - p(\nabla \cdot \vec{u}) - \vec{u} \cdot (\nabla p)$$

$$+ \vec{u} \cdot (\nabla \cdot T_\tau) + \tau_{ij} \frac{\partial u_j}{\partial x_i} + \vec{f} \cdot \vec{u} + \frac{\dot{Q}_s}{dx_1 dx_2 dx_3} \quad .$$

(3.73)

This equation is further modified by using the momentum equation. Multiplication of Eq. 3.49 with the velocity vector yields the second term on the left hand side of Eq. 3.73,

$$\rho \vec{u} \frac{D\vec{u}}{Dt} = \rho \frac{D}{Dt}\left(\frac{|\vec{u}|^2}{2}\right) = -\vec{u}\cdot(\nabla p) + \vec{u}\cdot(\nabla\cdot T_\tau) + \vec{u}\cdot\vec{f} , \qquad (3.74)$$

and Eq. 3.73 now reads

$$\rho\frac{De}{Dt} = \nabla\cdot(\lambda\nabla T) - p(\nabla\cdot\vec{u}) + \tau_{ij}\frac{\partial u_j}{\partial x_i} + \frac{\dot{Q}_s}{dx_1 dx_2 dx_3} , \qquad (3.75)$$

which is a widely used form of the first law of thermodynamics for fluid motion, the so-called thermal energy equation. In the case of a Newtonian fluid, the components of the stress tensor can be expressed by Eq. 3.51. This finally yields

$$\rho\frac{De}{Dt} = \nabla\cdot(\lambda\nabla T) - p(\nabla\cdot\vec{u}) + \mu\Phi + \frac{\dot{Q}_s}{dx_1 dx_2 dx_3} , \qquad (3.76)$$

where

$$\Phi = 2\left[\frac{\partial^2 u_1}{\partial x_1^2} + \frac{\partial^2 u_2}{\partial x_2^2} + \frac{\partial^2 u_3}{\partial x_3^2}\right] + \left(\frac{\partial u_2}{\partial x_1} + \frac{\partial u_1}{\partial x_2}\right)^2$$

$$+ \left(\frac{\partial u_3}{\partial x_2} + \frac{\partial u_2}{\partial x_3}\right)^2 + \left(\frac{\partial u_1}{\partial x_3} + \frac{\partial u_3}{\partial x_1}\right)^2 - \frac{2}{3}\left(\frac{\partial u_1}{\partial x_1} + \frac{\partial u_2}{\partial x_2} + \frac{\partial u_3}{\partial x_3}\right)^2 \qquad (3.77)$$

is the dissipation function.

Eq. 3.76 may be rearranged in order to get an expression based on the enthalpy h. The continuity equation (Eq. 3.40) can be rewritten to give

$$p(\nabla\cdot\vec{u}) = -\frac{p}{\rho}\frac{D\rho}{Dt} = \rho\frac{D}{Dt}\left(\frac{p}{\rho}\right) - \frac{Dp}{Dt} . \qquad (3.78)$$

Combining Eqs. 3.78 and 3.76 and neglecting the source term $\dot{Q}_s$ yields

$$\rho\frac{D}{Dt}\left(e + \frac{p}{\rho}\right) = \rho\frac{Dh}{Dt} = \frac{Dp}{Dt} + \nabla\cdot(\lambda\nabla T) + \mu\Phi . \qquad (3.79)$$

The dissipation term can usually be neglected. This term becomes important for high Mach number flows (e.g. reentry of a spacecraft into the earth's atmosphere). Assuming that the thermal conductivity and the specific heat capacity are constants and neglecting the usually small effect of the pressure term on temperature results in the following often used form of the thermal energy equation:

$$\rho c_p\frac{DT}{Dt} = \rho c_p\left(\frac{\partial T}{\partial t} + u_i\frac{\partial T}{\partial x_i}\right) = \lambda\frac{\partial^2 T}{\partial x_i^2} . \qquad (3.80)$$

3.1.4 Turbulent Flows

3.1.4.1 RANS Equations

The Navier-Stokes equations can be solved directly without averaging or using turbulence models if the grid spacing is fine enough to resolve the smallest eddies in a flow field. This approach is called direct numerical simulation (DNS). The size of the smallest eddies is proportional to the Kolmogorov length scale, which decreases with increasing Reynolds numbers. For a cylinder volume of 1.0 liters at least 10^{12} grid points are needed due to the highly turbulent flow, where the smallest eddies are in the range of 0.01 mm. The requirements for the application of DNS to in-cylinder processes, concerning computer speed and memory, exceed the power of today's processors by far. Up to now, DNS simulations have not been suitable for solving engineering problems and are limited to basic research applications with low Reynolds numbers and small geometric domains.

The large eddy simulation (LES) distinguishes between large and small eddies. Only the large eddies, which contain most of the energy and thus are much more important concerning the transport of the conserved quantities than the small ones, are resolved by the grid. The effect of the small eddies is described by sub-models. The separation of small and large eddies is obtained by filtering the flow field using an appropriate threshold value of the eddy size. Because the small eddies are described by so-called subgrid-scale Reynolds stress models, the Navier–Stokes equations are expressed in terms of averaged quantities, similar to the RANS equations described in the following. The numerical algorithms are usually specially adapted to the individual problem and the geometry. Although the large eddy simulation consumes less time and computational power than DNS, it is still not suitable for engineering applications. A more detailed description of DNS and LES is given in [6] for example.

Today, the computation of technical flow fields is performed using the so-called Reynolds averaged Navier-Stokes equations (RANS equations) in combination with an appropriate turbulence model. Following the basic approach of Reynolds (1895), the instantaneous values of the turbulent flow quantities are split into a mean and a fluctuating component (Reynolds decomposition):

$$u_i = \overline{u}_i + u_i' \, , \ \rho = \overline{\rho} + \rho' \, , \ T = \overline{T} + T' \, . \tag{3.81}$$

The overbar denotes the time-average, and the superscript ($'$) denotes the superimposed fluctuation, see Fig. 3.9. The time-average of a flow quantity, e.g. a velocity component, is

$$\overline{u}_i = \frac{1}{\Delta t} \int_t^{t+\Delta t} u_i(\vec{x}, t) dt \, . \tag{3.82}$$

The time interval Δt must be large compared to the relevant period of the fluctuations, but small enough to map the time dependence of an unsteady mean flow.

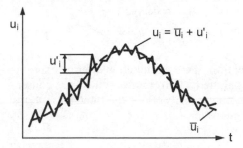

Fig. 3.9. Unsteady turbulent flow

The RANS equations are obtained by substituting the Reynolds decomposition terms into the instantaneous conservation equations, and then by averaging the entire equations over time. Now the turbulent flow is described by the conservation equations in terms of time average quantities. Due to the time-averaging, the information about the turbulent fluctuating quantities is lost. This effect is shown in additional terms in the momentum and energy equations, the Reynolds stresses, and the turbulent heat flux, which have to be described by a turbulence model in order to close the system of equations again.

In general, thermodynamic properties like μ, c_p, and λ can also fluctuate due to the fluctuation of pressure and temperature, but these fluctuations are usually small, and they are neglected for the treatment of turbulence in this book. Special attention must be given to the density fluctuations. Neglecting the density fluctuations does not mean that the density is constant. It simply expresses the fact that the effect of turbulent density variations is not included; the mean density variations however may be large. Should fluctuations in density be considered, the appropriate equations can be derived either by simple time-averaging (Reynolds averaging, Eq. 3.81), or by a mass-weighted time-averaging procedure (Favre-averaging). Both kinds of averaging are in use today and will be described in the following.

Substituting the Reynolds decomposition terms into the instantaneous continuity equation (here only shown for two-dimensional flow) and averaging over time (denoted by a long overbar) yields

$$\overline{\frac{\partial}{\partial t}(\bar{\rho}+\rho')+\frac{\partial}{\partial x_1}\left[(\bar{\rho}+\rho')(\bar{u}_1+u'_1)\right]+\frac{\partial}{\partial x_2}\left[(\bar{\rho}+\rho')(\bar{u}_2+u'_2)\right]}=0. \qquad (3.83)$$

By applying the following rules of averaging (e.g. [20, 11]),

$$\overline{f'}=0, \quad \overline{\bar{f}}=\bar{f}, \quad \overline{\bar{f}\,\bar{g}}=\bar{f}\,\bar{g}, \quad \overline{f'\bar{g}}=0, \quad \overline{f+g}=\bar{f}+\bar{g},$$

$$\overline{f\,g}=\bar{f}\,\bar{g}+\overline{f'\,g'}, \quad \overline{\frac{\partial f}{\partial s}}=\frac{\partial \bar{f}}{\partial s}, \qquad (3.84)$$

where f and g are two turbulent flow quantities, the continuity equation for a compressible flow becomes

$$\frac{\partial \overline{\rho}}{\partial t} + \frac{\partial}{\partial x_1}\left(\overline{\rho}\overline{u}_1\right) + \frac{\partial}{\partial x_2}\left(\overline{\rho}\overline{u}_2\right) + \frac{\partial}{\partial x_1}\left(\overline{\rho'u_1'}\right) + \frac{\partial}{\partial x_2}\left(\overline{\rho'u_2'}\right) = 0 .$$
(3.85)

Compared to the original continuity equation, there are additional terms expressing the fact that the mean flow quantities alone no longer express the mass conservation principle. For incompressible (constant density) flow, the fluctuation terms disappear, and the continuity equation reduces to

$$\frac{\partial \overline{u}_i}{\partial x_i} = 0 .$$
(3.86)

The momentum and energy equations are treated accordingly. Replacing the instantaneous values by the Reynolds decompositions, averaging over time, separating components, and dropping all terms that contain only one fluctuating quantity (these terms are zero), the x_1-momentum equation for a compressible flow in the absence of external forces becomes (here only shown for two-dimensional flow)

$$\frac{\partial}{\partial t}\left(\overline{\rho}\overline{u}_1 + \overline{\rho'u_1'}\right) + \frac{\partial}{\partial x_1}\left(\overline{\rho}\overline{u}_1^2\right) + \frac{\partial}{\partial x_2}\left(\overline{\rho}\overline{u}_1\overline{u}_2\right) + \frac{\partial}{\partial x_1}\left(\overline{\rho}\overline{u_1'^2}\right) + \frac{\partial}{\partial x_2}\left(\overline{\rho}\overline{u_1'u_2'}\right)$$

$$+ \frac{\partial}{\partial x_1}\left(\overline{\rho'u_1'^2}\right) + \frac{\partial}{\partial x_2}\left(\overline{\rho'u_1'u_2'}\right) + \frac{\partial}{\partial x_1}\left(2\overline{u}_1\,\overline{\rho'u_1'}\right) + \frac{\partial}{\partial x_2}\left(\overline{u}_1\overline{\rho'u_2'} + \overline{u}_2\,\overline{\rho'u_1'}\right)$$

$$= -\frac{\partial \overline{p}}{\partial x_1} + \frac{\partial \overline{\tau}_{x_1x_1}}{\partial x_1} + \frac{\partial \overline{\tau}_{x_2x_1}}{\partial x_2} .$$
(3.87)

The fluctuating parts of the viscous stress tensor disappear entirely upon time-averaging. The x_2- and x_3- momentum equations as well as the energy equation may be treated in the same manner. Compared to the original equations, there are again a lot of extra terms due to additional turbulent momentum and heat transfer, complicating the system of conservation equations.

In the case of incompressible flow however, the resulting momentum and energy conservation equations reduce to

$$\rho\left(\frac{\partial \overline{u}_j}{\partial t} + \overline{u}_i\frac{\partial \overline{u}_j}{\partial x_i}\right) = -\frac{\partial \overline{p}}{\partial x_j} + \frac{\partial}{\partial x_i}\left(\overline{\tau}_{ij} - \rho\overline{u_i'u_j'}\right) ,$$
(3.88)

$$\rho c_p\left(\frac{\partial \overline{T}}{\partial t} + \overline{u}_i\frac{\partial \overline{T}}{\partial x_i}\right) = -\frac{\partial}{\partial x_i}\left(\overline{q}_i + \rho c_p\overline{u_i'T'}\right) ,$$
(3.89)

where $j = 1, 2, 3$ indicates that there are three momentum equations, each for one coordinate direction. These equations are very similar to the original set of conservation equations. However, two additional terms have been added as a result of the averaging process. In the momentum equation, this is the so-called turbulent stress tensor,

$$\overline{\tau}_{ij,t} = -\rho \overline{u_i' u_j'} \tag{3.90}$$

(with the minus sign dropped it is called Reynolds stress tensor), and in the energy equation there is the turbulent heat flux,

$$\overline{\dot{q}}_{i,t} = \rho c_p \overline{u_i' T'} . \tag{3.91}$$

The Reynolds stress term is not a stress but an inertia effect due to additional momentum exchange caused by turbulence and is only referred to as a stress because of the way it appears in the equations. The same holds true for the turbulent heat flux: this term represents the increase of heat transfer due to the presence of turbulent eddies.

Altogether, is has been shown that, except for the turbulent stresses and the turbulent heat flux, the RANS equations for incompressible flow have the same form as the original conservation equations. In the case of fully compressible flow however, the equations are much more complicated.

For this reason, the conservation equations for compressible turbulent flows are often obtained by a mass-weighted time-averaging procedure according to A. Favre (1965) (e.g. Cebeci and Smith [4] and Oertel [16]). Favre-averaging results in conservation equations which are very similar to the original ones, and which contain only few additional terms due to the effect of turbulence.

Favre-averaging of a flow quantity, e.g. the velocity component u_i, is obtained by dividing the time-averaged product of density and velocity,

$$\overline{pu_i} = \frac{1}{\Delta t} \int_{t}^{t+\Delta t} (pu_i) dt , \tag{3.92}$$

by the time-averaged density:

$$\tilde{u}_i = \frac{\overline{\rho u_i}}{\overline{\rho}} . \tag{3.93}$$

The instantaneous flow quantities are again split into a mean value, denoted by $(\tilde{\ })$, and a fluctuating value, denoted by $('')$. This is done for all flow quantities except for static pressure and the density, which are only time-averaged:

$$\rho = \overline{\rho} + \rho' , \quad p = \overline{p} + p' , \quad u_i = \tilde{u}_i + u_i'' , \quad T = \tilde{T} + T'' . \tag{3.94}$$

In contrast to the fluctuation components of the simple Reynolds decompositions $(\overline{u_i'})$, the time average of the Favre-averaged fluctuation components $(\overline{u_i''})$ is not zero. Instead $\overline{\rho u_i''} = 0$, which can be shown by the following calculation:

$$\frac{\overline{\rho u_i}}{\overline{\rho}} = \frac{\overline{\rho \tilde{u}_i}}{\overline{\rho}} + \frac{\overline{\rho u_i''}}{\overline{\rho}} = \frac{\overline{\rho} \tilde{u}_i}{\overline{\rho}} + \frac{\overline{\rho u_i''}}{\overline{\rho}} = \tilde{u}_i + \frac{\overline{\rho u_i''}}{\overline{\rho}} . \tag{3.95}$$

Because $\tilde{u}_i = \overline{\rho u_i} / \overline{\rho}$ by definition, the last term in Eq. 3.95 is zero.

Using the following rules of averaging,

$$\overline{f+g} = \overline{f}+\overline{g}, \quad \overline{\rho'\tilde{f}} = 0, \quad \overline{\frac{\partial f}{\partial s}} = \frac{\partial \overline{f}}{\partial s},$$

(3.96)

the conservation equations can be derived. The continuity equation becomes

$$\frac{\partial \overline{\rho}}{\partial t} + \frac{\partial \overline{\left(\rho\left(\tilde{u}_i + u_i''\right)\right)}}{\partial x_i} = \frac{\partial \overline{\rho}}{\partial t} + \frac{\partial\left(\overline{\rho}\tilde{u}_i\right)}{\partial x_i} + \frac{\partial\overline{\left(\rho u_i''\right)}}{\partial x_i}$$

$$= \frac{\partial \overline{\rho}}{\partial t} + \frac{\partial\left(\overline{\rho}\tilde{u}_i\right)}{\partial x_i} = 0.$$

(3.97)

The general form of this equation is identical to that of the original continuity equation. The Navier-Stokes equations and the energy equation are treated accordingly (e.g. [16]). Again, the general form of the resulting equations is identical to the original form, except for an additional turbulent stress term in the momentum equations and an additional turbulent heat flux in the energy equation:

$$\overline{\rho}\frac{D\tilde{u}_j}{Dt} = \overline{\rho}\left(\frac{\partial \tilde{u}_j}{\partial t} + \tilde{u}_i\frac{\partial \tilde{u}_j}{\partial x_i}\right) = \frac{\partial\left(\overline{\rho}\tilde{u}_j\right)}{\partial t} + \frac{\partial\left(\overline{\rho}\tilde{u}_j\tilde{u}_i\right)}{\partial x_i}$$

$$= -\frac{\partial \overline{p}}{\partial x_j} + \frac{\partial}{\partial x_i}\left(\overline{\tau}_{ij} - \overline{\rho u_i'' u_j''}\right),$$

(3.98)

$$\overline{\rho}c_p\frac{D\tilde{T}}{Dt} = \overline{\rho}c_p\left(\frac{\partial \tilde{T}}{\partial t} + \tilde{u}_i\frac{\partial \tilde{T}}{\partial x_i}\right) = \frac{\partial\left(\overline{\rho}c_p\tilde{T}\right)}{\partial t} + \frac{\partial\left(\overline{\rho}c_p\tilde{T}\tilde{u}_i\right)}{\partial x_i}$$

$$= -\frac{\partial}{\partial x_i}\left(\overline{\tilde{q}}_i + c_p\overline{\rho T'' u_i''}\right).$$

(3.99)

The components of the stress tensor in Eq. 3.98 are obtained by using the Favre-averaged decomposition terms in Eq. 3.51, but in contrast to simple time-averaging the fluctuation terms do not disappear:

$$\overline{\tau}_{ij} = \left[\mu\left(\frac{\partial \tilde{u}_i}{\partial x_j} + \frac{\partial \tilde{u}_j}{\partial x_i}\right) - \delta_{ij}\frac{2}{3}\mu\frac{\partial \tilde{u}_i}{\partial x_i}\right] + \left[\mu\left(\overline{\frac{\partial u_i''}{\partial x_j}} + \overline{\frac{\partial u_j''}{\partial x_i}}\right) - \delta_{ij}\frac{2}{3}\mu\overline{\frac{\partial u_i''}{\partial x_i}}\right].$$

(3.100)

The heat flux in Eq. 3.99 is

$$\overline{\tilde{q}}_i = -\lambda\left(\frac{\partial \tilde{T}}{\partial x_i} + \overline{\frac{\partial T''}{\partial x_i}}\right).$$

(3.101)

For incompressible flows the mass-weighted time-average is equal to the simple time-average, and the Favre-averaged conservation equations are equal to the time-averaged ones, Eqs. 3.86, 3.88, 3.89.

Due to the presence of the Reynolds stresses and the turbulent heat flux, the conservation equations are not closed: they contain more variables than equations. Because it is impossible to derive a closed set of exact equations, the equations have to be closed by turbulence models. The following considerations are presented for incompressible flows, but they can also be applied to compressible flows with density fluctuations (e.g. [17]). The traditional modeling assumption, following J. Boussinesq (1877), is that the effect of turbulence in the momentum equation can be represented as increased viscosity. Hence, it is assumed that the Reynolds stress tensor can be expressed in analogy to the viscous stress tensor τ_{ij}, leading to the so-called eddy viscosity model,

$$\overline{\tau}_{ij,t} = -\rho\overline{u_i'u_j'} = \mu_t\left(\frac{\partial\overline{u}_i}{\partial x_j} + \frac{\partial\overline{u}_j}{\partial x_i}\right) - \frac{2}{3}\rho\delta_{ij}k\,, \tag{3.102}$$

where $\delta_{ij} = 1$ if $i = j$ and $\delta_{ij} = 0$ otherwise.

$$k = \frac{1}{2}\overline{u_i'u_i'} = \frac{1}{2}\left(\overline{u_1'u_1'} + \overline{u_2'u_2'} + \overline{u_3'u_3'}\right) \tag{3.103}$$

is the turbulent kinetic energy, and μ_t is the turbulent viscosity (eddy viscosity), which is referred to as viscosity, but is not a fluid property and is caused by turbulence. As shown in Eq. 3.103, the sum of the normal turbulent stresses must be equal to $2k$. The last term in Eq. 3.102 is added in order to guarantee that this holds true for $i = j$ (due to the continuity equation the first term on the right hand side of Eq. 3.102 then becomes zero). Because normal stresses behave like a pressure, the static pressure in the RANS equations simply has to be substituted by $p + (2/3)k$.

The turbulent heat flux is modeled in the same manner,

$$\overline{q}_{i,t} = -\lambda_t\frac{\partial\overline{T}}{\partial x_i} = -\rho c_p a_t\frac{\partial\overline{T}}{\partial x_i}\,, \tag{3.104}$$

where $a_t = \lambda_t/(\rho c_p)$ is the turbulent thermal diffusivity and λ_t is referred to as turbulent eddy conductivity, although it is not a fluid property. In analogy to laminar flows, the ratio of eddy viscosity to eddy conductivity, $Pr_t = \nu_t/a_t = \mu_t c_p/\lambda_t$, is called the turbulent Prandtl number. Since the turbulent flux terms in the momentum and the energy equation are caused by the same mechanism of time-averaged convection, it follows that their ratio, Pr_t, ought to be of order one, which is the Reynolds analogy for turbulent flows. The turbulent Prandtl number is commonly assumed to have a constant value of $Pr_t = 0.9$ or $Pr_t = 1.0$. Now λ_t in Eq. 3.104 can be calculated from $Pr_t = \mu_t c_p/\lambda_t$, and the problem of closing the RANS equations is reduced to the approximation of the turbulent viscosity by an appropriate turbulence model.

3.1.4.2 Turbulence Modeling

The task of a turbulence model is to provide a relation for the turbulent eddy viscosity of a flow field in order to close the RANS equations. In the RANS equations, all unsteadiness of the flow field is averaged out, treated as part of turbulence, and included in the turbulent stresses and fluxes. However, due to the complexity of turbulence, it has not been possible up to now to develop a single universal model capable of predicting the turbulent behavior of the Navier-Stokes equations for all kinds of turbulent flow fields. Hence, the models can only be regarded as approximations and not as universal laws.

According to the number of partial differential equations necessary for their description, turbulence models are classified as zero-equation models (algebraic models) and one- and two-equation models.

Prandtl Mixing-Length Model

The Prandtl mixing-length theory (1925) is the simplest and the oldest of all turbulence closure models. The mixing length l is the distance a turbulent eddy can travel in a turbulent flow field until it has completely mixed with its surroundings and lost its identity due to the dissipation of its energy. In order to estimate the mixing length, a liquid element with mean velocity $\bar{u}_1(x_2)$ in the main flow direction at the position x_2 is regarded, Fig. 3.10. Due to the turbulent fluctuating velocity component u'_2, this element may be transported to some position $(x_2 + l_1)$ or $(x_2 - l_1)$ below or above the original one. At these new positions, the velocity $\bar{u}_1(x_2)$ of the element differs from the mean velocity $\bar{u}_1 (x_2 \pm l_1) = \bar{u}_1 (x_2) \pm l_1 d\bar{u}_1/dx_2$ of the surroundings by the term

$$\Delta u_1 = u'_1 = \pm l_1 \frac{d\overline{u}_1}{dx_2},$$

(3.105)

which is regarded as fluctuation velocity. Hence, the fluctuation velocity is expressed using the time-averaged velocity again. Due to mass conservation, the fluctuation velocities normal to the mean flow direction must be of the same order:

$$|u'_1| = l_1 \left| \frac{d\overline{u}_1}{dx_2} \right|, \quad |u'_2| = l_2 \left| \frac{d\overline{u}_1}{dx_2} \right|.$$

(3.106)

According to Fig. 3.10, where $d\bar{u}_1/dx_2 > 0$, the product $u'_1 u'_2$ of both velocity fluctuations is always negative. If $d\bar{u}_1/dx_2 < 0$, $u'_1 u'_2 > 0$. Thus, the signs of both expressions are always different, and the expression for the turbulent stress becomes

$$\overline{\tau}_{x_1,t} = -\rho\overline{u'_1 u'_2} = \rho l^2 \left| \frac{d\overline{u}_1}{dx_2} \right| \frac{d\overline{u}_1}{dx_2} = \mu_t \frac{d\overline{u}_1}{dx_2},$$

(3.107)

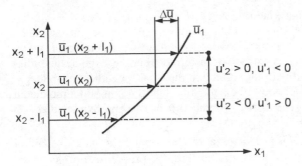

Fig. 3.10. Prandtl mixing-length model

where $l^2 = l_1 l_2$, and

$$\mu_t = \rho l^2 \left| \frac{d\overline{u}_1}{dx_2} \right| \tag{3.108}$$

is the eddy viscosity. The mixing length l must be determined experimentally. It depends mainly on the distance from a wall. For simple flows, the mixing length model has been shown to produce satisfying results, but this no longer holds true in the case of fully three-dimensional flows. In three-dimensional flows, more detailed models like the two-equation k-e model are used, which introduces two additional differential equations for the description of turbulence.

k-ε Model

The two-equation k-ε model has been published by Launder and Spalding in 1974 [14] and is still the standard turbulence model today. The turbulent viscosity μ_t has the same dimension as the molecular one: it is the product of velocity and length. Following the approach of Kolmogorov and Prandtl, μ_t can be expressed as

$$\mu_t = C_\mu \cdot \rho \cdot l \cdot q, \tag{3.109}$$

where C_μ is a model constant, and l and q are characteristic length scales and velocities. In the simple mixing length model $C_\mu \cdot q = l \cdot d\overline{u}_1/dx_2$ is used, and l is a prescribed function of the distance from a wall. In the model of Launder and Spalding, the characteristic velocity q is taken to be the square root of the turbulent kinetic energy, which is defined by Eq. 3.103. The choice of an expression describing the length scale is much more complicated. The most popular model is based on the fact that in flows where the production of turbulent kinetic energy equals its dissipation, the dissipation rate ε, the length scale l, and the turbulent kinetic energy k can be related by

$$\varepsilon \approx \frac{k^{3/2}}{l}. \tag{3.110}$$

In turbulent flows, turbulent kinetic energy is only dissipated in the smallest eddies, while only the large eddies are able to convert the kinetic energy of the main flow into turbulent kinetic energy (production). According to the energy cascade model, the turbulent kinetic energy is transferred from the large-scale eddies to the smaller ones, until it is finally dissipated in the smallest eddies (characteristic size associated with l), which loose their identity due to mixing on a molecular level with their surroundings. The dissipation rate of k can be approximated as follows.

The drag of an eddy of size l is proportional to $l^2 \overline{u'^2}$, where $\overline{u'^2}$ is the square of its resulting velocity relative to the surrounding, and l^2 is proportional to its cross-sectional area. The energy dissipated per unit time is proportional to $l^2 \overline{u'^3}$. Hence, the dissipation rate ε, which is the amount of turbulent kinetic energy that is dissipated per unit time and volume ($\sim l^3$), is

$$\frac{l^2 \overline{u'^3}}{l^3} \sim C \cdot \frac{k^{3/2}}{l} \approx \varepsilon , \tag{3.111}$$

where $u' = \sqrt{2k}$. In Eq. 3.110, the constant C is neglected, because it can be combined with other constants in the complete model.

Using Eq. 3.110 in Eq. 3.109 gives

$$\mu_t = C_\mu \cdot \rho \frac{k^2}{\varepsilon} . \tag{3.112}$$

The transport equations for k and ε can be derived from the Navier-Stokes equations for the fluctuating quantities (e.g. [11]). However, there are new unknown terms in the resulting equations, which have to be modeled. Finally, the following equations for the description of the turbulent kinetic energy and its dissipation rate are obtained:

$$\rho \frac{\partial k}{\partial t} + \rho \overline{u}_i \frac{\partial k}{\partial x_i} \approx \frac{\partial}{\partial x_i} \left(\frac{\mu_t}{C_k} \frac{\partial k}{\partial x_i} \right) + \mu_t \frac{\partial \overline{u}_j}{\partial x_i} \left(\frac{\partial \overline{u}_i}{\partial x_j} + \frac{\partial \overline{u}_j}{\partial x_i} \right) - \rho \varepsilon , \tag{3.113}$$

$$\rho \frac{\partial \varepsilon}{\partial t} + \rho \overline{u}_i \frac{\partial \varepsilon}{\partial x_i} \approx \frac{\partial}{\partial x_i} \left(\frac{\mu_t}{C_\varepsilon} \frac{\partial \varepsilon}{\partial x_i} \right) + C_1 \frac{\varepsilon}{k} \mu_t \frac{\partial \overline{u}_j}{\partial x_i} \left(\frac{\partial \overline{u}_i}{\partial x_j} + \frac{\partial \overline{u}_j}{\partial x_i} \right) - \rho C_2 \frac{\varepsilon^2}{k} . \tag{3.114}$$

The constants C_μ, C_k, C_ε, C_1, and C_2 are not universal and have to be adjusted to a specific problem. A set of values recommended by Jones and Launder [9] after examination of a large amount of experimental data is

$$C_\mu = 0.09, \ C_k = 1.0, \ C_\varepsilon = 1.3, \ C_1 = 1.44, \ C_2 = 1.92 . \tag{3.115}$$

In the case of compressible flow and interaction of the gas phase with sprays, additional terms appear in the equations. They are given in [2], for example.

Further Turbulence Models

The standard k-ε model is based on the assumption that the eddy viscosity is isotropic and thus the same for all Reynolds stresses, Eq. 3.102. Especially in fully three-dimensional flows, the assumption of isotropic turbulence results in significant drawbacks in the eddy viscosity model. Furthermore, the constants in the k-ε model are not universal. For these reasons, advanced turbulence models have been developed.

A modified k-ε model, the so-called RNG k-ε model, has been derived using the ideas from renormalization group (RNG) theory by Yakhot and Orszag [22, 23]. The basic idea of this approach is that the smallest scales of turbulence are systematically removed, and the remaining scales are resolved numerically. In the limit of all scales being removed, the RNG k-ε model is obtained. An important feature of the model is that the model constants are obtained explicitly in the derivation. Further on, the original k-equation is not modified, but the ε-equation is modified with the inclusion of an additional term, which changes dynamically with the rate of strain of turbulence. This results in a more detailed description of flows with rapid distortion and anisotropic large-scale eddies. The RNG k-ε model has been successfully applied to spray modeling [7, 1]. In general, the RNG k-ε model results in greater mixing in the jets relative to the standard model.

The most complex of the advanced turbulence models today are the so-called Reynolds stress models. Here, partial differential equations for each of the components of the Reynolds stress tensor are derived from the Navier-Stokes equations. These equations are quite complicated and contain new terms of the form $\overline{u_i' u_j' u_k'}$ and $\overline{u_i' p'}$, which have to be modeled again. The interested reader may refer to refs. [8] and [10] for more details. Altogether, the Reynolds stress models add at least six new partial differential equations to the Reynolds averaged conservation equations, which have to be solved. Today, research is still in progress, and lots of different approaches and modifications are being proposed. Although the Reynolds stress models have a larger potential to represent the three-dimensional anisotropic nature of turbulence than the standard two-equation models, their application to engineering problems has not always produced superior results. However, errors originated by the numerical solution of the complex system of equations are often larger than the improvement by the model. For these reasons, the k-ε model is still the standard turbulence model for CFD simulations of in-cylinder processes.

3.1.4.3 Velocity and Temperature Distribution in Boundary Layers

In order to solve the basic flow equations numerically, the boundary conditions at the wall must be taken into account. The shear stress and heat flux boundary conditions can be calculated if the velocity and temperature profiles in the usually very thin boundary layers are known. An accurate calculation of the boundary conditions is necessary, because this directly affects the prediction of efficiency, pollutant formation, and thermal load of engine components. One possibility is to solve the equations accurately right up to the wall. Then, the standard k-ε equations have to be modified in order to extend the model all the way to the wall. However, due to the extremely thin boundary layers and the strong velocity and temperature gradients at the wall, it is impossible to resolve the velocity and temperature profiles with the desired accuracy using standard grid spacing. Extremely fine grids with a sufficient resolution would considerably increase computational time and are usually not applicable in CFD calculations. Instead, it is possible to develop a semi-empirical solution of the velocity and temperature profiles in the boundary layer, so-called wall-functions, and to use these profiles as sub-models in CFD codes. In this case, the computed flow velocities and temperatures at the grid points closest to the wall are matched to the wall functions in order to get the correct velocities and temperatures in the boundary layer, which are then used to determine the shear stresses and heat fluxes.

The boundary layer is divided into an inner and an outer region, Fig. 3.11. While the outer region strongly depends on the flow conditions outside the boundary layer, the inner region, which has to be described by a sub-model in CFD codes, does not. The inner region is again divided into a so-called viscous wall layer directly at the wall and a so-called fully turbulent layer on top of the viscous sublayer. Both layers can be described by self-similar universal dimensionless velocity and temperature profiles, which are then used as standard models in CFD codes (e.g. [11, 6]).

The flow in a turbulent boundary layer is described by the turbulent boundary layer equations, which are directly derived from the Reynolds averaged conservation equations [3, 11]. Assuming a two-dimensional ($u_1 = u$, $u_2 = v$), steady, and incompressible flow over a flat plate with negligible pressure gradient parallel to the wall, the equations read

$$\frac{\partial \overline{u}}{\partial x} + \frac{\partial \overline{v}}{\partial y} = 0 , \tag{3.116}$$

$$\overline{u}\frac{\partial \overline{u}}{\partial x} + \overline{v}\frac{\partial \overline{u}}{\partial y} = \frac{1}{\rho}\frac{\partial \tau}{\partial y} , \tag{3.117}$$

$$\rho c_p \left(\overline{u}\frac{\partial \overline{T}}{\partial x} + \overline{v}\frac{\partial \overline{T}}{\partial y} \right) = \frac{\partial \dot{q}}{\partial y} , \tag{3.118}$$

where

$$\tau = \mu \frac{\partial \overline{u}}{\partial y} - \rho \overline{u'v'} = \left(\mu + \mu_t \right)\frac{\partial \overline{u}}{\partial y} = \rho \left(v + v_t \right)\frac{\partial \overline{u}}{\partial y} , \tag{3.119}$$

$$\dot{q} = \lambda \frac{\partial \overline{T}}{\partial y} - \rho c_p \overline{v'T'} = \left(\lambda + \lambda_t \right)\frac{\partial \overline{T}}{\partial y} = \rho c_p \left(a + a_t \right)\frac{\partial \overline{T}}{\partial y} . \tag{3.120}$$

In Eqs. 3.119 and 3.120, v_t and a_t are the turbulent viscosity and turbulent thermal diffusivity.

Near the wall $\bar{v} = 0$, and the term $\partial \bar{u}/\partial x$ can also be neglected. Hence, the momentum equation simplifies to $\partial \tau /\partial y = 0$. Integrating from $y = 0$ ($\tau(y = 0) = \tau_w$) to y results in $\tau = \tau_w = $ constant. Further on, the heat flux to the wall is also constant, which can be easily shown by using $\bar{v} = 0$ and neglecting $\partial \bar{T} /\partial x$ in the energy equation. Thus, the resulting shear stress, which is the sum of the molecular and the turbulent stresses, is constant near the wall:

$$\tau_w = \rho\left(v + v_t\right) d\bar{u} / dy \,. \tag{3.121}$$

In the viscous sublayer directly at the wall, turbulent velocity fluctuations are of minor importance, the flow is more or less laminar, and the molecular momentum transport is much greater than the turbulent one ($v_t \ll v$). Thus, the resulting shear stress is

$$\tau_w = \rho v d\bar{u} / dy \,, \tag{3.122}$$

and the integration yields

$$\bar{u}\left(y\right) = \frac{\tau_w}{\rho v} y \,. \tag{3.123}$$

Using the so-called shear velocity $u_\tau = \left(\tau_w / \rho\right)^{0.5}$ gives

$$\frac{\bar{u}}{u_\tau} = \frac{u_\tau y}{v} \,. \tag{3.124}$$

The left hand side of Eq. 3.124 is the dimensionless velocity $u^+ = \bar{u} / u_\tau$, while the right hand side is the dimensionless distance from the wall $y^+ = u_\tau y /v$. Hence, the universal velocity profile in the viscous sublayer is a linear relationship:

$$u^+ = y^+ \,. \tag{3.125}$$

Experiments show that the linear distribution holds up to $y^+ \approx 5.0$, which may be taken to be the limit of the viscous sublayer, Fig. 3.11.

In the fully turbulent layer on top of the viscous layer, the turbulent momentum transport is predominant ($v_t \gg v$), and

$$\tau_w = \rho v_{turb} \frac{du}{dy} = \rho l^2 \left(\frac{d\bar{u}}{dy}\right)^2 \,, \tag{3.126}$$

if Prandtl's mixing length model is utilized. Using $l = \kappa y$ in Eq. 3.126 gives

$$\frac{d\bar{u}}{dy} = \frac{1}{\kappa y} \sqrt{\frac{\tau_w}{\rho}} \,, \tag{3.127}$$

which can be transformed into

$$\frac{d\overline{u}}{u_\tau} = \frac{1}{\kappa}\frac{dy}{y}. \tag{3.128}$$

The integration finally yields the universal velocity profile of the fully turbulent sublayer,

$$u^+ = \frac{1}{\kappa}\ln y^+ + C, \tag{3.129}$$

where the Karman constant κ and the constant C must be experimentally determined and are typically set to $\kappa = 0.4$ and $C = 5.5$.

The transition region between the linear and logarithmic velocity profiles ($5 < y^+ < 30$) with no specified velocity profile is called the buffer layer, Fig. 3.11. This region is usually neglected since it has turned out that the two-layer model is already a good approximation.

Because the thermal boundary layer is very similar to a velocity boundary layer, the two-layer model can also be applied to Eq. 3.120 for a viscous and a turbulent sublayer [11],

$$\frac{\dot{q}_w}{\rho c_p} = \left(a + a_t\right)\frac{d\overline{T}}{dy}, \tag{3.130}$$

where a and a_t are the molecular and turbulent thermal diffusivities, and $\dot{q}_w$ = constant. Using the dimensionless distance from the wall, $y^+ = u_\tau y / \nu$, as well as the dimensionless boundary layer temperature,

$$T^+ = \frac{\left(\overline{T} - T_w\right)u_\tau}{\dot{q}_w / \rho c_p}, \tag{3.131}$$

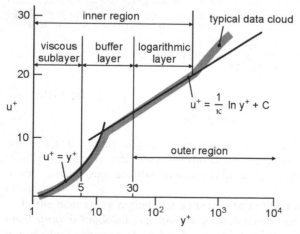

Fig. 3.11. Law of the wall: velocity profile in a turbulent boundary layer [12]

results in

$$\frac{dT^+}{dy^+} = \left(\frac{1}{Pr} + \frac{a_t}{v}\right)^{-1},$$ (3.132)

where $Pr = v/a$ is the Prandtl number. Neglecting the term (a_t/v) in the viscous sublayer and integrating Eq. 3.132 finally yields

$$T^+ = Pr\, y^+.$$ (3.133)

Neglecting the term $(1/Pr)$ in the fully turbulent layer gives

$$\frac{dT^+}{dy^+} = \frac{v}{v_t} Pr_t = \frac{1}{\kappa y^+} Pr_t,$$ (3.134)

where $Pr_t = v_t/a_t$ is the turbulent Prandtl number. Again, Prandtl's mixing length model, $v_t = \kappa^2 y^2\, d\bar{u}/dy$, or

$$\frac{v_t}{v} = \kappa^2 y^{+2} \frac{du^+}{dy^+},$$ (3.135)

is applied, where $du^+/dy^+ = (\kappa y^+)^{-1}$ is derived from Eq. 3.129. Integrating Eq. 3.135 finally yields the universal temperature profile in the fully turbulent sublayer:

$$T^+ = \frac{Pr_t}{\kappa}\left(\ln y^+ - \ln y^+_{visc}\right) + Pr\, y^+_{visc}.$$ (3.136)

In Eq. 3.136, $y^+_{visc} \approx 13.2$ is the thickness of the viscous sublayer and thus the beginning of the turbulent sublayer, $\kappa = 0.41$, and $Pr_t = 0.9$ [11].

In order to calculate the heat flux to the wall from Eq. 3.131, the values of the dimensionless temperature T^+ and the absolute temperature $\bar{T}$ at some known distant y^+ from the wall must be given. T^+ is known from Eqs. 3.133 and 3.136 for any position in the viscous and turbulent sublayers. $\bar{T}$ is the value corresponding to the first grid cell from the wall. However, the grid cells in CFD applications are often much greater than the boundary layer thickness, and in these cases y^+ exceeds the range of applicability of the logarithmic velocity profile (approx. $y^+ <$ 200), which is a well-known source of uncertainty. One approach to eliminate these uncertainties is the use of adaptive meshes near the wall, where the grid spacing at the cylinder wall is dynamically adjusted to the required resolution during the calculation (e.g. [15]), see also Chap. 5.

3.1.5 Application to In-Cylinder Processes

The conservation equations derived in the previous sections are valid for single-component continua. However, in the case of in-cylinder processes, the gas phase consists of several components (e.g. nitrogen, oxygen, fuel vapor, combustion products). Hence, there is a multi-component gas mixture with changing composition, and additional transport terms have to be added to the conservation equations that account for a change of species concentration due to mass diffusion and chemical reactions.

In addition to this, the injection of liquid fuel into the gas phase results in the presence of a two-phase flow, either in the combustion chamber itself (direct injection), or in the intake manifold. Due to aerodynamic drag, evaporation, heat exchange between liquid and gas etc., the fuel droplets (dispersed liquid phase) exchange momentum, mass, and energy with the gas phase. Hence, appropriate source terms have to be included in the conservation equations of the continuous gas phase as well. Details about the derivation of the additional source and transport terms are given in [2] for example.

Altogether, the conservation equations for mass, momentum, and energy for the gas phase have to be modified as follows. Because the gas phase consists of different species, there is no longer just one single mass conservation equation. The mass conservation for each species m becomes

$$\frac{\partial \rho_m}{\partial t} + \frac{\partial (\rho_m u_i)}{\partial x_i} = \frac{\partial}{\partial x_i}\left(\rho D \frac{\partial(\rho_m/\rho)}{\partial x_i}\right) + \dot{\rho}_{m,spray} + \dot{\rho}_{m,comb}, \tag{3.137}$$

where ρ_m is the species density. The first term on the right hand side accounts for species mass diffusion, the last two terms are due to spray and combustion. The total fluid density equation,

$$\frac{\partial \rho}{\partial t} + \frac{\partial(\rho u_i)}{\partial x_i} = \dot{\rho}_{spray}, \tag{3.138}$$

is obtained by summing Eq. 3.137 over all species m. The source term due to combustion does not appear in this equation, because chemical reactions do not change the total mass.

In the momentum equations the change of the gas phase momentum due to the interaction with the spray is included by the source term $f_{j,spray}$:

$$\frac{\partial(\rho u_j)}{\partial t} + \frac{\partial(\rho u_j u_i)}{\partial x_i} = -\frac{\partial p}{\partial x_j} + \frac{\partial \tau_{ij}}{\partial x_i} + f_j + f_{j,spray}. \tag{3.139}$$

In the energy equation,

$$c_p\left(\frac{\partial(\rho T)}{\partial t} + \frac{\partial(\rho T u_i)}{\partial x_i}\right) =$$

$$\lambda \frac{\partial^2 T}{\partial x_i^2} + \frac{\partial}{\partial x_i}\left(\rho D \sum_m h_m \frac{\partial(\rho_m/\rho)}{\partial x_i}\right) + \rho\varepsilon + \frac{\dot{Q}_{spray} + \dot{Q}_{comb}}{dx_1 dx_2 dx_3}, \tag{3.140}$$

the second term on the right hand side is due to enthalpy transfer caused by mass diffusion of the different species, the third term includes dissipation of turbulent kinetic energy, and the fourth term includes two source terms due to spray and combustion.

The various sub-models used in CFD codes in order to determine the different source terms in the conservation equations of the gas phase and to calculate the mass, momentum, and energy exchange between the dispersed liquid and the continuous gas phase are discussed in Chap. 4.

3.2 Description of the Disperse Phase

3.2.1 Spray Equation

The task of simulating the spray break-up consists in calculating for each time t and at all positions $\vec{x} = (x, y, z)$ the values of droplet radius r, droplet velocity $\vec{u}$, and droplet temperature T inside a spray consisting of millions of droplets. A statistical description of the spray can be given using the droplet probability distribution function (PDF) $f(\vec{x}, \vec{u}, r, T, t)$, which is defined in such a way that

$$f(\vec{x}, \vec{u}, r, T, t) d\vec{u} dr dT \qquad (3.141)$$

is the probable number of droplets per unit volume at time t and point $\vec{x}$, the velocities of which are in the range of $d\vec{u}$ around $\vec{u}$, while the droplet radius is in the interval dr around r, and the droplet temperature is in the interval dT around T. Thus, f is a function in a nine-dimensional space (three spatial and three velocity coordinates, one dimension for radius, temperature, and time). If further droplet properties like droplet deformation and deformation velocity are regarded, even more dimensions have to be considered. The time evolution of f can be described in differential form by a transport equation, the so-called spray equation, (Williams [21], Ramos [19]),

$$\frac{\partial f}{\partial t} + \nabla_{\vec{x}} \left(f \vec{u} \right) + \nabla_{\vec{u}} \left(f \vec{a} \right) + \frac{\partial}{\partial T} \left(f \frac{dT}{dt} \right) + \frac{\partial}{\partial r} \left(f \frac{dr}{dt} \right) = \dot{f}_{source}, \qquad (3.142)$$

which was derived using phenomenological considerations. In Eq. 3.142, dr/dt and dT/dt are the rates of change of droplet size and temperature (vaporization and heating), and $\vec{a} = (d\vec{u}/dt)$ is the acceleration of the liquid drop (drop drag). The effect of additional processes that could alter the number of droplets, like break-up processes, collision etc., is represented by the source term on the right hand side. Further information about its modeling is given in Amsden et al. [2] for example. Because the individual terms of the spray equation are again described by functions and differential equations, the spray equation becomes extremely complex and cannot be solved directly. The approach used today in order to obtain a satisfying approximation of the exact solution is the Stochastic-Parcel Technique, which is based in turn on the Monte-Carlo Method.

3.2.2 Monte-Carlo Method

The basic idea of the Monte-Carlo method is that it is possible to approximate an overall solution of a complex problem without knowing each of the numerous sub-solutions (often an infinite number) contributing to the overall solution. As an example, Fig. 3.12 shows the function $F(x)$, whose value y at each position x is known, but which cannot be integrated. The complex problem is to determine the area below the curve. Using the Monte-Carlo Method, a fixed number of discrete points is regarded, whose positions inside the known area $A = H \cdot L$ are randomly distributed. The procedure is now to detect for each point whether it is above or below the curve. If for example 70% of the discrete points are below the curve, the solution of the complex problem is that the area below the curve in Fig. 3.12 represents approximately 70% of A. Hence, using the Monte-Carlo-Method, the behavior of a few discrete points is taken to estimate the behavior of all points (in this example an infinite number) that would be necessary in order to get the exact solution. The more discrete points are used, the better the approximation.

3.2.3 Stochastic-Parcel Method

Applying the Monte-Carlo method to the calculation of sprays consisting of millions of droplets, only the behavior of a subset (discrete number) of all droplets is calculated in detail. It is assumed that the properties and behavior of all droplets in the spray can be approximately represented by these discrete droplets. Nevertheless, in order to have the correct fuel mass in the cylinder, the mass of all droplets must be regarded. This is why every representative droplet gets a number of further droplets with identical size, temperature and velocity components etc., who have exactly the same behavior and properties like the representative one. These groups of equal droplets are called parcels. Thus, the spray is represented by a stochastic system of a discrete number of parcels (Stochastic Parcel Method, Dukowicz [5]).

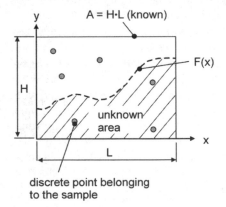

Fig. 3.12. Basic idea of the Monte-Carlo Method

3.2.4 Eulerian-Lagrangian Description

In order to calculate a spray penetrating into the gaseous atmosphere of the combustion chamber, two phases, the dispersed liquid and the continuous gas phase, must be considered. The gas phase is described using the Navier-Stokes equations in conjunction with a turbulence model, usually the well-known k-ε model. For each time t, this Eulerian formulation describes the behavior of the relevant flow parameters like velocity components, pressure, density, and temperature as a function of position (x,y,z) for the whole three-dimensional flow field. The Eulerian formulation is not appropriate for the description of the disperse phase (no continuum), and the Lagrangian description is used. The droplets are treated as individual mass points, the velocity of which is reduced by drag forces due to the relative velocity between gas and droplet, the mass of which is decreased by evaporation, etc. The behavior of both phases is coupled because momentum (droplet velocity is reduced, gas velocity is increased), energy (heat exchange), and mass (evaporated mass passes over to the gas phase) are exchanged. Source terms in the conservation equations of the gas phase enable the increase or decrease of momentum, energy, and mass in each grid cell. The effect of the gas phase on the dispersed liquid is accounted for by using the actual data (temperature, gas velocity etc.) of the grid cell the droplet is crossing at time t as a boundary condition. The numerous sub-models that describe the change of droplet momentum, energy, and mass are presented in Chap. 4. The gas phase source terms of each grid cell are then obtained by summing the rates of change of mass, momentum, and energy of all drops inside the cell at time t.

References

[1] Abraham J, Magi V (1997) Computations of Transient Jets: RNG k-ε Model Versus Standard k-ε Model. SAE paper 970885
[2] Amsden AA, O'Rourke PJ, Butler TD (1989) KIVA II: A Computer Program for Chemically Reactive Flows with Sprays. Los Alamos National Laboratory, LA-11560-MS
[3] Cebeci T, Bradshaw P (1988) Physical and Computational Aspects of Convective Heat Transfer. Springer-Verlag
[4] Cebeci T, Smith AMO (1974) Analysis of Turbulent Boundary Layers. Academic, New York
[5] Dukowicz JK (1980) A Particle-Fluid Numerical Model for Liquid Sprays. Journal of Computational Physics, vol 35, pp 229–253
[6] Ferziger JH, Peric M (1996) Computational Methods for Fluid Dynamics. Springer-Verlag
[7] Han Z, Reitz RD (1995) Turbulence Modeling of Internal Combustion Engines Using RNG k-ε Models. Combustion Science Technology, vol 106, pp 267–295
[8] Hanjalic K (1994) Advanced Turbulence Closure Models: Review of Current Status and Future Prospects. Int J Heat Fluid Flow, 15, pp 178–203

[9] Jones WP, Launder BE (1972) The Prediction of Laminarization with a Two-Equation Model of Turbulence. Int J Heat and Mass Transfer, vol 15, p 301

[10] Jones WP (1994) Turbulence Modelling and Numerical Solution Methods for Variable Density and Combusting Flows. In Libby PA and Williams FA (editors): Turbulent Reacting Flows. Academic Press, London, pp 309–374

[11] Kays WM, Crawford ME, Weigand B (2005) Convective Heat and Mass Transfer. McGraw-Hill

[12] Kundu PK (1990) Fluid Mechanics. Academic Press

[13] Landau LD, Lifshitz EM (1959) Fluid Mechanics. Addison-Wesley

[14] Launder BE, Spalding DB (1974) The Numerical Computation of Turbulent Flows. Comp Meth Appl Mech Eng 3, pp 269–289

[15] Lettmann H, Eckert P, Baumgarten C, Merker GP (2004) Assessment of Three-dimensional In-Cylinder Heat Transfer Models in DI Diesel Engines. 4th European Thermal Sciences Conference, Birmingham

[16] Oertel H, Böhle M (2002) Strömungsmechanik. Second edition, Vieweg-Verlag

[17] Peters N (2000) Turbulent Combustion. Cambridge University Press

[18] Potter MC, Wiggert DC (1997) Mechanics of Fluids. Second Edition, Prentice-Hall Inc

[19] Ramos JI (1989) Internal Combustion Engine Modeling. Hemisphere Publishing Corporation

[20] White FM (1991) Viscous Fluid Flow. McGraw-Hill Inc

[21] Williams FA. (1965) Combustion Theory: Fundamental Theory of Chemically Reacting Flow Systems. Addison-Wesley Publishing Company Inc, Massachusetts

[22] Yakhot V, Orszag SA (1986) Renormalization Group Analysis of Turbulence. I. Basic Theory. Journal of Scientific Computing, vol 1(1), pp 3–51

[23] Yakhot V, Orszag SA (1992) Development of Turbulence Models for Shear Flows by a Double Expansion Technique. Phys Fluids, vol 4(7), pp 1510–1520

4 Modeling Spray and Mixture Formation

4.1 Primary Break-Up

The primary break-up process provides the starting conditions for the calculation of the subsequent mixture formation inside the cylinder, and for this reason a detailed modeling of the transition from the nozzle flow into the dense spray is essential. Because the Lagrangian description of the liquid phase requires the existence of drops, the simulation of spray formation always begins with drops starting to penetrate into the combustion chamber. The task of a primary break-up model is to determine the starting conditions of these drops, such as initial radius and velocity components (spray angle), which are mainly influenced by the flow conditions inside the nozzle holes.

There are only very few detailed models for the simulation of primary break-up of high-pressure sprays. One reason is that the experimental investigation is extremely complicated because of the dense spray and the small dimensions. Thus, it is difficult to understand the relevant processes and to verify primary break-up models. On the other hand, it is now possible to simulate the flow inside high-pressure injectors, but because of different mathematical descriptions of the liquid phase inside (Eulerian description) and outside the nozzle (Lagrangian description), it is not possible to calculate the primary break-up directly, and models must be used.

Different classes of break-up models exist concerning the way the relevant mechanisms like aerodynamic-induced, cavitation-induced and turbulence-induced break-up are treated. The simpler the model, the less input data is required, but the less the nozzle flow is linked with the primary spray and the more assumptions about the upstream conditions have to be made. This results in a significant loss of quality concerning the prediction of structure and starting conditions of the first spray near the nozzle. On the other hand, an advantage of the simpler models is that their area of application is wider because of the more global modeling. Furthermore, detailed models often require a complete CFD simulation of the injector flow as input data. This results in an enormous increase of computational time, but the close linking of injector flow and spray guarantees the most accurate simulation of the primary break-up process and its effect on spray and mixture formation in the cylinder that is possible today.

It must be pointed out that all kind of models have their special field of application. Depending on the available input data, the computational time, the relevant break-up processes of the specific configuration as well as the required accuracy of the simulation, the appropriate model has to be chosen.

4.1.1 Blob-Method

The simplest and most popular way of defining the starting conditions of the first droplets at the nozzle hole exit of full-cone diesel sprays is the so-called blob-method. This approach was developed by Reitz and Diwakar [114, 118]. The blob method is based on the assumption that atomization and drop break-up within the dense spray near the nozzle are indistinguishable processes, and that a detailed simulation can be replaced by the injection of big spherical droplets with uniform size, which are then subject to secondary aerodynamic-induced break-up, see Fig. 4.1. The diameter of these blobs equals the nozzle hole diameter D (mono-disperse injection) and the number of drops injected per unit time is determined from the mass flow rate. Although the blobs break up due to their interaction with the gas, there is a region of large discrete liquid particles near the nozzle, which is conceptually equivalent to a dense core. Assuming slug flow inside the nozzle hole, the conservation of mass gives the injection velocity $U_{inj}(t)$ of the blobs

$$U_{inj}(t) = \frac{\dot{m}_{inj}(t)}{A_{hole}\rho_l},$$

(4.1)

where $A_{hole} = \pi D^2/4$ is the cross-sectional area of the nozzle hole, ρ_l is the liquid density, and $\dot{m}_{inj}(t)$ is the fuel mass flow rate (measurement).

If there are no measurements about the injected mass flow, the Bernoulli equation for frictionless flow can be used in order to calculate an upper limit of the initial velocity,

$$U_{inj,max} = \sqrt{\frac{2\Delta p_{inj}}{\rho_l}},$$

(4.2)

where Δp_{inj} is the difference between the sac hole and combustion chamber pressures. Because the flow is not frictionless, $U_{inj,max}$ is reduced by energy losses. According to measurements of Schugger et al. [127], Walther et al. [145], and Meingast et al. [86] the flow velocity at the nozzle hole exit is about 70%–90% of the Bernoulli velocity.

In order to define the velocity components of each blob, the spray cone angle ϕ must be known from measurements or has to be estimated using semi-empirical relations (e.g. Hiroyasu and Arai [53]). The direction of the resulting velocity U_{inj} of the primary blob inside the 3D spray cone is randomly chosen by using two random numbers ξ_1 and ξ_2 in the range of [0, 1] in order to predict the azimuthal angle φ and the polar angle ψ in the spherical coordinate system, see Fig. 4.2:

$$\varphi = 2\pi\xi_1,$$

(4.3)

$$\psi = \frac{\phi}{2}\xi_2.$$

(4.4)

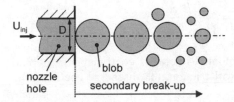

Fig. 4.1. Blob-method

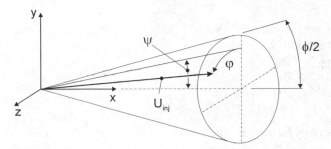

Fig. 4.2. 3D spray cone angle and coordinate system

Kuensberg et al. [70] have developed an enhanced blob-method that calculates an effective injection velocity and an effective blob diameter dynamically during the entire injection event taking the reduction of nozzle flow area due to cavitation into account. The given mass flow and the nozzle geometry (length L, diameter D, radius of inlet edge r) are used as input parameters for a one-dimensional analytical model of the nozzle hole flow. During injection, this model determines for every time step whether the nozzle hole flow is turbulent or cavitating.

The static pressure p_1 at the vena contracta (point 1 in Fig. 4.3), which can be estimated using the Bernoulli equation for frictionless flow from point 0 to point 1,

$$p_1 = p_0 - \frac{\rho_l}{2} u_{vena}^2 , \qquad (4.5)$$

must be lower than the vapor pressure p_{vap} in the case of cavitating flow and higher in the case of turbulent flow. In Eq. 4.5, the inlet pressure p_0 and the velocity u_{vena} at the smallest flow area are unknown. The inlet pressure p_0 is estimated using again the Bernoulli equation

$$p_0 = p_2 + \frac{\rho_l}{2} \left(\frac{2(p_0 - p_2)}{\rho_l} \right) = p_2 + \frac{\rho_l}{2} \left(\frac{u_{mean}}{C_d} \right)^2 , \qquad (4.6)$$

where $u_{mean} = \dot{m}_{inj} / (A_{hole}\, \rho_l)$ is the average velocity inside the hole assuming slug flow and

$$C_d = \frac{\dot{m}_{inj}}{\dot{m}_{Bernoulli}} = \frac{\rho_l A_{hole} u_{mean}}{\rho_l A_{hole} u_{Bernoulli}} = \frac{u_{mean}}{\left(2(p_0 - p_2)/\rho_l\right)^{0.5}} \qquad (4.7)$$

is the discharge coefficient. By using tabulated inlet loss coefficients (k_{inlet}) and the Blasius or the laminar equation for wall friction, the discharge coefficient is given by

$$C_d = \left(k_{inlet} + f \cdot L/D + 1\right)^{-0.5}, \qquad (4.8)$$

$$f = \max\left(0.316 \cdot Re^{-0.25}, 64/Re\right). \qquad (4.9)$$

Assuming a flat velocity profile and using Nurick's expression for the size of the contraction at the vena contracta [92],

$$A_{vena} = A_{hole} C_c, \quad C_c = \left[\left(\frac{1}{C_{c0}}\right)^2 - 11.4\frac{r}{D}\right]^{-0.5}, \qquad (4.10)$$

where $C_{c0} = 0.61$. Mass conservation gives the velocity at the smallest flow area:

$$u_{vena} = \frac{\dot{m}_{inj}}{A_{hole} C_c \rho_l} = \frac{u_{mean}}{C_c}. \qquad (4.11)$$

At the beginning and end of injection, the flow is usually turbulent, the blob size equals the nozzle hole diameter D, and the injection velocity is calculated using Eq. 4.1. During the main injection phase, the flow is usually cavitating, see Fig. 4.3. In this case, the effective cross-sectional area of the nozzle hole exit A_{eff} is smaller than the geometrical area A_{hole} resulting in a decrease of the blob diameter,

$$D_{eff} = \sqrt{\frac{4A_{eff}}{\pi}}, \qquad (4.12)$$

while the injection velocity is increased. A momentum balance from the vena contracta (point 1) to the hole exit (point 2), together with the conservation of mass,

$$\dot{m}_{inj} = \rho_l u_{vena} A_{hole} C_c = \rho_l u_{eff} A_{eff}, \qquad (4.13)$$

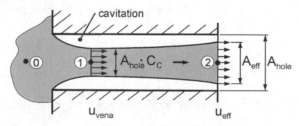

Fig. 4.3. One-dimensional cavitating nozzle hole flow

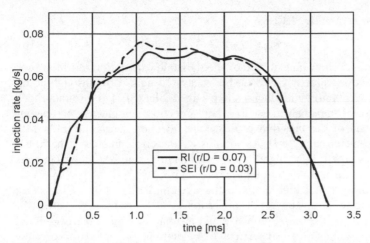

Fig. 4.4. Injection profiles for sharp-edged (SEI) and round-edged inlet (RI), [70]

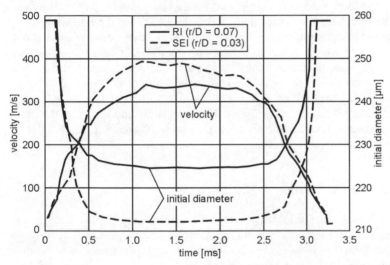

Fig. 4.5. Dynamic calculation of blob diameter and injection velocity, data from [70]

gives the injection velocity

$$u_{eff} = \frac{A_{hole}}{\dot{m}_{inj}}\left(p_{vap} - p_2\right) + u_{vena},$$ (4.14)

where

$$u_{vena} = \frac{\dot{m}_{inj}}{\rho_l A_{hole} C_c}.$$ (4.15)

The effective flow area in Eq. 4.12 is

$$A_{eff} = \dot{m}_{inj} / \left(u_{eff} \rho_l \right) . \tag{4.16}$$

Figures 4.4 and 4.5 show an example of the dynamic calculation of initial blob diameter and injection velocity based on the measured mass flow through a six-hole injector ($D = 259$ μm) with either a sharp-edged inlet (SEI) or a round-edged inlet (RI). Because of the small mass flows and injection velocities at the beginning and at the end of the injection process, the injection starts and ends with blobs having a diameter equal to D. As soon as cavitation is predicted, the initial diameter decreases rapidly. The highest velocities and the smallest blob diameters are predicted during the main injection process.

Compared to the original blob-method, the dynamic calculation of blob size and injection velocity during the whole injection event introduces the effect of cavitation by decreasing the initial blob size and estimating a more realistic initial velocity. However, only the passive effect of cavitation, the reduction of flow area, is considered. The increase of turbulence and break-up energy due to cavitation bubble implosions is not included.

Altogether, the blob method is a simple and well-known method of treating the primary break-up in Eulerian-Lagrangian CFD codes. As far as there is no detailed information about the composition of the primary spray, and measurements about the spray cone angle are available, it is the best way to define the initial starting conditions for the liquid entering the combustion chamber. Nevertheless, this method does not represent a detailed physical and satisfying modeling of the relevant processes during primary break-up. The most important disadvantage is that the influence of the 3D nozzle hole flow on 3D spray angle and drop size distribution cannot be mapped and that the promotion of primary break-up by turbulence and by implosions of cavitation bubbles outside the nozzle is not regarded at all.

4.1.2 Distribution Functions

This method assumes that the fuel is already fully atomized at the nozzle exit and that the distribution of drop sizes can be described by mathematical functions. In this case, a distribution of droplet sizes is injected. In high-pressure sprays, neither the droplet sizes nor their distribution in the dense spray near the nozzle could be quantified experimentally up to now. Thus the droplet size distribution must be guessed and iteratively adjusted until the measured drop sizes in the far field of the nozzle are similar to the simulated ones. This of course does not represent a detailed modeling of the relevant processes during primary break-up, but can be used as an alternative to the mono-disperse injection of the blob-method.

Full-Cone Sprays

Martinelli et al. [84], for example, assume an initial distribution of drop sizes at the nozzle with the Sauter mean diameter (SMD) given by the correlation

$$SMD = A_2 \cdot \left[12\pi \cdot \frac{\sigma}{\rho_g \cdot U_{inj}^2} \right],$$
(4.17)

where A_2 is independent of the nozzle geometry and of order one. Equation 4.17 accounts for the effect that the initial mean drop size decreases when the chamber pressure increases.

According to Levy et al. [74], the droplet's diameter D at the nozzle exit should be sampled from a χ^2-law in order to get a good agreement between measured and simulated downstream drop size distributions,

$$P(D) = \frac{1}{6\bar{D}^4} \cdot D^3 \cdot e^{-D \cdot \bar{D}},$$
(4.18)

where $\bar{D} = SMD / 6$. Further analytical functions to match size distributions in diesel sprays are published in Long et al. [82], Simmons [131], and Levebvre [73].

Hollow-Cone Sprays

In the case of a hollow-cone spray, an annular liquid sheet is produced at the nozzle orifice forming a free cone-shaped liquid sheet that disintegrates into droplets. The simulation of the liquid phase starts at the point of sheet disintegration, when the first droplets are formed. Two functions are commonly used in order to sample the initial drop sizes, the χ^2 and the Rosin-Rammler distribution. In the case of the χ^2 distribution, the volume distribution (cumulative distribution) V is given by

$$V = 1 - \exp\left(-\frac{D}{\bar{D}}\right)\left[1 + \frac{D}{\bar{D}} + \frac{1}{2}\frac{D^2}{\bar{D}} + \frac{1}{6}\frac{D^3}{\bar{D}}\right],$$
(4.19)

where V is the fraction of the total volume contained in drops of diameter less than D, and $\bar{D} = SMD/3$ is a characteristic mean drop size. The corresponding volume distribution (dV/dD) is shown in Fig. 4.6.

In the case of the Rosin-Rammler distribution, the cumulative volume distribution V is given by

$$V = 1 - \exp\left(-\left(\frac{D}{\bar{D}}\right)^q\right),$$
(4.20)

and the corresponding volume distribution is

$$\frac{dV}{dD} = \frac{qD^{q-1}}{\bar{D}^q} \exp\left(-\frac{D^q}{\bar{D}}\right),$$
(4.21)

where D is the size of the individual droplets, q is the distribution parameter ($q \approx 3.5$, [49]), and

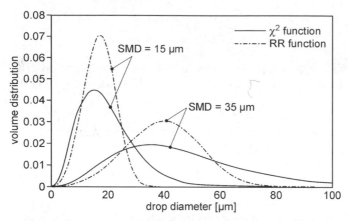

Fig. 4.6. Comparison of the Rosin-Rammler (RR) and χ^2 size distributions [49]

$$\bar{D} = SMD \cdot \Gamma\left(1 - q^{-1}\right). \tag{4.22}$$

SMD is the Sauter mean diameter of the distribution and Γ is the gamma function:

$$\Gamma(x) = \int_0^\infty e^{-t} t^{x-1} dt. \tag{4.23}$$

Figure 4.6 shows a comparison between the χ^2 and the Rosin-Rammler distribution. For a given Sauter mean diameter, the χ^2 distribution yields larger standard deviation and includes more large drops, while the Rosin-Rammler distribution contains more droplets with sizes closer to the SMD value. The larger droplets of the χ^2 distribution will evaporate slower and result in larger tip penetration due to the higher momentum. In the case of pressure-swirl atomizers, the investigations of Han et al. [49] have shown that the measured droplet sizes are better represented by the Rosin-Rammler distribution.

Maximum Entropy Formalism

Instead of defining the shape of the distribution function a priori by mathematical distribution functions, it is possible to predict the function using the so-called Maximum Entropy Formalism (MEF). Due to the use of a physically based criterion, the entropy of the distribution given by Shannon [130], it is possible to estimate the most probable distribution function. For example, if the mean diameter of a spray just after primary break-up is known, either from theoretical calculation or from data extrapolation, and if mass, energy, and momentum have to be conserved (constraints), there is still an infinite number of possible distribution functions that may fit it. Using the Shannon entropy, the one with the largest entropy can be chosen. This distribution is regarded as the most probable one.

Thus, the MEF is a method of statistical inference that provides the least biased estimate of a probability distribution, consistent with a set of constraints that express the available information about the relevant phenomena.

The idea of entropy (information uncertainty) of a probability distribution was introduced by Shannon [130], who showed that for a set of n states, each of them having the probability P_i, the uncertainty of the probability distribution is given by

$$S(P_i) = -k\sum_{i=1}^{n} P_i \ln P_i , \qquad (4.24)$$

where k is a constant. Because of its similarity to the thermodynamic entropy, S is called Shannon's entropy. The constraints mathematically express the mean values of information available. If $f_k(D)$ is a function describing a droplet property (e.g. mass or momentum) and its known mean value over all droplets is F_k, the corresponding constraint written for all the size classes of the spray spectrum is

$$\sum_{i=1}^{n} P_i f_k(D_i) = F_k . \qquad (4.25)$$

In the case of mass conservation ($f_k(D_i) = m_i$), the mass m before atomization must be equal to the sum of all droplet masses after atomization:

$$m = \sum_{i=1}^{n} m_i = \frac{\pi}{6} \rho_l N \sum_{i=1}^{n} D_i^3 P_i . \qquad (4.26)$$

Hence, it is assumed that evaporation can be neglected during break-up. In Eq. 4.26, n is the number of drop size classes, P_i is the probability that a droplet is in the size class i, D_i is the arithmetic mean diameter of the drops in size class i, and N is the total number of droplets inside the control volume, see Fig. 4.7. The inclusion of the mean spray diameter D_{30} (e.g. known from experiment) via the relation $m = (\pi/6)\rho_l N(D_{30})^3$ finally results in the mass conservation constraint for MEF:

$$\sum_{i=1}^{n} \frac{D_i^3}{D_{30}^3} P_i = 1 . \qquad (4.27)$$

Further constraints may be derived from momentum and energy conservation, e.g. conservation of kinetic energy in main spray direction. One general constraint, which always has to be satisfied, is the normalization constraint,

$$\sum_{i=1}^{n} P_i = 1 . \qquad (4.28)$$

The problem of finding the most likely size distribution for the given mean diameter D_{30} is expressed mathematically by a maximization problem: the maximum of the function defined in Eq. 4.24 has to be calculated subject to the constraints.

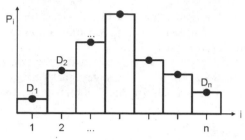

Fig. 4.7. Droplet size distribution: definition of relevant parameters

This problem is usually solved using the method of Lagrange multipliers. The solution is an exponential distribution of the form

$$P_i = \exp\left(-\lambda_0 - \lambda_1 f_1(d_i)...- \lambda_m f_m(d_i)\right), \tag{4.29}$$

with as many Lagrange multipliers (λ_0, λ_1, λ_2,...λ_m) as constraints. To evaluate the Lagrange multipliers, the constraints (Eqs. 4.25 and 4.28) are used. This results in a system of non-linear equations that has to be solved [1]. Further information about the MEF is given, for example, in [128, 25, 2].

The MEF has been used by several authors in order to reconstruct drop size distributions. For example, Dobre and Bolle [28] used the MEF in order to predict the drop size distribution of an ultrasonic atomizer, and Cousin et al. [26] predicted drop size distributions in sprays from pressure-swirl atomizers. Gavaises and Arcoumanis [39] also applied the MEF to hollow-cone sprays and determined the droplet size distribution at every stage of liquid film atomization and, in the case of the subsequent droplet break-up, using only mass conservation and Eq. 4.28 as constraints. Arcoumanis et al. [7] estimated droplets sizes from secondary break-up processes in full-cone diesel sprays. Further applications of the MEF are summarized in [10].

4.1.3 Turbulence-Induced Break-Up

Huh and Gosman [56] have published a phenomenological model of turbulence-induced atomization for full-cone diesel sprays, which is also used to predict the primary spray cone angle. The authors assume that the turbulent forces within the liquid emerging from the nozzle are the producers of initial surface perturbations, which grow exponentially due to aerodynamic forces and form new droplets. The wavelength of the most unstable surface wave is determined by the turbulent length scale. The turbulent kinetic energy at the nozzle exit is estimated using simple overall mass, momentum, and energy balances.

The atomization model starts with the injection of spherical blobs the diameter of which equal the nozzle hole diameter D. Initial surface waves grow due to the relative velocity between gas and drop (Kelvin-Helmholtz (KH) mechanism) and break up with a characteristic atomization length scale L_A and time scale τ_A. The effects of turbulence are introduced by postulating that, on the one hand, the characteristic atomization length scale L_A is proportional to the turbulence length scale L_t,

$$L_A = C_1 L_t = C_2 L_w, \tag{4.30}$$

where $C_1 = 2.0$, $C_2 = 0.5$, and L_w is the wavelength of surface perturbations determined by turbulence, and that, on the other hand, the characteristic atomization time scale τ_A is a linear combination of the turbulence time scale τ_t (from nozzle flow) and the wave growth time scale τ_w (KH model),

$$\tau_A = C_3 \tau_t + C_4 \tau_w = \tau_{spontaneous} + \tau_{exponential}, \tag{4.31}$$

where $C_3 = 1.2$ and $C_4 = 0.5$ [55]. The spontaneous growth time is due to jet turbulence, while the exponential one is caused by the KH wave growth mechanism. The wave growth time scale τ_w provided by the KH instability theory applied to an infinite plane is

$$\tau_w = \left[\frac{\rho_l \rho_g}{\left(\rho_l + \rho_g\right)^2} \left(\frac{U_{inj}}{L_w}\right)^2 - \frac{\sigma}{\left(\rho_l + \rho_g\right)L_w^3} \right]^{-1} \tag{4.32}$$

for an inviscid liquid. The turbulent length and time scales L_{t0} and τ_{t0} at the time the blob leaves the nozzle are related to the average turbulent kinetic energy k_0 and the average energy dissipation rate ε_0 at the nozzle exit:

$$L_t = C_\mu \frac{k^{1.5}}{\varepsilon}, \quad L_{t0} = C_\mu \frac{k_0^{1.5}}{\varepsilon_0} \tag{4.33}$$

$$\tau_t = C_\mu \frac{k}{\varepsilon}, \quad \tau_{t0} = C_\mu \frac{k_0}{\varepsilon_0}, \tag{4.34}$$

where $C_\mu = 0.09$ is a constant given in the k-ε-model [71]. In the above equations k_0 and ε_0 are estimated as follows [55]:

$$k_0 = \frac{U_{inj}^2}{8L/D} \left[\frac{1}{C_d^2} - K_c - \left(1 - s^2\right) \right] \tag{4.35}$$

$$\varepsilon_0 = K_\varepsilon \frac{U_{inj}^3}{2L} \left[\frac{1}{C_d^2} - K_c - \left(1 - s^2\right) \right], \tag{4.36}$$

where C_d is the discharge coefficient, $K_\varepsilon = 0.27$ is a model constant, $K_c = 0.45$ and $s = 0.01$ are the form loss coefficient and the area ratio at the contraction corner (both values for sharp-edged entry), and L is the nozzle hole length.

In order to predict the primary spray cone angle, Huh and Gosman [56] assume that the spray diverges with a radial velocity L_{A0}/τ_{A0}. The combination of the radial and axial velocities gives the spray cone angle ϕ:

$$\tan\left(\frac{\phi}{2}\right) = \frac{L_{A0}/\tau_{A0}}{U_{inj}}.$$ (4.37)

The direction of the resulting velocity of the primary blob inside the 3D spray cone is randomly chosen, see also Sect. 4.1.1.

From the atomization length and time scales, Eqs. 4.30 and 4.31, the break-up rate of the primary blob and the size of the new secondary drops are derived (new drops are regarded as secondary ones). The break-up rate of a primary blob is set proportional to the atomization length and time scale with an arbitrary constant,

$$\frac{d}{dt}\left(d_{drop}(t)\right) = k_1 \frac{L_A(t)}{\tau_A(t)},$$ (4.38)

where $k_1 = 0.05$. Analogously to the KH model (Sect. 4.2.4), the primary blob radius is reduced by break-up. The values of the atomization length and time scales in Eq. 4.38 are time-dependent because outside the nozzle the internal turbulence of parent drops decays with time as they travel downstream. Hence, the actual values of turbulent kinetic energy and dissipation must be corrected. Assuming isotropic turbulence and negligible diffusion as well as convection and production of turbulent kinetic energy during this time, the simplified k-ε equations are

$$\frac{dk(t)}{dt} = -\varepsilon(t)$$ (4.39)

and

$$\frac{d\varepsilon(t)}{dt} = -C_\varepsilon \frac{\varepsilon(t)^2}{k(t)},$$ (4.40)

where $C_\varepsilon = 1.92$. These equations can be solved analytically. Eq. 4.39 can be rewritten as $\varepsilon(t)dt = -dk(t)$, and Eq. 4.40 gives $d\varepsilon(t)/\varepsilon(t) = -C_\varepsilon(\varepsilon(t)/k(t))dt$, which results in

$$\frac{d\varepsilon(t)}{\varepsilon(t)} = C_\varepsilon \frac{dk(t)}{k(t)}.$$ (4.41)

With the initial values k_0 and ε_0 at the time the blobs leave the nozzle, the integration from ε_0 to ε and k_0 to k gives the time-dependent dissipation rate

$$\varepsilon(t) = \varepsilon_0 \left(\frac{k(t)}{k_0}\right)^{C_\varepsilon}.$$ (4.42)

Using Eq. 4.41 in Eq. 4.39 yields

$$\frac{dk(t)}{k(t)^{C_\varepsilon}} = \frac{-\varepsilon_0}{k_0^{C_\varepsilon}} dt ,$$

(4.43)

and integrating from k_0 to k and from t_0 to t finally gives the time-dependent turbulent kinetic energy of the drop

$$k(t) = \left(\frac{k_0^{C_\varepsilon}}{\varepsilon_0 t (C_\varepsilon - 1) + k_0} \right)^{\frac{1}{C_\varepsilon - 1}} .$$

(4.44)

The values of $k(t)$ and $\varepsilon(t)$ are used in Eqs. 4.33 and 4.34 in order to calculate the time-dependent turbulent length and time scales

$$L_t(t) = L_{t0} \left(1.0 + \frac{0.0828 \cdot t}{\tau_{t0}} \right)^{0.457}$$

(4.45)

and

$$\tau_t(t) = \tau_{t0} + 0.0828 \cdot t ,$$

(4.46)

which again are inputs in Eqs. 4.30 and 4.31 in order to calculate $L_A(t)$ and $\tau_A(t)$ and to predict the break-up rate of the primary drop, Eq. 4.38. In Eq. 4.32 the actual relative velocity between drop and gas is used instead of U_{inj}.

The probability density function for the diameter d_{drop} of the secondary drops is assumed to be proportional to the turbulence energy spectrum [52],

$$\Phi(d_{drop}, t) = C \cdot \frac{(k / k_e(t))^2}{\left(1 + (k / k_e(t))^2 \right)^{11/6}} ,$$

(4.47)

and as inversely proportional to the atomization time scale as follows:

$$P(d_{drop}, t) = C \cdot \frac{\Phi(d_{drop}, t)}{\tau_A(d_{drop}, t)} .$$

(4.48)

The wave number k is the inverse of the droplet diameter d_{drop}. Further on, $k_e(t) = 0.75/L_t(t)$, and the constant C is determined by the normalization condition

$$\int_0^\infty P(d) dd_{drop} = 1 .$$

(4.49)

Each time the diameter of the primary drop is reduced by break-up, the diameter of the secondary drops is determined using Eq. 4.47.

The model of Huh and Gosman [56] predicts the spray cone angle of steady-flow single-hole experiments reasonably well. However, the effects of cavitation are not included. Instead it is assumed that the turbulence at the nozzle hole exit completely represents the influence of the nozzle characteristics on the primary spray break-up. The use of this model is limited to sprays from non-cavitating turbulent nozzle hole flows.

4.1.4 Cavitation-Induced Break-Up

Arcoumanis et al. [7] have developed a primary break-up model for full-cone diesel sprays that takes cavitation, turbulence, and aerodynamic effects into account. In order to link the spray characteristics with the nozzle hole flow, the authors use a one-dimensional sub-model to estimate input data like effective hole area A_{eff}, injection velocity U_{inj}, and turbulent kinetic energy k. The initial droplet diameter is set equal to the effective hole diameter (blob-method), and the first break-up of these blobs is modeled using the Kelvin-Helmholtz mechanism (see Sect. 4.2.4) in the case of aerodynamic-induced break-up, the model of Huh and Gosman [56] for turbulence-induced break-up, and a new phenomenological model in the case of cavitation-induced break-up. This cavitation-induced break-up model will be described in the following. It assumes that the cavitation bubbles are transported to the blob surface by the turbulent velocity inside the liquid and either burst on the surface or collapse before reaching it. For both cases, a characteristic time scale is calculated, and the smaller one causes break-up.

Bubble collapse: in order to estimate the collapse-time, the cavitation bubbles are lumped together into a single big artificial bubble occupying the same area as all the small ones, Fig. 4.8:

$$R_{cav} = \sqrt{r_{hole}^2 - r_{eff}^2} \quad , \quad r_{eff} = \sqrt{A_{eff} / \pi} \ . \tag{4.50}$$

Although the collapse-time of cavitation bubbles is directly dependent on their size (smaller bubbles collapse earlier) the radius of this artificial bubble, which is always bigger than the radius of any single bubble present in the nozzle hole flow, is used in order to estimate the atomization time from the Rayleigh theory of bubble dynamics [18]:

$$\tau_{coll} = 0.9145 \cdot R_{cav} \cdot \sqrt{\frac{\rho_l}{\rho_{g,bubble}}} \ . \tag{4.51}$$

Bursting of bubbles: although cavitation structures are usually located along the nozzle hole walls, the artificial bubble is placed in the center of the liquid and then transported to its surface with a turbulent velocity $u_{turb} = (2 \cdot k/3)^{0.5}$, resulting in a burst time of

$$\tau_{burst} = \frac{r_{hole} - R_{cav}}{u_{turb}} \ . \tag{4.52}$$

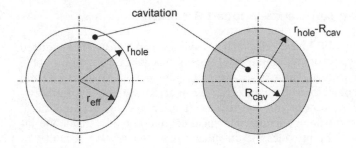

Fig. 4.8. Estimation of the artificial bubble size

The resulting time scale τ_A of atomization is assumed to be the smaller one of τ_{coll} and τ_{burst}, and the length scale of atomization is given by the correlation

$$L_A = 2\pi \left(r_{hole} - R_{cav} \right). \tag{4.53}$$

From L_A and τ_A, the force acting on the jet surface at the time of collapsing or bursting of a cavitation bubble with radius R_{cav} can be estimated from dimensional analysis as

$$F_{total} = C \cdot CN \cdot m_{jet} \cdot \frac{L_A}{\tau_A^2}, \tag{4.54}$$

where m_{jet} is the mass of the blob to be atomized, $CN = (p_{back} - p_{vap})/(0.5\rho_l U^2_{inj})$ is the dynamic cavitation number, and $C = 0.9$ is an empirical constant. The maximum size of the new droplets is calculated according to the condition $F_{total} = F_{surf}$ with the surface tension force $F_{surf} = 2\pi r_{blob}\sigma$ opposing the break-up process. The exact size of the new droplets is sampled from a distribution function. The spray angle is calculated as proposed in Huh and Gosman (Eq. 4.37):

$$\tan\left(\frac{\phi}{2}\right) = \frac{L_A / \tau_A}{U_{inj}}. \tag{4.55}$$

Although the complete aerodynamic, turbulence, and cavitation-induced atomization model of Arcoumanis et al. [7] is quite extensive and takes account of all important effects, the cavitation model is based on strong simplifications. The use of a single artificial bubble for example does not represent the dynamic behavior of a multitude of small bubbles, and the influence of bubble-collapse energy on droplet sizes and spray angle is not included. Although the authors have validated this model against experimental data and have shown its suitability for the simulation of high-pressure diesel injection, the use of a detailed model of bubble dynamics also taking bubble-collapse energy into account would definitely improve the quality of the model.

4.1.5 Cavitation and Turbulence-Induced Break-Up

Nishimura and Assanis [91] have presented a cavitation and turbulence-induced primary break-up model for full-cone diesel sprays that takes cavitation bubble collapse energy into account. Discrete cylindrical ligaments with diameter D and volume equal to that of a blob with diameter D are injected, see Fig. 4.9. Each cylinder contains bubbles, according to the volume fraction and size distribution at the hole exit, computed from a phenomenological cavitation model inside the injector. This sub-model also provides the turbulent kinetic energy k_{flow} and the injection velocity U_{inj}.

The bubbles collapse outside the nozzle, and the energy $E_{bu} = \Delta V_{bubble}\, p_{back}$ is released, resulting in an increase of turbulent kinetic energy $k_{bu} = \Sigma E_{bu}/m_{cylinder}$. The reduction of bubble volume V during collapse is calculated using the Rayleigh theory of bubble dynamics [18]. Assuming isotropic turbulence, the turbulent velocity inside the liquid cylinder can be determined as

$$u_{turb} = \sqrt{\frac{2 \cdot \left(k_{bu} + k_{flow}\right)}{3}}. \tag{4.56}$$

The authors assume that the velocity fluctuations inside the cylinder induce a deformation force

$$F_{turb} = surface \cdot dynamic\ pressure = \pi \cdot D \cdot \frac{2}{3} D \cdot \frac{\rho_l}{2} \cdot u_{turb}^2 \tag{4.57}$$

on its surface, and that it breaks up if the sum of F_{turb} and the aerodynamic drag force $F_{aero} = \pi (2/3) D_{hole}^2\, 0.5 \rho_g\, U_{rel}^2$ is no longer compensated by the surface tension force $F_{surf} = \pi D \sigma$. In this case, the diameter of the original cylinder is reduced until $F_{turb} + F_{aero} = F_{surf}$ again.

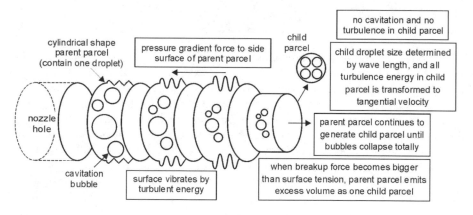

Fig. 4.9 Primary break-up model of Nishimura and Assanis [91]

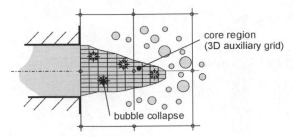

Fig. 4.10 Primary break-up model based on locally resolved break-up rates

The surplus volume (only liquid) forms a parcel of spherical child droplets the diameters of which are estimated by the Kelvin-Helmholtz break-up theory, see Sect. 4.2.4, but u_{turb} is used instead of the relative velocity between gas and liquid in order to predict the surface wave growth. Furthermore, it is assumed that the turbulent kinetic energy contained within the mass of the child droplets is transformed into kinetic energy normal to the original path giving the spray cone angle. The velocity in the original direction is not changed. At the end of bubble collapse, the remaining parent cylinder is transformed into a cylindrical drop and is subject to the secondary aerodynamic-induced break-up like all the child droplets produced before.

The authors have successfully validated their model against experimental data. Because of the detailed modeling of the relevant processes like the increase of break-up energy due to the bubble collapse energy and its direct effect on drop sizes and spray angle, the model is well suited for the simulation of cavitation and turbulence-induced primary break-up. However, it assumes axis-symmetry and is therefore not capable of predicting the influence of asymmetric nozzle hole flows on the 3D structure of primary sprays.

Von Berg et al [143] have published a turbulence and cavitation-induced primary break-up model for full-cone diesel sprays that releases droplets from a coherent core, Fig. 4.10. The sizes and velocity components of the droplets are calculated based on locally resolved turbulence scales and break-up rates on the core surface. The model uses detailed information from a 3D turbulent cavitating nozzle flow simulation as input data. Firstly, this data is utilized in order to estimate the erosion of the jet and the 3D shape of a core region. An auxiliary grid in the core region allows to realize the desired resolution, which cannot be provided by the coarse grid of the combustion chamber. Secondly, a sub-model, based on the simplified Rayleigh theory of cavitation bubble dynamics [18, 143] is used in order to determine the increase of turbulence due to the oscillations of collapsing bubbles within the core. A source term in the k-ε model of each core cell allows to increase the local turbulent kinetic energy and to modify the turbulent time and length scales until the fluid reaches the core surface. Thirdly, an advanced version of the turbulence-induced break-up model of Huh and Gosman [56] is applied locally at the surface elements, which represent the sources of droplet production. The turbulent length scale determines the atomization length scale L_A and the droplet size $r_{drop} = L_A/2$, while the break-up time scale τ_A is a linear combination

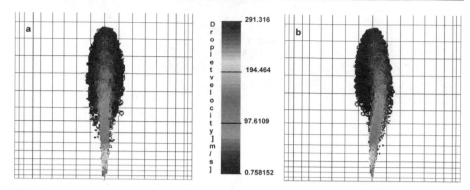

Fig. 4.11. Appearance of spray cone under perpendicular views and droplet velocities. **a** front view, **b** side view [143]

of the aerodynamic (Kelvin-Helmholtz-model) and the turbulent time scale. From the local atomization lengths and time scales, the local break-up rates are calculated. The model finally delivers the initial droplet sizes and velocities as well as the initial spray angle.

The model has been successfully validated. In the case of an asymmetric distribution of cavitation within the nozzle hole, the model is able to represent the experimentally observed spray asymmetry: due to better atomization, smaller droplets and an increased spray angle are produced on the cavitation side, Fig. 4.11. The most important features of this primary break-up model are the detailed modeling of the effect of bubble collapse outside the nozzle, the very close linking of nozzle hole flow and spray formation using locally resolved flow properties, and the ability to simulate the influence of asymmetric nozzle flow on the 3D spray structure. This model is well suited for the simulation of high-pressure diesel injection, and because of the very close linking of nozzle hole flow and spray formation, it offers the opportunity to optimize the injection system. However, a complete CFD simulation of the nozzle flow is required as input data.

Another detailed cavitation and turbulence-induced primary break-up model for full-cone sprays has been developed by Baumgarten et al. [12, 13, 14]. The model is based on energy and force balances and predicts all starting conditions needed for the simulation of further break-up and mixing processes. The input data is extracted from a CFD calculation of the nozzle flow. This guarantees a very close linking of nozzle characteristics and spray formation. The model simplifies the complex distribution of bubbles and liquid inside the hole and defines two zones (Fig. 4.12): a liquid zone (zone 1) and a mixture zone (zone 2), consisting of liquid and bubbles and surrounding zone 1. The circumferential thickness L_{cav} (φ), Fig. 4.13, and the average void fraction $\alpha_2 = (\rho - \rho_l)/(\rho_{vap} - \rho_l)$ of zone 2 as well as further input data like the flow velocity w, the turbulent kinetic energies (k_1, k_2) and dissipation rates (ε_1, ε_2), and the mass flows $\dot{m}_1$ and $\dot{m}_2$ of zones 1 and 2 are extracted from a CFD calculation of the nozzle flow. The indices "vap" and "l" indicate vapor and liquid, and ρ is the average density of zone 2.

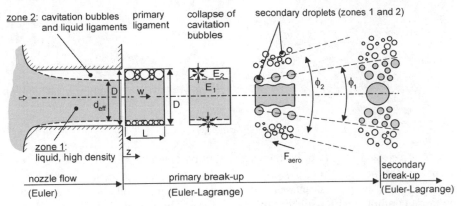

Fig. 4.12. Primary break-up model of Baumgarten et al. [13, 14]

The model starts with the injection of cylindrical primary ligaments. Each ligament consists of two zones the distribution of which equals that at the nozzle hole exit. The diameter of the primary ligaments is the hole diameter D, and the length L is equal to the effective diameter d_{eff} of the liquid zone.

The size distribution of the cavitation bubbles leaving the nozzle is unknown and must be modeled. Because the stochastic parcel method is used, all bubbles inside a primary ligament have the same size, but from ligament to ligament the sizes differ. The bubble sizes are sampled from a Gaussian distribution (mean value: 10 μm, standard deviation: 10 μm, [13, 14]). The volume of pure vapor and the number of bubbles inside a primary ligament can be calculated from the known average void fraction and size of zone 2. The use of a detailed model of bubble dynamics [13, 14] gives the collapse energy that is released in zone 2 and the collapse-time t_{coll} of the bubbles, which is used as the break-up time of the primary ligament. A part of the cavitation energy, E_{cav1}, is absorbed from zone 1 and contributes to the break-up of this zone. The amount of energy absorbed from zone 1 depends on the distribution of the cavitation zone and is calculated using geometrical relations [13]. Further on, the turbulent kinetic energy produced inside the injection holes is reduced due to dissipation until break-up occurs. The total break-up energies $E_1 = E_{turb1} + E_{cav1}$ and $E_2 = E_{turb2} + E_{cav2}$ of each zone at the time of break-up are the sums of the particular turbulence and cavitation fractions. The disintegration of the zones into secondary droplets with a velocity component normal to the spray axis (spray angles ϕ_1, ϕ_2, Fig. 4.12) is calculated separately for each zone. The new droplets are exposed to the aerodynamic forces of the secondary break-up.

Break-up of the cavitation zone: It is assumed that the energy E_2 is transformed into surface energy $E_{\sigma2}$ (formation of n_2 new droplets, σ: surface tension fuel/gas) and into kinetic energy E_{kin2} normal to the main direction (velocity component v_{r2} normal to the spray axis),

$$E_{\sigma2} = n_2 \sigma \pi d_2^2 , \qquad (4.58)$$

$$n_2 = \frac{6m_2}{\pi d_2^3 \rho_l}, \tag{4.59}$$

$$E_{kin2} = n_2 \frac{1}{2} \frac{\pi}{6} d_2^3 \rho_l v_{r2}^2, \tag{4.60}$$

$$\phi_2 = 2\arctan\left(v_{r2}/v_{ax}\right), \tag{4.61}$$

where $v_{ax} \approx w$. The ratio of both energies is unknown and must be modeled. It is assumed that the ratio $\kappa = E_{\sigma2}/E_2$ or $(1-\kappa) = E_{kin2}/E_2$ is only dependent on the cylinder pressure, because an increase of cylinder pressure results in a faster bubble collapse combined with a higher collapse energy. The model gives reasonable results, if the calculation of the energy ratio κ_{1bar} for a cylinder pressure of 1 bar is performed first and the additional collapse energy due to the rise of pressure is completely added to the kinetic energy (spray angle),

$$E_{\sigma2} = \kappa_{1bar} \left[E_{turb2} + \frac{1bar}{P_{chamber}(bar)} E_{cav2} \right], \tag{4.62}$$

where $\kappa_{1bar} = 0.15$. Although the angle ϕ_2 between the spray axis and the direction of the new droplets can now be determined, the velocity vector in the plane perpendicular to the spray axis must be modeled in order to specify the exact direction of droplet motion inside the 3D-spray cone, see Fig. 4.13. Because the thickness $L_{cav}(\varphi)$ of the cavitation zone directly effects the distribution of collapse energy, the probability $P(\varphi)$ that new droplets will be created at a certain position φ is assumed to be proportional to $L_{cav}(\varphi)$. The new droplets are assumed to

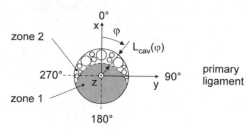

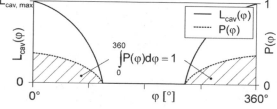

Fig. 4.13. 3D spray angle, example: asymmetric flow [14]

move radial outwards with velocity v_{r2}. Altogether, this results in a larger spray divergence at the cavitation side. The model is thus capable of producing asymmetric primary sprays according to the nozzle hole flow. The high break-up energy per unit mass (high collapse energy, low density) results in small droplets and a large spray divergence near the nozzle. Consequently, zone 2 forms the outer spray region, which appears optically dense but does not contain much mass.

Break-up of the liquid zone: Outside the nozzle, zone 1 forms the extremely dense core in which energy is dissipated due to chaotic collision, break-up, and coalescence processes. The chaotic nature of these dissipation processes is modeled by using an efficiency η_1, which is sampled randomly for each ligament between 0 and 1 and which reduces the effective break-up energy of zone 1 to $E_{1,eff} = \eta_1 E_1$. As in the model of Nishimura and Assanis [91], it is assumed that all break-up energy is present as turbulent kinetic energy the velocity fluctuations $u_{turb} = (2E_{1,eff}/3m_1)$ of which cause a disruptive turbulence force F_{turb}. Mass is split off the liquid cylinder of zone 1 until the surface tension force F_σ is equal to F_{turb}:

$$F_{turb} = dynamic\ pressure \cdot surface\ area = \frac{\rho_l}{2} u_{turb}^2 d_{eff} L , \qquad (4.63)$$

$$F_\sigma = 2\sigma \left(L + d_{eff} \right). \qquad (4.64)$$

The remaining cylinder with diameter

$$d_{cyl} = \frac{4\sigma L}{\rho_l u_{turb}^2 L - 4\sigma} \qquad (4.65)$$

is transformed into a spherical drop, and its turbulent kinetic energy is used to determine its spray angle (radial velocity). The disintegration of the split mass into droplets with a velocity component normal to the main spray direction is modeled in the same way as for zone 2, but the direction φ of the child parcel in the x-y-plane is sampled from a uniform distribution between 0° and 360°. Altogether this results in a symmetric full-cone spray (angle ϕ_1, Fig. 4.12). It forms the dense inner spray region with high momentum in the axial direction and is responsible for the spray characteristics in the far field of the nozzle. Because of the small break-up energy per unit mass, the average droplet diameter is larger and the spray angle is smaller than in zone 2.

Cavitation bubble dynamics: It is assumed that the bubbles leaving the nozzle experience a sudden rise in pressure from the vapor pressure p_{vap} to the pressure p_∞ of the gas inside the chamber and that the bubbles collapse inside zone 2. In order to calculate the time-dependent development of the bubble radius $R(t)$ during the collapse process, the equation of Herring [50] and Trilling [141] (see Prosperetti and Lezzi [108]),

$$\left(1 - \frac{2\dot{R}(t)}{a}\right) R(t)\ddot{R}(t) + \frac{3}{2}\left(1 - \frac{4\dot{R}(t)}{3a}\right)\dot{R}(t)^2$$

$$= \frac{1}{\rho_\infty}\left(p_{vap} - \frac{2\sigma}{R(t)} - \frac{4\mu}{R(t)}\dot{R}(t) - p_\infty\right), \tag{4.66}$$

is used. The model also accounts for the effect of compressibility (accoustic approximation, a: average sound speed of zone 2) as well as surface tension σ and viscosity μ. $R(t)$ is the bubble radius, $\dot{R}(t)$ and $\ddot{R}(t)$ are the time-dependent velocity and acceleration of the bubble wall, and ρ_∞ is the density of the fluid around the bubble (zone 2). The kinetic energy of the fluid surrounding the bubble can be estimated as (Knapp et al. [64]):

$$E(t) = 2\pi\rho_\infty \dot{R}(t)^2 R(t)^3. \tag{4.67}$$

In Fig. 4.14 the curves of radius and kinetic energy for the collapse of a representative bubble are shown. The liquid surrounding the bubble is accelerated towards the bubble center, and the kinetic energy grows. At the end of the collapse, the liquid is decelerated again. The begin of deceleration is regarded as the break-up time t_{coll} of the primary ligament, because from this time on the kinetic energy E_{max} of the collapse process is used for the disintegration process (formation of droplets and radial velocity). E_{max} is the amount of energy that is released during the collapse of a single bubble.

Turbulence-modeling outside the nozzle: The turbulent kinetic energy k_0 and the dissipation rate ε_0 of each zone at the nozzle exit are known from the calculation of the nozzle flow. From the time the ligament leaves the nozzle until break-up, the turbulent kinetic energy is reduced by dissipation. The simplified (no diffusion and production) 0-dimensional k-ε model gives (see Sect. 4.1.3):

$$k(t) = \left(\frac{k_0^{C_\varepsilon}}{\varepsilon_0 t\left(C_\varepsilon - 1\right) + k_0}\right)^{\frac{1}{C_\varepsilon - 1}}. \tag{4.68}$$

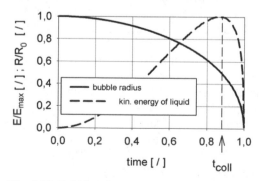

Fig. 4.14. Bubble radius and kinetic energy during collapse

Using this equation, the actual values $k(t_{coll})$ and E_{turb} available for break-up are estimated.

Using their primary break-up model in combination with the Kelvin-Helmholtz model for secondary break-up, the authors have shown that experimentally observed asymmetric spray structures due to asymmetric nozzle hole flows as well as penetration, spray angle, and droplet sizes could be well predicted. Figs. 4.15 and 4.16 show the results for an asymmetric nozzle hole flow resulting in an asymmetric distribution of cavitation and liquid zone inside the primary ligament as shown in Fig. 4.13. Diesel fuel (rail pressure: 65 MPa) is injected in compressed air (5 MPa, 25°C).

In Fig. 4.15, the results of a statistical analysis (frequency distribution) of the relevant primary break-up parameters during the disintegration of the primary ligaments are shown. As expected, zone 2 has the highest break-up energy per unit mass because of the collapse of cavitation bubbles and its low mass, Fig. 4.15a. This results in larger primary spray angles and smaller droplets compared to zone 1 (Figs. 4.15b and 4.15c). The frequency distribution of the formation of new droplets at different circumferential positions φ in the x-y-plane, see also Fig. 4.13, is, as implied by the model, proportional to the thickness of the cavitation zone in case of zone 2, and equally distributed in case of zone 1, Fig. 4.15d. The combination of the diagrams gives the following picture of the 3D primary spray: zone 1 forms a dense full-cone spray consisting of big droplets (spray center). Zone 2 forms the outer spray region, whose asymmetry is directly dependent on the asymmetric distribution of the cavitation zone inside the nozzle holes.

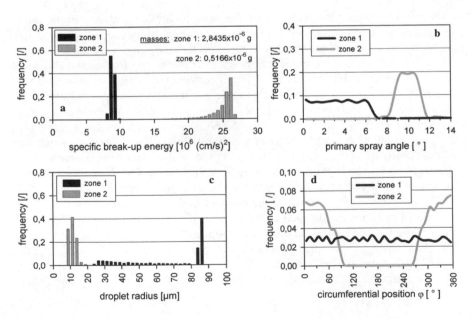

Fig. 4.15. Statistical analysis of relevant parameters during primary break-up [12]

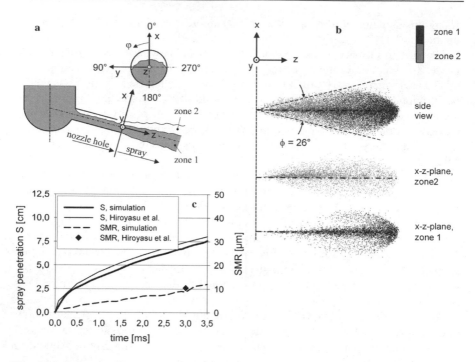

Fig. 4.16. Influence of the new primary break-up model on the dilute spray [12]

Figure 4.16 shows the influence of the primary break-up model on the dilute spray. In Fig. 4.16b two slices of the spray containing only droplets of zone 1 or only droplets of zone 2 are presented. As expected, the droplets of zone 2 lead to the strong and asymmetric spray divergence near the nozzle (main mass of zone 2 on the cavitation side), while zone 1 forms a symmetric spray with small divergence near the nozzle, which is responsible for the spray in the far field of the nozzle due to its high momentum (macroscopic spray angle, penetration). Global spray parameters like spray angle (Fig. 4.16b, side view), temporal development of spray penetration, and Sauter mean radius (SMR) (Fig. 4.16c) show feasible values compared to the ones estimated using the well-known correlations of Hiroyasu et al. [53]. The slight increase of SMR over time is a consequence of the collision model (droplet coalescence) and the low ambient temperatures.

The most important features of the primary break-up model of Baumgarten et al. [12, 13, 14] are the detailed consideration of cavitation, the close linking of nozzle hole flow and spray, and the ability to simulate the influence of asymmetric nozzle flows on the 3D spray structure. This model is well suited for the simulation of high-pressure diesel injection, but, similar to the model of von Berg et al. [143], a complete CFD simulation of the nozzle hole flow is needed as input data.

4.1.6 Sheet Atomization Model for Hollow-Cone Sprays

In direct injection spark ignition (DISI) engines, nozzles producing a hollow-cone spray are usually used in order to achieve maximum dispersion of the liquid phase at moderate injection pressures of about 5 to 10 MPa. Compared to the boundary conditions in the case of diesel engines, the backpressures and temperatures are small and the use of full-cone sprays would result in poor mixture formation and increased wall impingement. Hollow-cone sprays are typically characterized by small droplet diameters, effective fuel-air mixing, reduced penetration, and thus high atomization efficiencies. Fig. 4.17 shows two nozzle concepts, an inwardly opening pressure-swirl atomizer and an outwardly opening nozzle. In case the of a swirl-atomizer, the fuel passes through tangentially arranged swirl ports and gets a rotational motion inside the swirl chamber, Fig. 4.18. The centrifugal motion of the liquid forms a hollow air core.

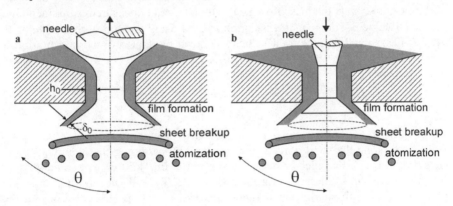

Fig. 4.17. a Inwardly opening pressure-swirl atomizer, **b** outwardly opening nozzle

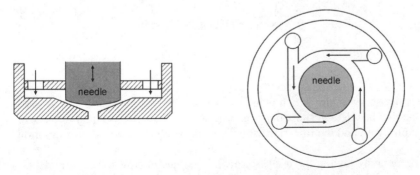

Fig. 4.18. Schematic illustration of the tangentially arranged swirl ports

Because the area of the swirl chamber reduces to a nozzle, the rotational motion is further increased. The liquid passes through the nozzle and forms a free cone-shaped liquid sheet inside the combustion chamber, which becomes thinner because of the conservation of mass as it departs from the nozzle and subsequently disintegrates into droplets.

Schmidt et al. [125] have developed a model which is well suited to describe the primary break-up of hollow-cone sprays. First, a zero-dimensional approach is used to represent the internal injector flow and to determine the velocity U_{inj} at the nozzle hole exit. The rotational motion of the fuel inside the swirl chamber creates an air core surrounded by a liquid film, Fig. 4.17a, the thickness δ_0 of which is related to the mass flow rate by

$$\dot{m} = \pi \rho_l u \left(D - \delta_0 \right), \tag{4.69}$$

where u is the axial velocity component at the nozzle exit, D is the hole diameter at the exit, and $\dot{m}$ is the mass flow rate, which must be measured experimentally. The axial velocity component u is approximated using the approach of Han et al. [49]. It is assumed that the total velocity U_{inj} can be related to the injection pressure $\Delta p = p_{rail} - p_{back}$,

$$U_{inj} = C_d \sqrt{\frac{2\Delta p}{\rho_l}}, \tag{4.70}$$

where C_d is the discharge coefficient of the swirl injector. The value of C_d is derived by treating the swirl ports as nozzles, and by assuming that the major pressure drop through the injector occurs at the ports. Eq. 4.70 is then an expression for the coefficient of discharge for the swirl ports. Typical values of C_d for single phase nozzles with sharp inlet corners are 0,78 or less [76], depending on the amount of momentum loss due to friction and reduction of cross-sectional area. C_d must be less than one in order to conserve energy. On the other hand C_d must be large enough to conserve mass and to avoid the prediction of negative air core sizes. These restrictions are satisfied using the following expression for C_d:

$$C_d = \max \left\{ 0.7, \ \frac{4\dot{m}}{\pi D^2 \rho_l u} \right\}. \tag{4.71}$$

The axial velocity u is

$$u = U_{inj} \cos(\theta), \tag{4.72}$$

where θ is the spray angle, which is assumed to be known. In the case of an outwardly opening nozzle, the sheet thickness at the nozzle orifice is equal to the size of the annular gap, and the injection velocity is determined from the conservation of mass.

In the near spray region, the transition from the injector flow to the fully developed spray is modeled as a three-step mechanism, see Figs. 4.17 and 4.19, consisting of film formation, sheet break-up, and, finally, disintegration into droplets.

Fig. 4.19. Break-up of the liquid sheet into ligaments and droplets

First, a liquid film with initial thickness h, spray angle θ, axial velocity u, and total velocity U_{inj} is formed, which penetrates into the combustion chamber. The liquid sheet becomes thinner because of the conservation of mass as it departs from the nozzle. Further on, Kelvin-Helmholtz instabilities are assumed to develop and grow on the film surface resulting in a first break-up and in the formation of ring-shaped ligaments. These ligaments finally break up into droplets.

Schmidt et al. [125] have shown that when the film exits the injector as a free liquid sheet, the tangential velocity (which is present in the case of a swirl atomizer) is completely transformed into radial velocity, and the trajectory approaches a straight line. The angle θ from the centerline is determined by the ratio of radial and axial velocity. Hence, at the point of film break-up and drop formation, the swirl velocity of the drops can be neglected. This is a very convenient approach because the swirl velocity at the nozzle orifice is usually unknown. Nevertheless, because the swirl velocity is responsible for the formation of the spray half-cone angle θ, the spray cone angle must be known from experimental investigations in this case. Alternatively, the detailed liquid film movement model of Gavaises et al. [39] could be used at this point in order numerically to predict the spray angle, the reduction of the film thickness, and the decay of the swirl velocity during the film disintegration process.

The disintegration of the liquid film is modeled in direct analogy to the Kelvin-Helmholtz model for diesel sprays, see Sect. 4.2.4. The model, which is presented in detail in Senecal et al. [129], assumes that a two-dimensional, viscous and incompressible liquid sheet with thickness $2h$ moves with velocity U through a quiescent, inviscid, and incompressible gas. A spectrum of infinitesimal disturbances

$$\eta(t) = R\left(\eta_0 \exp[ikx + \omega t]\right) \tag{4.73}$$

is imposed on the sheet surface producing pressure and velocity fluctuations in both the liquid and the gas. The amplitudes of these disturbances grow due to the liquid-gas interaction. In Eq. 4.73, η_0 is the initial wave amplitude, $k = 2\pi/\lambda$ is the wave number, and $\omega = \omega_r + i\omega_i$ is the complex wave growth rate. It is assumed that the most unstable disturbance with growth rate $\omega_r = \Omega$ causes sheet break-up. Hence, a dispersion relation $\omega = \omega(k)$ is derived, from which the most unstable wave can be deduced. In ref. [45], it has been shown that two modes of film disturbances exist, both of which satisfy the liquid-governing equations subject to the boundary conditions at the upper and lower interfaces of the sheet. In the case of the first mode (sinuous mode), the waves at both surfaces are exactly in phase, while in the case of the second mode (varicose mode), the waves are 180 degrees out of phase. Senecal et al. [129] have shown that the sinuous mode dominates the growth of the varicose waves for low velocities, and that both modes become indistinguishable for high flow velocities U. Hence, only the sinuous waves are regarded in the further analysis, resulting in the dispersion relation [129]

$$\omega^2 \left[\tanh(kh) + \frac{\rho_l}{\rho_g} \right] + \left[4v_l k^2 \tanh(kh) + 2i \frac{\rho_l}{\rho_g} kU \right] +$$

$$4v_l k^4 \tanh(kh) - 4v_l^2 k^3 l \tanh(lh) - \frac{\rho_l}{\rho_g} U^2 k^2 + \frac{\sigma k^3}{\rho_l} = 0 \ , \tag{4.74}$$

where $l^2 = k^2 + \omega/v_l$, μ_l is the liquid viscosity, and ρ_l and ρ_g are the liquid and gas densities. Furthermore, Senecal et al. [129] have shown that a simplified version of the dispersion relation,

$$\omega_r = -2v_l k^2 + \sqrt{4v_l^2 k^4 + \frac{\rho_l}{\rho_g} U^2 k^2 - \frac{\sigma k^3}{\rho_l}} \ , \tag{4.75}$$

can be used if the following three assumptions are valid: first of all, an order of magnitude analysis using typical values from the inviscid solutions shows that the terms of second order viscosity can be neglected. Second, it can be shown that above a critical Weber number of $W_e = 27/16$ (based on the relative velocity, the gas density, and the sheet half-thickness) short waves, the growth rate of which are independent of the sheet thickness, grow on the sheet surface. Finally, the gas-to-liquid density ratio has to be small ($\rho_g/\rho_l \ll 1$). All these assumptions are valid in the case of typical pressure-swirl injectors used in modern DISI engines.

   If the unstable waves on the sheet surface have reached a critical amplitude, the sheet is assumed to break up into primary ligaments. Because the wave growth is independent of the sheet thickness, the break-up time τ_b and the corresponding break-up length L_b can be formulated based on an analogy with the break-up of cylindrical liquid jets (e.g. Reitz [114]),

$$\eta_b = \eta_0 \exp(\Omega \tau_b), \quad \tau_b = \frac{1}{\Omega} \ln\left(\frac{\eta_b}{\eta_0} \right), \tag{4.76}$$

where Ω is obtained by numerically maximizing Eq. 4.75 as a function of k. Assuming that the velocity U_{inj} of the liquid sheet is not reduced until break-up occurs yields the break-up length

$$L_b = U_{inj} \tau_b = \frac{U}{\Omega} \ln\left(\frac{\eta_b}{\eta_o} \right), \tag{4.77}$$

where the value of $\ln(\eta_b / \eta_0)$ is taken to be 12 as suggested by Dombrowski and Hoper [29]. It must be noted that the quantity U in Eq. 4.75 is the relative velocity between liquid and gas while U_{inj} in Eq. 4.77 is the absolute velocity of the sheet. However, it is assumed in the numerical implementation that the relative velocity is equal to the absolute velocity because the liquid is usually injected into quiescent gas.

The diameter of the primary ligaments is obtained from a mass balance. It is assumed that every tear in the sheet forms a primary ligament. The resulting diameter is given by

$$d_{ligament} = \sqrt{\frac{16h_b}{K_s}} .$$
(4.78)

K_s is the wave number of the fastest growing surface wave. Because the sheet thins as it departs from the nozzle, the position of first the break-up determines the ligament diameter. The sheet half-thickness h_b at the point of break-up is

$$h_b = \frac{h_0 \left(D - \tau_b \right)}{2L_b \sin \left(\theta \right) + D - \tau_b} .$$
(4.79)

The initial half-thickness of the film at the nozzle orifice is approximately $h_0 \approx 0.5 \delta_0 \cos(\theta)$.

The subsequent disintegration of the ligament into drops is calculated using another stability analysis, which is based on an analogy to Weber's result for the growth of waves on cylindrical, viscous liquid columns. The wave number for the fastest growing wave on the ligament is

$$K_{ligament} = \frac{1}{d_{ligament}} \left(0.5 + \frac{3\mu_l}{2\sqrt{\rho_l \sigma d_{ligament}}} \right)^{-1/2} .$$
(4.80)

As proposed in Dombrowski and Johns [30], it is assumed that break-up occurs if the amplitude of the wave is equal to the radius of the ligament, and that one drop is formed per wavelength. A mass balance gives the drop diameter:

$$d_{drop} = \sqrt[3]{\frac{3\pi d_{ligament}^2}{K_{ligament}}} .$$
(4.81)

The starting conditions of the drops formed at the point of break-up are used as input for the CFD calculation of spray formation. The parcels are injected into a hollow-cone with a mean spray angle θ from the spray axis. The exact trajectory of each discrete parcel is determined by randomly distributing the angle over a range $\theta \pm \Delta\theta$, where $\Delta\theta \approx 10$ deg. is the dispersion angle. In the numerical implementation, it is assumed that the sheet does not interact with the gas phase. The sheet is represented by dormant parcels, which are injected with velocity U_{inj} (Eq. 4.70), and do not experience any drag nor undergo any break-up, evaporation, or collision until their distance from the nozzle orifice is equal or greater than the break-up length as given by Eq. 4.77. At the point of break-up, the parcels are given a size which is sampled from a Rosin-Rammler distribution with SMD = d_{drop} (Eq. 4.81). From now on, the parcels are treated as normal and are subject to aerodynamic drag forces as well as break-up and collision processes. Usually the secondary break-up is modeled using the Taylor-Analogy (TAB) or the Droplet Deformation and Break-up (DDB) model, see Sects. 4.2.2 and 4.2.3.

As the injection starts, some amount of fuel that had been trapped in the tangential slots from the previous injection event flows out with low velocity and nearly zero swirl and forms a kind of solid-cone spray with narrow spray cone angle and large drops, the so-called pre-spray. As the fuel velocity inside the injector increases, the angular momentum and the centrifugal forces increase too, and the liquid inside the swirl chamber forms a hollow-cylinder structure. This structure is then transformed into a hollow-cone spray as it leaves the nozzle. Hence, the development of the spray can be divided into two phases: the very short and transient phase at the beginning of injection and the steady-state phase corresponding to the largest part of the injection duration. While the LISA model can be used to predict the spray behavior during the steady-state phase, a simple approach of Chryssakis et al. [23] can be used in order to model the pre-spray: A solid-cone injection (blob method) is performed and the cone angle is gradually increased until the steady-state hollow-cone spray angle is reached, using a linear profile. At some point, which must be estimated from experimental investigations, the full-cone injection switches into a hollow-cone injection, while the cone angle is still increasing according to the given profile.

4.2 Secondary Break-Up

Secondary break-up is the disintegration of already existing droplets into smaller ones due to the aerodynamic forces that are induced by the relative velocity u_{rel} between droplet and surrounding gas. These forces result in an instable growing of waves on the droplet surface or of the whole droplet itself, and finally lead to its disintegration. The surface tension force on the other hand tries to keep the droplet spherical and counteracts the deformation force. This behavior is expressed by a non-dimensional number, the gas phase Weber number

$$We_g = \frac{\rho_g u_{rel}^2 d}{\sigma},$$
(4.82)

which represents the ratio of aerodynamic and surface tension forces. The smaller the droplet diameter d, the bigger the surface tension force and the bigger the critical relative velocity needed for break-up. From experimental investigations, it is known that, depending on the Weber number, different break-up modes and break-up mechanisms of droplets exist. A detailed description is given in Hwang et al. [58] and Krzeczkowski [69], for example. The models used in order to simulate secondary break-up processes in full-cone as well as hollow-cone fuel sprays are described in the next sections.

4.2.1 Phenomenological Models

Arcoumanis et al. [7] distinguish between seven different droplet break-up modes, which are all described using semi-empirical relationships for the resulting droplet sizes and break-up times,

$$t_{bu} = \tau_{break} \frac{d}{u_{rel}} \cdot \sqrt{\frac{\rho_l}{\rho_g}},$$
(4.83)

where τ_{break} is given in Table 4.1. Some of them appear within the same range of the Weber number.

The product droplet sizes are sampled from distribution functions, the Sauter mean diameters (SMD) of which are estimated using the following phenomenological relations. According to Arcoumanis et al. [7], the SMD of the first three modes is

$$SMD = \frac{4d_{drop}}{4 + 0.5\left(1 + 0.19\sqrt{We_g}\right)},$$
(4.84)

while for the chaotic and catastrophic regimes the correlation

Table 4.1. Break-up modes and break-up times of droplets [7]

Break-up mode	Break-up time τ_{break} [/]	Weber number [/]
Vibrational	$\frac{\pi}{4}\left[\frac{\sigma}{\rho_l d^3} - 6.25\frac{\mu_l}{\rho_l d^2}\right]^{-0,5}$	$We_g \approx 12$
Bag	$6\left(We_g - 12\right)^{-0.25}$	$12 \leq We_g \leq 18$
Bag-and-Streamer	$2.45\left(We_g - 12\right)^{0.25}$	$18 \leq We_g \leq 45$
Chaotic	$14.1\left(We_g - 12\right)^{-0.25}$	$45 \leq We_g \leq 100$
Sheet Stripping	$14.1\left(We_g - 12\right)^{-0.25}$	$100 \leq We_g \leq 350$
Wave Crest Stripping	$0.766\left(We_g - 12\right)^{0.25}$	$350 \leq We_g \leq 1000$
Catastrophic	$0.766\left(We_g - 12\right)^{0.25}$	$1000 \leq We_g \leq 2760$
	5.5	$We_g > 2760$

$$SMD = 6.2 \frac{\sigma}{\rho_g u_{rel}^2} \cdot \sqrt[4]{\frac{\rho_l}{\rho_g}} \cdot \sqrt[2]{\frac{\mu_l}{\rho_l d_{drop} u_{rel}}} \cdot We_g \qquad (4.85)$$

is used. In the case of stripping break-up, small product drops are stripped from the parent ones, the size of which decreases continuously and can be predicted subtracting the mass

$$\frac{dm_{strip}}{dt} = 12\rho_l \left(\frac{\rho_g}{\rho_l}\right)^{\frac{1}{3}} \cdot \left(\frac{\mu_g}{\mu_l}\right)^{\frac{1}{6}} \cdot \mu_l^{1/2} \cdot u_{rel}^{1/2} \cdot d_{drop}^{3/2} \qquad (4.86)$$

leaving the parent droplet. This mass disintegrates into small product droplets with volume mean diameter $D_{30} = 0.2\ d_{drop}$. Altogether, this phenomenological modeling results in a multitude of different regimes with different correlations for break-up time and product droplet size. This sub-division into many different break-up modes is necessary because in contrast to detailed models, phenomenological correlations are not able to describe the break-up processes of several Weber number classes. In the following, the more detailed break-up models usually used in CFD codes today will be presented.

4.2.2 Taylor-Analogy Break-Up Model

The Taylor Analogy Break-up model (TAB model), which was proposed by O'Rourke and Amsden [96], is based on an analogy between a forced oscillating spring-mass system and an oscillating drop that penetrates into a gaseous atmosphere with a relative velocity u_{rel}, see Fig. 4.20. The force F initiating the oscillation of the mass m corresponds to the aerodynamic forces deforming the droplet and thus making its mass oscillate. The restoring force $F_{spring} = k \cdot x$ is analogous to the surface tension force, which tries to keep the drop spherical and to minimize its deformation. The damping force $F_{damping} = d \cdot \dot{x}$ corresponds with the friction forces inside the droplet due to the dynamic viscosity μ_l of the liquid. The second order differential equation of motion for the damped spring-mass-system is

$$\ddot{x} = \frac{F}{m} - \frac{k}{m} x - \frac{d}{m} \dot{x}, \qquad (4.87)$$

where x is the displacement of the mass from the idle state. According to the analogy, the coefficients in Eq. 4.87 have to be replaced by

$$\frac{F}{m} = C_F \frac{\rho_g u_{rel}^2}{\rho_l r}, \quad \frac{k}{m} = C_k \frac{\sigma}{\rho_l r^3}, \quad \frac{d}{m} = C_d \frac{\mu_l}{\rho_l r^2}, \qquad (4.88)$$

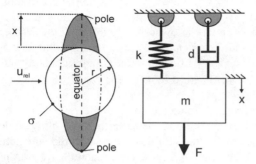

Fig. 4.20. Taylor-Analogy break-up model

and x is the displacement of the droplet's equator from its equilibrium position, see Fig. 4.20. C_F, C_k, C_d, and C_b are model constants. In Eq. 4.88, r is the droplet radius in idle state (spherical drop), ρ_g and ρ_l are the gas and liquid densities, and σ is the appropriate surface tension. Using the dimensionless displacement $y = x /(C_b \cdot r)$ the equation of motion becomes

$$\ddot{y} = \frac{C_F}{C_b} \frac{\rho_g}{\rho_l} \frac{u_{rel}^2}{r^2} - C_k \frac{\sigma}{\rho_l r^3} y - C_d \frac{\mu_l}{\rho_l r^2} \dot{y} . \qquad (4.89)$$

Assuming a constant relative velocity u_{rel}, which is satisfied in the numerical solution process during a given time interval, the equation of motion can be solved analytically [96]:

$$y(t) = \frac{C_F}{C_k C_b} We_g + e^{-\frac{t}{t_d}} \cdot \left[A \cdot cos\,\omega t + \frac{1}{\omega t_d} \cdot B \cdot sin\,\omega t \right] , \qquad (4.90)$$

where

$$A = \left(y_0 - \frac{C_F}{C_k C_b} We_g \right), \quad B = \left(\dot{y}_0 t_d + y_0 - \frac{C_F}{C_k C_b} We_g \right),$$

$$We_g = \frac{\rho_g u_{rel}^2 r}{\sigma} , \quad \frac{1}{t_d} = \frac{C_d \mu_l}{2\rho_l r^2} , \quad \omega^2 = C_k \frac{\sigma}{\rho_l r^3} - \frac{1}{t_d^2} ,$$

$$y_0 = y_{t=0} , \quad \dot{y}_0 = (dy / dt)\big|_{t=0} .$$

In contrast to other definitions, the Weber number of the gas phase is calculated using the droplet radius and not its diameter. Equation 4.90 describes the dimensionless time-dependent oscillation of the droplet equator. Although there are several possible modes of oscillation that can result in a drop break-up, the TAB model only describes the fundamental mode corresponding to the lowest order spherical harmonic, which is the most important one [96]. If the blob-method is used as primary break-up model, $y_0 = \dot{y}_0 = 0$ are usually used as initial conditions. Because the experimentally determined critical Weber number resulting in a first drop break-up is $We_{g,crit} \approx 6$, break-up is not allowed to occur below this Weber number. Furthermore, it is assumed that break-up occurs if a deformation of $x \geq 0.5 \cdot r$ is reached, resulting in $C_b = 0.5$ and $y \geq 1.0$. The remaining model constants are $C_k = 8.0$, $C_d = 5.0$ and $C_F = 1/3$ [96]. Using Eq. 4.90, the break-up time of a droplet can be calculated.

The TAB model can also be used to determine the spray angle. This is important if the model is combined with the blob method, because then the spray angle must not be specified independently, and the initial blobs can be injected with zero spray cone angle into the computational domain. It is assumed that the new droplets retain the velocity $\vec{u}$ of the old drop, and that they get an additional velocity component

$$\left| \vec{v}_n \right| = C_v \left| \dot{x} \right| = C_v C_b r \left| \dot{y} \right|, \quad C_v \approx 1.0 \tag{4.91}$$

normal to the original path of the old drop. This normal velocity is the deformation velocity of the old droplet at the time of break-up. This approach results in an angle of

$$\tan\left(\Phi / 2\right) = \left| \vec{v}_n \right| / \left| \vec{u} \right| \tag{4.92}$$

between old and new path. The exact direction of in $\vec{v}_n$ the plane normal to $\vec{u}$ must be sampled randomly. However, this method of modeling the spray angle is somewhat imprecise and is usually not used in CFD applications.

The Taylor-Analogy break-up results in a complete disintegration of the old drop into a number of smaller new ones. However, this model gives no information about their number and size. These values are estimated using an energy balance (kinetic energy due to oscillation, surface energy, [96]). Because the velocity in the old direction (before break-up) is not changed during break-up, this kinetic energy is not included in the analysis. Before break-up, the energy of the drop is the sum of its minimum surface energy

$$E_{surf,old} = 4\pi r^2 \sigma \tag{4.93}$$

and the energy in oscillation (kinetic energy) and distortion (remaining surface energy), which is

$$E_{osc,old} = \frac{\pi}{5} \rho_l r^5 \left(\dot{y}^2 + \omega^2 y^2 \right) \tag{4.94}$$

for the fundamental mode of droplet oscillation. Because in reality there is energy in other modes as well, the model constant K is implemented, and the resulting energy before break-up is

$$E_{old} = K \frac{\pi}{5} \rho_l r^5 \left(\dot{y}^2 + \omega^2 y^2 \right) + 4\pi r^2 \sigma . \tag{4.95}$$

After break-up, it is assumed that the new droplets are spherical and do not oscillate ($y_0 = \dot{y}_0 = 0$). Hence, the energy after break-up is the sum of the surface energies

$$E_{surf,new} = 4\pi r^2 \sigma \frac{r}{SMR} \tag{4.96}$$

and the kinetic energy

$$E_{kin,new} = \frac{1}{2} \cdot \frac{4}{3}\pi r^3 \rho_l v_n^2 = \frac{\pi}{6}r^5 \rho_l \dot{y}^2 \tag{4.97}$$

of the product drops due to an additional new velocity component v_n (Eq. 4.91) normal to the path of the old drop before break-up:

$$E_{new} = 4\pi r^2 \sigma \cdot \frac{r}{SMR} + \frac{\pi}{6}r^5 \rho_l \dot{y}^2 . \tag{4.98}$$

Equating E_{new} and E_{old} and using $y = 1$ and $\omega^2 = 8\sigma/(\rho_l r^3)$ finally yields

$$\frac{r}{SMR} = 1 + \frac{8K}{20} + \frac{\rho_l r^3}{\sigma}\dot{y}^2\left(\frac{6K-5}{120}\right), \tag{4.99}$$

where $K = 10/3$ is determined from experiments and $\dot{y}$ is the deformation velocity at the time of break-up. For each break-up event, the radius of the product drops is chosen randomly from a χ-square distribution with a Sauter mean radius SMR as predicted by Eq. 4.99. Finally, the number of product drops can be predicted using the mass conservation constraint.

The TAB model is generally known to underpredict droplet sizes of full-cone diesel sprays (Tanner et al. [138], Liu et al. [79]) and to underestimate penetration if it is combined with the blob-method (Park et al. [104], Allocca et al. [4]). Today, the TAB model has lost its leading position with regard to the break-up prediction of diesel sprays. However, the TAB model is used in order to predict droplet deformation (independent of break-up), which is needed to calculate the dynamic drag coefficient of the droplets in a spray, Liu et al. [79]. In contrast to diesel spray simulations, the TAB model is the most important secondary break-up model in the case of hollow-cone gasoline sprays, see Sect. 4.3.4.

A further development of the TAB model is the ETAB model (Enhanced TAB model), proposed by Tanner [138]. The dynamics of the TAB model have been left unchanged. Besides a modified calculation of the new droplet radius after break-up, the most important improvement is that an initial oscillation $\dot{y}_0 \neq 0$ is chosen for the spherical drops ($y_0 = 0$) emerging from the nozzle. The deformation velocity is chosen in a way that the droplets first elongate in the direction of flight, see Fig. 4.21, return to their spherical shape and finally are flattened and break up. This increases the lifetime of the blobs, simulates the dense fragmented liquid core near the nozzle, and allows a more realistic representation of the dense core as well as the calculation of larger and more realistic droplet sizes within the spray.

The value of $\dot{y}_0$ is chosen in such a way that the first break-up of the blobs occurs at the experimentally obtained break-up length

$$L_b = U_{inj} \cdot t_{bu} = C \cdot D_{nozzle} \sqrt{\frac{\rho_l}{\rho_g}} \tag{4.100}$$

of the dense fragmented core, where $C = (3.3...15.8)$ [21, 19, 136] is an arbitrary constant representing the influence of the nozzle. Tanner [138] chooses $C = 5.5$. Using $y(t_{bu}) = 1$ and assuming an inviscid liquid, the expression for $\dot{y}_0$ can be derived from Eq. 4.90,

$$\dot{y}_0 = \left[1 - \frac{C_F}{C_k C_b} We_g \left(1 - \cos\left(\omega t_{bu}\right)\right)\right] \frac{\omega}{\sin\left(\omega t_{bu}\right)}, \tag{4.101}$$

with t_{bu} from Eq. 4.100. This method is only used for the first break-up of the blobs. The subsequent break-up of secondary droplets is again calculated according to the original TAB-model ($y_0 = \dot{y}_0 = 0$).

The second difference between TAB and ETAB-model is a modified calculation of the new droplet radius after break-up. Again, $y = 1$ is used as the break-up condition, and $We > 6$ must be fulfilled, but the product droplet size is related to the break-up time and not determined by an energy balance. In order to derive the new break-up law, it is asumed that the rate of product droplet generation (dn/dt) is proportional to the number of product droplets,

$$\frac{dn(t)}{dt} = 3K_{bu}n(t), \tag{4.102}$$

where $3K_{bu}$ is a constant of proportionality. Using the mass conservation, the number of product droplets that are formed by a break-up occurring at time t is given by

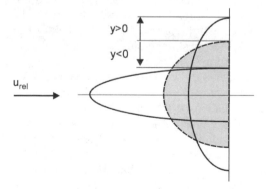

Fig. 4.21. ETAB model: an initial droplet deformation velocity results in an elongation of the drop in the direction of flight

$$n(t) = \frac{m_0}{m_{new}(t)} \quad , \quad \frac{dn(t)}{dt} = -\frac{m_0}{m_{new}(t)^2} \cdot \frac{dm_{new}}{dt} , \tag{4.103}$$

where m_0 is the mass of the drop before break-up and m_{new} is the mass of one new droplet after break-up. Substituting $dn(t)/dt$ and $n(t)$ in Eq. 4.102 using Eq. 4.103 gives

$$\frac{dm_{new}(t)}{dt} = -3 \cdot K_{bu} \cdot m_{new}(t) . \tag{4.104}$$

Using $m_{new} = (4/3)\pi r^3_{new} \rho_l$ in Eq. 4.104 yields $dr^3_{new}/r^3_{new} = -3K_{bu}\, dt$. Thus, the radius of the new droplets depends on the time span $t = t_{bu}$ after which the old drop breaks up. The longer the time span, the smaller the product droplets. The break-up time of a given drop with radius R_0 is calculated by the TAB-model. At the time of break-up, new droplets are formed and the time is set to zero again. Integrating the left hand side from $r_{new} = R_0$ to $r_{new} = R_{new}$ (R_{new}: radius of the new droplets if break-up occurs at time $t = t_{bu}$) and the right hand side from $t = 0$ to $t = t_{bu}$ finally gives the relation between new product droplet size R_{new}, old drop size R_0, and break-up time t_{bu}:

$$\frac{R_{new}}{R_0} = e^{-K_{bu} \cdot t_{bu}} . \tag{4.105}$$

The break-up constant K_{bu} depends on the break-up regime (bag or stripping break-up) and thus on the Weber number of the drop before break-up,

$$K_{bu} = \begin{cases} (1/4.5) \cdot \omega & if\ We_g \le We_t & (bag\ break\text{-}up) \\ (1/4.5) \cdot \omega \cdot \sqrt{We} & if\ We_g > We_t & (stripping\ break\text{-}up) . \end{cases} \tag{4.106}$$

We_t is the regime-dividing Weber number, which is set to $We_t = 80$ [138]. Figure 4.22 shows the ratio of product and parent drop radii for the ETAB and the standard TAB model as a function of the Weber number for inviscid spherical droplets with $y_0 = \dot{y}_0 = 0$. The break-up time is only dependent on the Weber number and

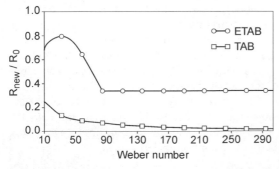

Fig. 4.22. Ratio of product and parent drop radii for the ETAB and the standard TAB model

does not depend on whether the TAB or the ETAB model is used. Compared to the TAB model, which generally predicts excessively small droplet sizes (e.g. [138, 79, 4]), the calculation of product droplet sizes according to the ETAB model results in bigger drops and thus a more realistic drop size distribution.

As in the standard TAB model, after break-up of a parent drop, the product droplets are initially supplied with a velocity component

$$\left|\vec{v}_n\right| = C_v\left|\dot{x}\right| = C_v C_b r\left|\dot{y}\right| \tag{4.107}$$

normal to the path of the parent drop, but in contrast to the original TAB model, where $C_v \approx 1$, this constant is determined from an energy balance consideration to be

$$C_v^2 = \frac{5}{4}C_{D,droplet} + \frac{18}{We_g}\left(1 - \frac{R_0}{R_{new}}\right). \tag{4.108}$$

Typical Reynolds numbers of $Re \approx 500$, drop drag coefficients of $C_D \approx 0.5$, Weber numbers of $We_g > 80$, and droplet radii of $R_{new}/R_0 \approx 0.7$ result in $C_v \approx 0.72$. Hence, in contrast to the original TAB model, the normal velocity component is slightly reduced.

The ETAB model was developed at a time when detailed primary break-up models were not available. The modifications of the TAB model were introduced in order to correct the shortcomings of the blob-method by the secondary break-up model. This way of modeling the spray break-up can of course not compete with the combination of a more detailed primary break-up model and a standard secondary break-up model.

4.2.3 Droplet Deformation and Break-Up Model

Ibrahim et al. [59] proposed a droplet deformation and break-up (DDB) model, which assumes that the liquid drop is deformed due to a pure extensional flow from an initial spherical drop of radius r to an oblate spheroid with an ellipsoidal cross section with major semi axis a and minor semi axis b, Fig. 4.23. The break-up mechanism of the DDB model is very similar to that of the TAB model and can be regarded as an alternative to this model. Again, the whole drop is deformed by aerodynamic forces. The air velocity distribution and air pressure distribution at any point on the surface of an initially spherical liquid drop, which is exposed to a steady air stream, are not uniform. The air velocity has a maximum at the equator of the drop and equals zero at its poles (stagnation points), resulting in low static pressures at the equator and high ones at the poles. This external pressure field causes the drop to deform from its undisturbed spherical shape and to become flattened to form an oblate ellipsoid normal to the airflow, Fig. 4.23. As the velocity at the equator increases, the pressure difference increases accordingly, which results in a further flattening, and finally the drop becomes disk-shaped.

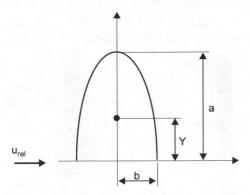

Fig. 4.23. Droplet deformation and breakup model (DDB model), schematic diagram of the deforming half drop

In contrast to the TAB model, the DDB model of Ibrahim et al. [59] does not use the droplet equator to describe the deformation, but regards the motion Y of the center of mass of the half-drop and conserves the droplet's volume as it distorts. In the case of a spherical drop (no deformation), the distance Y between the center of the half-drop and that of the complete drop is $Y = 4r/(3\pi)$. The model is based on the conservation of energy for a distorting drop. Assuming that the drop experiences no heat exchange with its surroundings, the energy equation of the half-drop is

$$\frac{dE}{dt} = -\frac{dW}{dt}.$$ (4.109)

E is the internal energy consisting of the kinetic energy ($dE_{kin}/dt = F \cdot v = (m \cdot dv/dt) \cdot v$) and the potential energy ($dE_{surf}/dt = \sigma dA/dt$),

$$\frac{dE}{dt} = \frac{2}{3}\pi r^3 \rho_l v \left(\frac{dv}{dt}\right) + \frac{1}{2}\sigma\frac{dA}{dt}.$$ (4.110)

In Eq. 4.110, $v = (dY/dt)$ and $a = (dv/dt)$ are the velocity and the acceleration of the center of mass of the deforming half-drop, and A is its surface area approximated by

$$A \approx 2\pi \left(a^2 + b^2\right).$$ (4.111)

W is the work done by pressure and viscous forces,

$$\frac{dW}{dt} = \frac{dW_{pressure}}{dt} + \Phi.$$ (4.112)

The work done by pressure is

$$\frac{dW_{pressure}}{dt} = -\frac{1}{2}pA_p\frac{dY}{dt} \approx -\frac{\pi}{4}r^2\rho_g u_{rel}^2\frac{dY}{dt},$$ (4.113)

where the pressure is approximated by the static pressure $p \approx 0.5\rho_g u^2_{rel}$ of the stagnation point, and the projected area of the drop is approximated by $A_p \approx \pi r^2$. The viscous dissipation due to the extensional flow is [59]

$$\Phi = \frac{8}{3}\pi r^3 \mu_l \left(\frac{1}{Y}\frac{dY}{dt}\right)^2. \tag{4.114}$$

During the deformation process, the drop volume is conserved, $(4/3)\pi a^2 b = (4/3)\pi r^3$ or $b = r^3/a^2$. Using $a = (3\pi/4)\cdot y$ and $b = r^3/a^2$ in Eq. 4.111 gives

$$\frac{dA}{dt} = \frac{9\pi^3}{4}Y\frac{dY}{dt}\left[1-2\left(\frac{3\pi Y}{4r}\right)^{-6}\right]. \tag{4.115}$$

Using Eqs. 4.110, 4.112, 4.113, 4.114 and 4.115 finally yields

$$\frac{2}{3}\pi r^3 \rho_l \left(\frac{dY}{dt}\right)\left(\frac{d^2Y}{dt^2}\right) + \frac{9\pi^3}{8}\sigma\cdot Y\left[1-2\left(\frac{3\pi\cdot Y}{4r}\right)^{-6}\right]\left(\frac{dY}{dt}\right)$$
$$+\frac{8}{3}\pi r^3 \mu_l \left(\frac{1}{Y}\frac{dY}{dt}\right)^2 = \frac{\pi r^2 \rho_g u^2_{rel}}{4}\left(\frac{dY}{dt}\right). \tag{4.116}$$

Defining $y* = Y/r$ and $t* = t\cdot u_{rel}/r$ to nondimensionalize Eq. 4.116 yields the equation of motion of the DDB model [81]

$$\frac{d^2 y*}{dt*^2} = \frac{3}{8K} - \frac{27\pi^2}{16KWe_g}y*\left[1-2\left(\frac{3\pi}{4}y*\right)^{-6}\right] - \frac{8N}{K Re_g}\left(\frac{1}{y*}\right)^2\frac{dy*}{dt*}, \tag{4.117}$$

where $N = \mu_l/\mu_g$, $K = \rho_l/\rho_g$, $We_g = \rho_g u^2_{rel} d/\sigma$, $Re_g = \rho_g u_{rel} d/\mu_g$, and $d = 2r$.

As in the TAB model, it is assumed that at time $t = 0$ (new droplet is formed by break-up of the old one or by injection), the droplets start with zero deformation ($Y = 4/(3\pi)$) and zero deformation velocity ($dY/dt = 0$). The break-up criterion is derived by noting that near break-up the kinetic energy as well as the viscous dissipation are negligible:

$$y*_{break} - 2\left(\frac{4}{3\pi}\right)^6\left(\frac{1}{y*_{break}}\right)^5 = \frac{6We_g}{27\pi^2}. \tag{4.118}$$

At the time of break-up, $y*_{break}$ reaches its maximum value. Hence, the second term on the left hand side of Eq. 4.118 can be neglected:

$$y*_{break} = \frac{6We_g}{27\pi^2} \quad \text{and} \quad \left(\frac{a}{r}\right)_{break} = \frac{3\pi}{4}y*_{break} \approx \frac{We_g}{6\pi}. \tag{4.119}$$

The break-up time of the droplet can be calculated combining Eqs. 4.119 and 4.117. The droplet sizes after break-up have to be estimated using further assumptions like the energy-based modeling used in the TAB model, for example.

4.2.4 Kelvin-Helmholtz Break-Up Model

The Kelvin-Helmholtz model (KH model) was proposed by Reitz [114]. The model is based on a first order linear analysis of a Kelvin-Helmholtz instability growing on the surface of a cylindrical liquid jet with initial diameter r_0 that is penetrating into a stationary incompressible gas with a relative velocity u_{rel}. Both the liquid and the gas are assumed to be incompressible, and the gas is assumed to be inviscid. Furthermore, it is assumed that due to the turbulence generated inside the nozzle hole the jet surface is covered with a spectrum of sinusoidal surface waves with an infinitesimal axisymmetric displacement $\eta = \eta_0 \cdot e^{\omega t}$ ($\eta \ll r$) causing small axisymmetric fluctuating pressures as well as axial and radial velocity components in both liquid and gas. These surface waves grow because of aerodynamic forces due to the relative velocity between liquid and gas (shear flow waves), Fig. 4.24. The motion of liquid and gas are described by the linearized Navier-Stokes equations for both phases. The solution is found by transforming the equations of motion into stream and potential functions.

The analysis, which is described in detail in Reitz and Bracco [117], yields a dispersion equation relating the growth rate ω (increase of amplitude per unit time) of a perturbation to its wavelength $\lambda = 2\pi/k$:

$$
\omega^2 + 2v_l k^2 \omega \left(\frac{I_1'(kr_0)}{I_0(kr_0)} - \frac{2kl}{k^2+l^2} \frac{I_1(kr_0)}{I_0(kr_0)} \frac{I_1'(lr_0)}{I_1(lr_0)} \right)
$$
$$
= \frac{\sigma k}{\rho_l r_0^2} \left(1 - r_0^2 k^2\right) \left(\frac{l^2-k^2}{l^2+k^2}\right) \frac{I_1(kr_0)}{I_0(kr_0)} \tag{4.120}
$$
$$
+ \frac{\rho_g}{\rho_l} \left(u_{rel} - \frac{i\omega}{k} \right)^2 k^2 \left(\frac{l^2-k^2}{l^2+k^2}\right) \frac{I_1(kr_0) K_0(kr_0)}{I_0(kr_0) K_1(kr_0)} ,
$$

where I_0 and I_1 are modified Bessel functions of the first kind, K_0 and K_1 are modified Bessel functions of the second kind, $k = 2\pi/\lambda$ is the wave number, σ is the surface tension, $l^2 = k^2 + \omega/v_l$, $v_l = \mu_l/\rho_l$ (kinematic viscosity), and the prime indicates differentiation.

The numerical solution of the dispersion function shows that there is a single maximum in the wave growth rate curve $\omega = \omega(k)$. It is assumed that the wave with the highest growth rate $\omega = \Omega$ will finally be sheared off the jet and form new droplets. Curve fits were generated from the numerical solutions to Eq. 4.120 for the growth rate Ω of the fastest growing and thus most unstable surface wave,

$$\Omega\left[\frac{\rho_l r_0^3}{\sigma}\right]^{0.5} = \frac{0.34 + 0.38 \cdot We_g^{1.5}}{(1+Z)(1+1.4\cdot T^{0.6})},$$ (4.121)

and the corresponding wavelength Λ,

$$\frac{\Lambda}{r_0} = 9.02\frac{\left(1+0.45\cdot Z^{0.5}\right)\left(1+0.4\cdot T^{0.7}\right)}{\left(1+0.865\cdot We_g^{1.67}\right)^{0.6}},$$ (4.122)

where

$$Z = \frac{\sqrt{We_l}}{Re_l}, \; T = Z\sqrt{We_g}, \; We_g = \frac{\rho_g r_0 u_{rel}^2}{\sigma}, \; We_l = \frac{\rho_l r_0 u_{rel}^2}{\sigma}, \; Re_l = \frac{\rho_l r_0 u_{rel}}{\eta_l}.$$

Z and T are the Ohnesorge number and the Taylor number, and r_0 is the radius of the undisturbed jet. These curve fits are shown in Figs. 4.25 and 4.26.

Reitz [114] applied this theory to the break-up modeling of liquid droplets with radius r. Again, waves grow on the drop surface with growth rate Ω and wavelength Λ. Because the new child drops are formed from the surface waves that are sheared off the parent drops, it is assumed that the size of the new droplets is proportional to the wavelength Λ,

$$r_{new} = B_0 \cdot \Lambda,$$ (4.123)

where $B_0 = 0.61$ is a constant, the value of which is fixed. A new parcel containing product drops of size r_{new} is created and added to the computations. In contrast to the TAB model, the parent drop does not perform a complete break-up, but continuously looses mass while penetrating into the gas. This results in a shrinking radius whose rate of reduction at a certain time t depends on the difference between the actual value of droplet radius r and an equilibrium droplet size (which is equal to the child droplet radius r_{new}) as well as on the value of a characteristic time span τ_{bu}, (Reitz [114], Ricart et al. [119]):

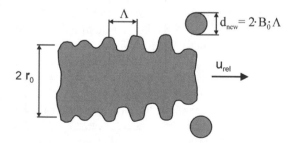

Fig. 4.24. Schematic illustration of the Kelvin-Helmholtz model

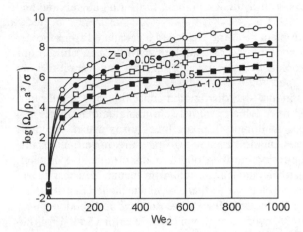

Fig. 4.25. Growth rate of the most unstable surface wave versus We_g as a function of Ohnesorge number Z [114].

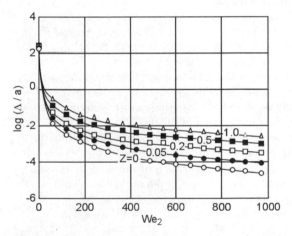

Fig. 4.26. Wavelength of the most unstable surface wave versus We_g as a function of Ohnesorge number Z [114].

$$\frac{dr}{dt} = -\frac{r - r_{new}}{\tau_{bu}}, \quad \tau_{bu} = 3.788 \cdot B_1 \frac{r}{\Lambda \cdot \Omega}. \tag{4.124}$$

If the KH model is used in combination with the blob method, the influence of the nozzle hole flow on primary break-up is not modeled satisfactorily. In this case, B_1 is an adjustable model constant including the influence of the nozzle hole flow like turbulence level and nozzle design on spray break-up. Values between $B_1 = 1.73$ and $B_1 = 60$ are proposed in the literature [107, 114]. A higher value of B_1 leads to reduced break-up and increased penetration, while a smaller value on the other hand results in increased spray disintegration, faster fuel-air mixing, and reduced penetration.

The number of new child droplets per time step can be calculated from the decrease of radius of the parent drop (Eq. 4.124) and the size of the new child drops (Eq. 4.125). The new child droplets become parent drops in a new parcel and are subject to further break-up.

The KH model predicts a bimodal size distribution consisting of a small number of big parent drops, the radius of which is slowly shrinking, and an increasing number of small child droplets. Although stripping break-up is one of the most important break-up mechanisms in the case of high-pressure injection, experiments [58] have shown that the formation of strong bimodal droplet size distributions is unrealistic, and that another important mechanism, the sudden disintegration of the complete drop into droplets with diameter much bigger that the KH-child droplets is important near the nozzle. For this reason, the KH-model is usually combined with the Rayleigh-Taylor model, which is described in the following section.

4.2.5 Rayleigh-Taylor Break-Up Model

The Rayleigh-Taylor model (RT model) is based on the theoretical work of Taylor [139], who investigated the instability of the interface between two fluids of different densities in the case of an acceleration (or deceleration) normal to this interface. If the two fluids are liquid and gas, the interface is stable when the acceleration is directed into the liquid, and instable disturbances can grow if the acceleration is directed into the gas. Regarding droplet and gas moving with velocity u_{rel} relative to each other, the deceleration of the drop (in the forward direction) due to drag forces can also be treated as an acceleration of the drop in the direction of the airflow (backward direction). Thus, instable waves can grow on the back side of the drop, see Fig. 4.27. The disintegration of the drop is induced by the inertia of the liquid if drops and ligaments leaving the nozzle with high velocities are strongly decelerated by the aerodynamic drag force

$$F_{aero} = \pi r^2 c_D \frac{\rho_g u_{rel}^2}{2} . \tag{4.125}$$

Dividing the drag force by the mass of the drop, the acceleration of the interface can be found,

$$a = \frac{3}{8} c_D \frac{\rho_g u_{rel}^2}{\rho_l r} , \tag{4.126}$$

where c_D is the drag coefficient of the drop. Using a linear stability analysis (Chang [20]) and neglecting liquid viscosity (Bellmann and Pennington [15]), the growth rate Ω and the corresponding wavelength Λ of the fastest growing wave are (Su et al. [136], Chan et al. [19], Patterson and Reitz [107]):

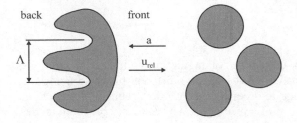

Fig. 4.27. Rayleigh-Taylor instability on a liquid drop

$$\Omega = \sqrt{\frac{2}{3\sqrt{3}\sigma} \frac{\left[a\left(\rho_l - \rho_g\right)\right]^{3/2}}{\rho_l + \rho_g}} \qquad (4.127)$$

and

$$\Lambda = C_3 2\pi \sqrt{\frac{3\sigma}{a\left(\rho_l - \rho_g\right)}}. \qquad (4.128)$$

In many applications, the gas density is neglected because it is much smaller than the liquid density. The break-up time $t_{bu} = \Omega^{-1}$ is found to be the reciprocal of the frequency of the fastest growing wave. At $t = t_{bu}$ the drop disintegrates completely into small droplets whose radius $d_{new} = \Lambda$ is assumed to be proportional to the wavelength. The drop is only allowed to break up if Λ is smaller than its diameter. The number of new drops is calculated using the mass conservation principle [107]. In contrast to this method, Su et al. [136] determine the number of new droplets as the ratio of maximum diameter of the deformed drop to Λ, and the size of the new droplets is calculated using mass conservation.

The adjustable constant C_3 is introduced in order to allow a modification of the effective wavelength. Similar to the constant B_1 in the KH model, it includes the unknown effects of initial conditions like turbulence and cavitation inside the nozzle hole on the secondary break-up. C_3 changes the size of the new product droplets and the likelihood of break-up events. By increasing its value, break-up is reduced (break-up is only allowed if $\Lambda < d_{drop}$), and the size of the new droplets is increased. Patterson and Reitz [107] use values in the range of $C_3 = 1.0$–5.33.

It has been observed that drag-deceleration and shear flow induced instabilities are simultaneous phenomena in the droplet break-up process (Hwang et al. [58]). The drag-deceleration induced RT break-up only results in effective and rapid disintegration near the nozzle, where the very high relative velocities between drop and gas result in a strong deceleration. This break-up mechanism describes the experimentally observed catastrophic break-up mode (Table 4.1). Further downstream, the shear flow induced KH break-up becomes the dominant process. Hence, the RT model is always used in combination with a second break-up model, usually the KH model.

4.3 Combined Models

Because a single break-up model is usually not able to describe all relevant classes of break-up processes and break-up regimes of engine sprays, the use of combined models, consisting of a combination of at least two different break-up models, is becoming more and more popular in order to improve the accuracy of prediction. Usually such a model is composed of a primary and a secondary break-up model. The most important hybrid models for full-cone and hollow-cone sprays are presented in the following.

4.3.1 Blob-KH/RT Model (Full-Cone-Sprays)

The experimental investigations of Hwang et al. [58] revealed that the break-up mechanism in the catastrophic break-up regime consists of a series of characteristic processes. The aerodynamic force on the drop flattens it into the shape of a liquid sheet, and the decelerating sheet breaks into large-scale fragments by means of RT instability. KH waves with a much shorter wavelength originate at the edges of the fragments, and these waves break up into micrometer-size drops as illustrated in Fig. 4.28.

If the blob-method is utilized in order to inject initial drops into the numerical grid, usually two secondary break-up models are used. The first one describes the relevant processes of the spray disintegration near the nozzle (part of the primary break-up that is not correctly described by the blob-method), and the second one is responsible for the remaining spray region. In the case of the KH-RT model, both models are implemented in CFD codes in a competing manner. Both models are allowed to grow unstable waves simultaneously, and if the RT-model predicts a break-up within the actual time step, the disintegration of the whole drop according to the RT mechanism occurs. Otherwise the KH model will produce small child droplets and reduce the diameter of the parent drop.

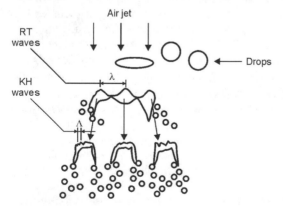

Fig. 4.28. Schematic diagram of break-up mechanisms in the catastrophic break-up regime [58]

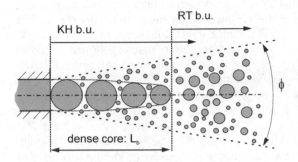

Fig. 4.29. Combined blob-HK/RT model

However, the reduction of droplet size by the RT model is too fast if it is applied to drops just leaving the nozzle hole. Thus, the model is applied to spray break-up beyond a certain distance from the nozzle, the break-up length L_b of the dense fragmented core (Eq. 4.100), and only KH stripping break-up is allowed to occur near the nozzle (Chan et al. [19]), as shown in Fig. 4.29. Hence, the secondary break-up process can be adjusted by modifying the break-up length L_b as well as the constants of the KH model and the RT model. Compared to the single use of the KH model, a faster disintegration of big drops is achieved, and an increased evaporation as well as a reduced penetration are calculated allowing a better matching of experimental data. Another positive effect of the combination of RT and KH model is that the RT model counteracts the formation of a strong bimodal droplet size distribution, because RT break-up always results in a number of equally sized droplets with intermediate size.

The KH-RT model is the most popular of all hybrid models used today. It has been successfully validated against experimental data and used by many authors in order to predict the disintegration process of high-pressure diesel sprays (e.g. [19, 135]).

4.3.2 Blob-KH/DDB Model (Full-Cone Sprays)

The concept of the Blob-KH/DDB model is similar to that of the KH-RT model. The blob-method is used in order to inject initial drops into the numerical grid, and the subsequent aerodynamic-induced break-up is described by a combination of DDB and KH model. In contrast to the KH-RT model, the KH model is not the relevant mechanism in the far field of the nozzle but is used together with the DDB-mechanism in a competing manner in order to predict the spray disintegration within the break-up length (Park and Lee, [106]). The authors also refer to the experiments of Hwang et al. [58] and argue that within this break-up length considerable flattening of the liquid drops was observed, and that the disintegration is affected by both wave instability (KH) and droplet deformation (here described by the DDB-model). Both models are allowed to deform the drop and to grow unstable waves simultaneously. If the DDB-model predicts a break-up within the actual time step, the disintegration of the whole drop according to the DDB mechanism occurs. Otherwise the KH model will produce small child droplets and reduce the diameter of the parent drop. Park and Lee [106] argue that beyond the break-up length the droplets are only affected by the deformation of the droplet (DDB model) since the relative gas/liquid velocities are low. This assumption does not conform with the KH-RT model, where the relevant break-up process beyond the break-up length is the KH-mechanism. However, it must be pointed out that although in both cases Eq. 4.100 is used for the break-up length, the constant C is different (e.g. $C = 30$ in the KH-DDB model, [106]; $C = 14$ in the KH-RT model, [19]). This results in different break-up lengths and thus in a different sub-division of the spray. Park and Lee [106] have validated their model against experimental data from high-pressure diesel sprays under atmospheric conditions. However, the small influence of the KH break-up mechanism beyond the break-up length is somewhat unexpected and may also be due to the low ambient gas densities used in the experiments.

4.3.3 Further Combined Models (Full-Cone Sprays)

Park et al. [105] have compared the prediction accuracy of different combined models using the Blob/KH model and the turbulence-induced model of Huh and Gosman [56] as primary break-up models, and the RT, DDB, and TAB model as secondary break-up models. Except for the KH-RT model, the models are not implemented as competing models, but there is a clear switching criterion from primary to secondary break-up. It is assumed that the primary break-up model switches to the secondary one if the drop diameter becomes less than 95% of the maximum diameter of the injected blobs. Park and Lee [105] have validated the models against experimental data from high-pressure diesel sprays under atmospheric conditions. The results concerning spray tip penetration and axial SMD distribution agree well with experimental data in the case of the Blob/KH-RT model

Table 4.2. Hybrid break-up model of Rotondi et al. [122]

Primary break-up	
Low-medium injection pressure	KH-model
High injection pressure	Turbulence-cavitation-aerodynamic-induced model

Secondary break-up		
Vibrational	$12 < We_g < 16$	TAB-model
Bag	$16 < We_g < 45$	DDB-model
Chaotic	$45 < We_g < 100$	DDB-model + KH-model
Stripping	$100 < We_g < 1000$	KH-model
Catastrophic	$We_g > 1000$	KH-model + RT-model

as well as the turbulence-RT model and the turbulence-DDB model, while other model combinations result in less accurate predictions.

Further hybrid models have been developed by Baumgarten et al. [12, 13], see Sect. 4.1.5, and Rotondi et al. [122], for example. The last named authors developed a hybrid model for the atomization of diesel sprays that also distinguishes between primary and secondary break-up. The primary break-up considers the effects of jet turbulence and cavitation (models of Huh and Gosman [56] and Arcoumanis et al. [7]), and, for the secondary break-up, different models are used as the droplet Weber number changes, see Table 4.2.

4.3.4 LISA-TAB Model (Hollow-Cone Sprays)

The most popular combined model for hollow-cone sprays is the combination of the LISA-model for primary break-up with the TAB model for the secondary break-up. Schmidt et al. [125] used the LISA-TAB model in order to simulate the spray development of a pressure-swirl injector and achieved a good agreement between experimental measurements and numerical predictions in the case of atmospheric ambient conditions and non-evaporating sprays. The injector characteristics are given in Table 4.3. The model was validated using experimental data of drop size, penetration, and mass flux distribution as well as photographic comparison. As an example, Figs. 4.30 and 4.31 show the measured and predicted SMD and penetration lengths for two different injection pressures. The pre-spray, which is produced at the beginning of injection due to the missing rotation and the low injection velocity, is modeled by using the blob-method and injecting a full-cone spray with a narrow cone angle. After a short time span, which must be determined experimentally, the full-cone injection is replaced by the hollow-cone LISA model. The pre-spray is responsible for the large droplets at the beginning of injection in Fig. 4.30.

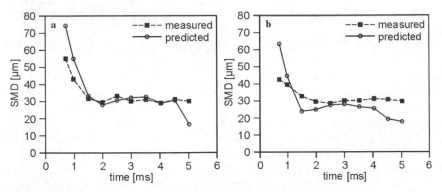

Fig. 4.30. Measured and predicted SMD. **a** case 1, **b** case 2, drop sizes averaged over a plane 39 mm downstream of the injector [125]

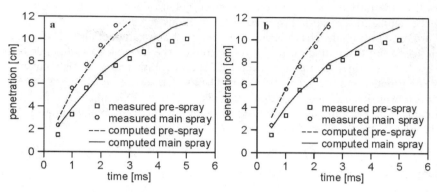

Fig. 4.31. Measured and predicted penetration. **a** case 1, **b** case 2 [125]

Table 4.3. Injection parameters for sprays shown in Figs. 4.30 and 4.31, data from [125]

Injection parameter	Quantity	
	Case 1	Case 2
Spray half-cone angle θ [°]	23	23
Dispersion angle Δθ [°]	10	10
Fuel delivery per injection [mm³]	56.8	56.8
Nozzle Diameter [μm]	560	560
Injection pressure [MPa]	4.76	6.12
Injection duration [ms]	3.86	3.4
Liquid density [g/cm³]	0.77	0.77

Senecal et al. [129] have used the LISA-TAB model to predict the spray formation in the case of outwardly opening nozzles. Due to the pintle, these injectors do not produce a pre-spray. Furthermore, one less equation is needed to initialize the calculation, since the initial thickness of the sheet at the nozzle orifice is determined by the size of the annular gap. The injection velocity is calculated from the conservation of mass.

Stiesch et al. [135] also combined the LISA sheet atomization model with the TAB model in order to simulate the non-evaporating spray development in a pressure bomb and to validate the model at ambient as well as elevated gas pressures and densities. The authors used a pressure-swirl injector with a half-cone angle of 27 degrees, a dispersion angle of 10 degrees, and a nozzle diameter of 0.458 mm. The injection pressure was 4.93 MPa. Figs. 4.32 and 4.33 show a comparison of computational and experimental results for three different timings after start of injection and for two different backpressures. While the photographs are side views of the complete spray, the calculated spray images represent spray slices containing the spray axis. Both series of pictures suggest that the hybrid LISA-TAB model is capable of predicting the spray behavior very well. Especially the recirculating vortex, which starts to form at the spray edges, is accurately predicted,

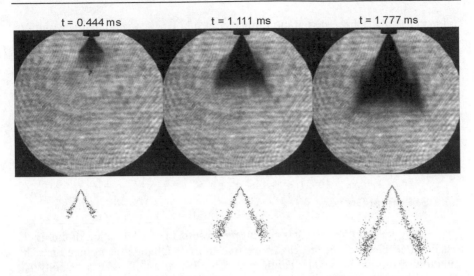

Fig. 4.32. Measured and calculated spray images, gas pressure: 101 kPa, [135]

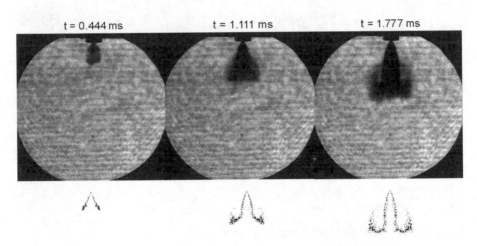

Fig. 4.33. Measured and calculated spray images, gas pressure: 366 kPa, [135]

and also the influence of an increase of backpressure (reduced penetration, narrower cone angle and a more distinct vortex at the spray edge) is predicted correctly.

4.3.5 LISA-DDB Model (Hollow-Cone Sprays)

Instead of using the TAB model, Park et al. [104] have combined the LISA sheet atomization model with the Droplet Deformation and Break-up (DDB) model in order to calculate the disintegration process of a hollow-cone spray. The authors have validated the model against experimental data concerning spray penetration, droplet sizes, spray shape, and mean velocity distribution in the case of atmospheric sprays and injection pressures of 5 and 7 MPa. Because the break-up mechanism of the DDB model is very similar to that of the TAB model, there are no significant differences regarding the prediction accuracy of both hybrid models.

4.4 Droplet Drag Modeling

4.4.1 Spherical Drops

The relative velocity between gas and droplet results in a deceleration of the liquid and in an acceleration of the gas phase due to the exchange of momentum. The equation of motion of a spherical drop with radius r moving with a velocity u_{rel} relative to the gas is

$$F_{drag} = \frac{\rho_g}{2} u_{rel}^2 c_D A_f = \rho_l \frac{4}{3} \pi r^3 \frac{d^2 x}{dt^2} , \qquad (4.129)$$

where c_D is the drag coefficient, $A_f = \pi r^2$ is the frontal area of the spherical drop, and x is the coordinate along the droplet trajectory. The drag coefficient is given by that of a rigid sphere [6],

$$c_{D,sphere} = \begin{cases} \dfrac{24}{Re_g} \left(1 + \dfrac{Re_g^{2/3}}{6} \right) & Re \leq 1000 \\ 0.424 & Re > 1000 \end{cases} , \qquad (4.130)$$

where $Re_g = 2r \cdot u_{rel} \rho_g / \mu_g$.

4.4.2 Dynamic Drag Modeling

When a liquid drop enters a gas stream with a sufficiently large Weber number, it deforms and is no longer spherical as it interacts with the gas. This has been observed experimentally by many researchers, e.g. [80, 111]. Hence, the drag coefficient should be a function of its Reynolds number as accounted for in Eq. 4.130 and of its oscillation amplitude as well. Based on these observations, Liu et al. [79] use the TAB model, Fig. 4.34, in order to predict the droplet distortion y (Eq. 4.90) and then modify the drag coefficient by relating it empirically to the magnitude of the drop deformation. Since the drag coefficient of a distorting drop should lie between that of a rigid sphere, Eq. 4.130 (lower limit), and that of a disk, which is about 3.6 times higher (upper limit), a simple linear expression is used for the dynamic drag coefficient:

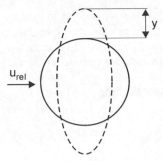

Fig. 4.34. The dynamic drag model accounts for drop distortion

$$c_D = c_{D,sphere}(1 + 2.632 \cdot y).$$ (4.131)

Hence, the drag coefficients of a spherical drop and a disk are recovered in the undeformed ($y = 0$) and the fully deformed limits ($y = 1$).

Compared to the use of the standard rigid sphere drag coefficients, the consideration of dynamic drag results in increased overall drag coefficients and a more realistic calculation of droplet deceleration. Compared to the standard model, the increased droplet deceleration results in reduced penetration and larger droplets due to the shorter time span with high relative velocity between gas and liquid.

It must be pointed out that the TAB model is only used for the calculation of drop distortion and does not influence the break-up modeling. If, for example, the Kelvin-Helmholtz model is used in order to predict drop break-up, then $y > 1$ does not result in any break-up.

Liu et al. [81] have found that their TAB drop drag model [79] significantly underestimates drop drag effects for high-speed drops and have presented a modification based on the DDB model [59]. According to the DDB model, the drops are deformed due to a pure extensional flow from an initial sphere shape to an oblate spheroid of ellipsoidal cross section with major semi axis a and minor semi axis b, Fig. 4.35. The dynamic drag coefficient is again calculated using Eqs. 4.130 and 4.131, but the dimensionless deformation of the droplet surface is given by

$$y = \min\left\{1, \left(\frac{a}{r} - 1\right)\right\},$$ (4.132)

such that the drag coefficients of a spherical drop and a disk were also recovered in the undeformed and the fully deformed limits ($a = 2r$ or $y = 1$, [59]), respectively. The drops undergo significant flattening, which changes the frontal area exposed to the airflow. Because the flattening occurs prior to a significant mass loss due to break-up [81], the calculation of the frontal area is also modified and given as $A_f = \pi a^2$ instead of $A_f = \pi r^2$.

The authors have extended their model in order to describe the dynamic drag after wall impingement. Before wall impingement, the drop's frontal area normal to the airflow is the largest cross-sectional area of the deforming drop ($A_f = \pi a^2$).

After wall impingement, drops with $We_{in} < 80$ (We_{in}: Weber number of incoming drops before wall impingement, $We_{in} = 2r\rho_l v_n^2 /\sigma$, v_n: drop velocity component normal to the wall) bounce off the wall, and it is assumed that they have an initial rotation which may be maintained or enhanced by the strong turbulence in the combustion chamber. In this case, the largest cross-sectional area is not always normal to the direction of drop motion, and it is assumed that the drops rotate stochastically. A random number ξ from the interval [0, 1] is used to modify the drop's frontal area

$$A_f^* = \pi a\left(b+\xi\cdot(a-b)\right) \tag{4.133}$$

and drag coefficient

$$c_D^* = c_{D,sphere}\left(1+2.632\cdot\xi\cdot y\right) \tag{4.134}$$

after they have impacted and rebounded from the wall. After wall impingement, the drops are assumed to start with an initial deformation $y = 1$, which can decrease again if the droplet velocity slows down and reduces the aerodynamic forces. Because the droplet is rotating, the value of ξ cannot be constant for the rest of its lifetime. Hence, ξ is only calculated for a time interval equal to that of the droplet's characteristic oscillation time

$$\tau_{rot} = B\cdot\sqrt{\frac{\rho_l\cdot r^3}{\sigma}}, \tag{4.135}$$

where r is the undisturbed droplet radius and B is a constant. Thereafter, a new value of ξ is sampled. The drop rotation effect after wall impingement is most important for large drops with high drop-gas relative velocities, but the Weber numbers (We_{in}) of which are less than 80.

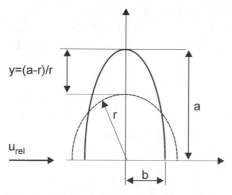

Fig. 4.35. Schematic diagram of the deforming half drop (DDB model)

4.5 Evaporation

In addition to the break-up of the spray and the mixing processes of air and fuel droplets, the evaporation of liquid droplets also has a significant influence on ignition, combustion, and formation of pollutants. The formation of fuel vapor due to evaporation is a prerequisite for the subsequent chemical reactions. The evaporation process determines the spatial distribution of the equivalence ratio, and thus strongly affects the timing and location of ignition. The energy for evaporation is transferred from the combustion chamber gas to the colder droplet due to conductive, convective, and radiative heat transfer, resulting in diffusive and convective mass transfer of fuel vapor from the boundary layer at the drop surface into the gas, Fig. 4.36. This again affects temperature, velocity, and vapor concentration in the gas phase. Hence, there is a strong linking of evaporation rate and gas conditions, and, for this reason, there must always be a combined calculation of heat and mass transfer processes.

In order to describe the evaporation process mathematically, the following assumptions are usually made: the radiative heat transfer is neglected because it is small compared to the convective one. Because it is not feasible to resolve the flow field around all the droplets of a spray, the evaporation modeling is based on averaged flow conditions and average transfer coefficients around the droplets. The droplets are usually assumed to be of spherical shape. Deformation, break-up, collisions, and other interactions of droplets are neglected during the calculation of evaporation. Further on, the droplet's interior is usually assumed to be well mixed. For this reason, there are no spatial gradients of the relevant quantities like liquid temperature, concentration of fuel components, boiling temperatures, and critical temperatures, heat of evaporation etc. inside the droplet, and only a dependence on time is possible. Furthermore, the solubility of the surrounding gas in the liquid and the effect of surface tension on the vapor pressure are neglected.

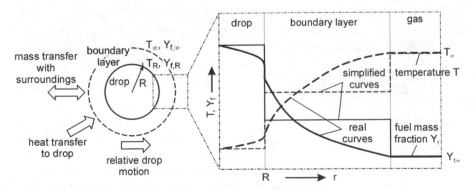

Fig. 4.36. Schematic view of drop vaporization

In order to determine the transport processes at the gas/liquid interface (mass and energy fluxes), phase equilibrium is assumed. It is presumed that the phase transition (liquid to vapor) is much faster than the vapor transport from the surface into the surrounding gas. Further on it is assumed that even if the conditions in the gas phase or inside the droplet change (e.g. temperature rise), phase equilibrium is always immediately reached.

The concentration of fuel vapor and thus also the properties of the gas mixture in the boundary layer are strongly dependent on radius, Fig. 4.36. In order to get representative values for the calculation of the diffusive mass transport, simplified vapor concentration curves are customarily used.

4.5.1 Evaporation of Single-Component Droplets

Although real fuels consist of a multitude of different components that influence the evaporation process (more volatile but less ignitable components evaporate first, components with higher molecular weight evaporate later), the standard approach today is to use a single-component model fuel. Usually tetradecane (n-$C_{14}H_{30}$) is used in order to represent the relevant properties of diesel, and octane is used for gasoline.

The temperature change of the liquid droplet can be obtained from an energy balance. The total heat flux

$$\dot{Q}_{drop} = \dot{Q}_{heating} + \dot{Q}_{evap} , \tag{4.136}$$

$$\dot{Q}_{heating} = m_{drop} c_{p,l} \frac{dT_{drop}}{dt} , \tag{4.137}$$

$$\dot{Q}_{evap} = \Delta h_{evap} \frac{dm_{evap}}{dt} , \tag{4.138}$$

transferred from the hot gas to the liquid droplet results in an increase of droplet temperature (heating) and in evaporation. In Eqs. 4.136–4.138, m_{drop} and T_{drop} are the droplet mass and temperature, $c_{p,l}$ is the specific heat capacity of the liquid fuel, Δh_{evap} is the enthalpy of evaporation, and m_{evap} is the mass that evaporates in the time interval dt. Using Eqs. 4.137 and 4.138 in Eq. 4.136 and solving for the temperature change yields

$$\frac{dT_{drop}}{dt} = \frac{1}{m_{drop} c_{p,l}} \left(\frac{dQ_{drop}}{dt} - \Delta h_{evap} \frac{dm_{evap}}{dt} \right) . \tag{4.139}$$

In order to solve Eq. 4.139, expressions for the quantities dQ_{drop}/dt and dm_{evap}/dt must be derived. A part of the total convective heat transferred from the gas to the liquid due to the temperature difference $\Delta T = T_{drop} - T_\infty$ (T_∞: temperature of the surrounding gas outside the boundary layer) is needed in order to heat up the evaporated mass of fuel that is transported from the drop surface ($T_{drop} = T_R$) into the gas (T_∞), see Fig. 4.37. A differential element of the gas atmosphere around the droplet (somewhere between $r = R$ and $r = \infty$) has to be regarded, and an energy balance considering both kinds of heat fluxes, Fig. 4.37, must be written down and solved. After several mathematical integrations and transformations, an equation for the heat transferred to the drop is obtained:

$$\dot{Q}_{drop} = \lambda_g \pi d_{drop} \left(T_\infty - T_R\right) \frac{\varsigma}{e^\varsigma - 1} Nu \,, \tag{4.140}$$

where

$$Nu = \frac{\alpha \cdot d_{drop}}{\lambda_g} \tag{4.141}$$

and

$$\varsigma = \frac{\dot{m}_{evap} c_{p,vap}}{Nu \lambda_g \pi d_{drop}}. \tag{4.142}$$

These equations are derived and discussed in [32] and [33] and are finally utilized in the droplet evaporation model of Borman and Johnson [16], which again is used in most spray and combustion models. The variable ς is a dimensionless correction factor taking account of the reduced heat transfer due to the simultaneous mass transfer from the drop into the gas atmosphere. In Eqs. 4.140–4.142, λ_g is the thermal conductivity of the gas mixture at the drop surface, $c_{p,vap}$ is the specific heat capacity of the fuel vapor (both calculated using the 1/3rd rule of Hubbard et al. [54], see also Eqs. 4.190 and 4.191, and d_{drop} is the droplet diameter. The appropriate Nusselt number that includes the effect of a relative velocity between droplet and gas has been proposed by Ranz and Marshall [112],

$$Nu = 2.0 + 0.6 \, Re^{1/2} \, Pr^{1/3} \,, \tag{4.143}$$

$$Re = \rho_\infty u_{rel} d_{drop} / \mu_g \,, \tag{4.144}$$

$$Pr = \mu_g c_p / \lambda_g \,. \tag{4.145}$$

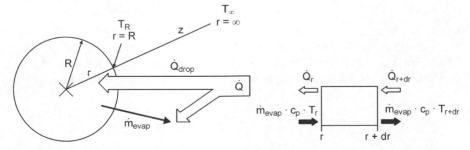

Fig. 4.37. Definition of coordinate system and heat fluxes

The properties of the gas phase inside the boundary layer are again calculated using the $1/3^{rd}$ rule.

Next, an expression for the fuel vapor mass flow due to evaporation in Eq. 4.139 must be derived. For the case of a one-dimensional binary diffusion, Fick's law can be used on a mass basis [142],

$$\frac{\dot{m}_A}{4\pi r^2} = Y_A\left(\frac{\dot{m}}{4\pi r^2}\right) - \rho D_{AB}\frac{dY_A}{dr}, \tag{4.146}$$

where $\dot{m}_A$ is the mass flow of species A, $\dot{m} = \dot{m}_A + \dot{m}_B$ is the total mixture flow rate, Y_A is the mass fraction, and D_{AB} is the binary diffusivity. The first term on the right hand side is the mass flow of species A associated with bulk flow, and the second one is the mass flow of species A associated with molecular diffusion. If it is assumed that species B (gas around the droplet) is insoluble in the liquid A, and that outside the boundary layer the concentration of A is constant, there will be no net transport of B in the boundary layer gas, and $\dot{m}_B = 0$. Thus, Eq. 4.146 can be expressed as

$$-\frac{\dot{m}_A}{4\pi\rho D_{AB}}\frac{dr}{r^2} = \frac{dY_A}{1-Y_A}. \tag{4.147}$$

Assuming that $\rho D_{AB} = \text{const}$, and integrating from $r = R$ to $r = \infty$ and from $Y_{A,R}$ to $Y_{A,\infty}$ gives

$$\dot{m}_A = 4\pi R\rho D_{AB}\ln\left(\frac{1-Y_{A,\infty}}{1-Y_{A,R}}\right) = 4\pi R\rho D_{AB}\ln(1+B), \tag{4.148}$$

where B is the Spalding transfer number,

$$B = \left(\frac{Y_{A,R}-Y_{A,\infty}}{1-Y_{A,R}}\right). \tag{4.149}$$

Using Eq. 4.148, the evaporating mass flow in Eq. 4.139 can be expressed as

$$\dot{m}_{evap} = 2d_{drop}\pi\rho D\ln\left(\frac{1-Y_{f,\infty}}{1-Y_{f,R}}\right), \tag{4.150}$$

where $d_{drop} = 2r$, D is the diffusivity of fuel vapor in air, $Y_{f,\infty}$ is the fuel vapor mass fraction outside the boundary layer, and $Y_{f,R}$ is the fuel vapor mass fraction at the droplet surface. The effect of an increased mass transport due to a relative velocity between droplet and surrounding gas is expressed by the Sherwood number:

$$\dot{m}_{evap} = d_{drop}\pi\rho D\ln\left(\frac{1-Y_{f,\infty}}{1-Y_{f,R}}\right)Sh. \tag{4.151}$$

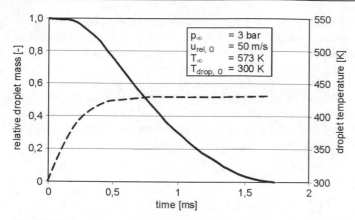

Fig. 4.38. Temperature and mass histories of an evaporating decane droplet

The Sherwood number has a value of $Sh = 2.0$ if no relative velocity is present. The appropriate Sherwood number, which includes the effect of a relative velocity between droplet and gas, has been proposed by Ranz and Marshall [112],

$$Sh = 2.0 + 0.6\, Re^{1/2}\, Sc^{1/3}\,, \tag{4.152}$$

where

$$Sc = \mu_g / \left(\rho_g D \right). \tag{4.143}$$

The properties of the gas phase inside the boundary layer are calculated using the $1/3^{\text{rd}}$ rule. Assuming equilibrium and using Raoult's law (e.g. [142]), the fuel vapor fraction in the boundary layer can be calculated as

$$Y_{f,R} = \frac{p_{vap}\left(T_l\right)}{p_{cyl}}\,\frac{MW_f}{MW_{mix,R}}\,, \tag{4.154}$$

where $p_{vap}(T_l)$ is the saturated vapor pressure belonging to the droplet temperature, and MW_f and $MW_{mix,R}$ are the molecular masses of fuel and gas mixture at the droplet surface. The vapor pressure can be determined from the Clausius-Clapeyron equation,

$$p_{vap}\left(T_l\right) = p_0 \exp\left[\frac{h_{fg}}{RT_0}\left(1 - \frac{T_0}{T_l}\right)\right], \tag{4.155}$$

where (p_0, T_o) is a known point on the vapor pressure curve.

As an example, Fig. 4.38 shows the temperature and mass histories of an evaporating decane droplet as predicted by the above model. The initial and boundary conditions are given in the figure. It can be seen that the droplet temperature increases at first due to the convective heat transfer from the hot gas to the liquid, while the evaporated mass is small. This causes an increased saturated vapor pressure at the droplet surface, which again results in increased diffusive mass transport into the gas. Due to the enhanced diffusion an increased amount of the energy (transferred from the gas to the liquid) is needed for evaporation, and a quasi-steady state is reached where the droplet temperature stays constant until it is completely vaporized.

It must be pointed out that the use of single-component model fuels is the standard approach today because of its simplicity and low consumption of computational time. Nevertheless, research work today concentrates on more sophisticated evaporation models, especially with respect to modeling more realistic multi-component fuels and to replace the single-component fuel calculations, see next section.

4.5.2 Evaporation of Multi-Component Droplets

The basic challenge when describing the evaporation process of fuel droplets is the choice of an appropriate reference fuel that represents the relevant behavior of the fuel with sufficient accuracy. For example, in standard CFD simulations, tetradecane (n-$C_{14}H_{30}$) is normally used to represent the relevant properties of diesel. However, diesel consists of more than 300 different components, and it is obvious that a single reference fuel cannot predict all of the relevant sub-processes during evaporation, ignition, and combustion with sufficient accuracy. Hence, multi-component model fuels are desirable. The goal is to achieve a high degree of accuracy while keeping the consumption of computational time small.

Initial studies on multi-component fuels concentrated on binary fuels consisting of two discrete hydrocarbons. Both discrete components were implemented in CFD codes using different fuel properties and transport equations for each of the two hydrocarbons. For example, Kneer [65] investigated the evaporation behavior of mixtures consisting of heptane and dodecane as well as hexane and tetradecane. Due to an increased evaporation of the more volatile component, especially during the initial stage of evaporation, these investigations have shown a significantly larger fuel vapor concentration. Further investigations have been published by Jin and Borman [60] and Klingsporn [63], for example. The investigations show that the overall mixture formation process, and thus also ignition and combustion, may be significantly influenced by the composition of the fuel vapor, and that it may be important to describe the multi-component character of fuels more accurately. However, the computational effort describing mixtures of ten and more components is enormous. For example, Rosseel and Sierens [121] developed a model including ten discrete components. According to the authors, the consumption of computational time for simulating the evaporation of a single droplet was twenty times higher than that of a single-component fuel.

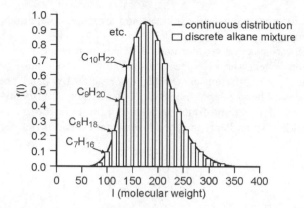

Fig. 4.39. Continuous and discrete representation of multi-component fuels

Tamin and Hallett [137] have shown a possibility of introducing the multi-component character of fuels without modeling and numerically solving the many differential equations needed to represent a blend of ten or more discrete components. Their approach is based on so-called continuous thermodynamics, which provides a relatively simple and elegant description of mixtures consisting of a multitude of different components. Although the modeling effort is larger compared to a discrete two-component model, the advantage of the continuous thermodynamic approach becomes apparent if more components need to be represented. In this approach, which has been used for gasoline and diesel (e.g. [78, 101, 137, 149, 110]), the composition of the fuel is described by a continuous distribution function.

The distribution function characterizes the distribution $f(I)$ of a macroscopic property I of the mixture, for example the molar mass, the boiling temperature, the number of carbon atoms, etc. In the case of hydrocarbon fuels, the molar mass is used as distribution variable. As an example, Fig. 4.39 shows a typical distribution of the different alkanes of diesel and the corresponding continuous distribution function, which has been used to model diesel fuel [77]. The continuous thermodynamic model only needs to solve one set of equations (as in the case of one single-component fuel) plus two additional equations for the first and the second moment of the distribution function, describing its mean value and variance. The main idea of the continuous thermodynamics approach is the description of the relevant fuel properties needed to determine the evaporation process like boiling and critical temperatures, heat conductivity, heat of evaporation etc. as a function of the distribution variable I. If the distribution of the different components inside the droplet changes during the evaporation process (the more volatile components evaporate first), the distribution function and thus the fuel properties also change. This kind of approach ensures that the effect of the different fuel components on the time-dependent evaporation process is accounted for. It should be mentioned at this point that this condition is only satisfied if the hydrocarbons belong to a particular family of molecules, e.g. alkanes. Hence, it is not possible to model components belonging to different groups, like alkanes and aromatic compounds, with a single distribution function.

Besides the distribution variable, the correct choice of an appropriate mathematical distribution function is necessary. On the one hand, this function must be able to describe the mixture with sufficient accuracy, but on the other hand, the mathematical description should be as simple as possible. In order to describe hydrocarbon mixtures, the so-called gamma-function is used [24, 137, 100]. This distribution function is also known as Pearson type III distribution and is defined as [102]:

$$f(I) = \frac{(I-\gamma)^{\alpha-1}}{\beta^{\alpha}\Gamma(\alpha)} \exp\left(-\frac{(I-\gamma)}{\beta}\right),$$ (4.156)

where α and β and are parameters describing the shape of the curve, and γ is responsible for the displacement of the origin of the curve, Fig. 4.40.

The quantity I is the distribution variable (molar mass of the fuel components), and $\Gamma(\alpha)$ is given by

$$\Gamma(\alpha) = \int_0^{\infty} e^{-t} t^{\alpha-1} dt.$$ (4.157)

Because Eq. 4.157 is the normalized gamma-function, the area below the curve is unity:

$$F(I) = \int_0^{\infty} f(I) dI = 1.$$ (4.158)

Thus, the first and the second moment of the distribution are given by

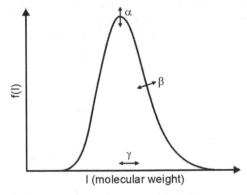

Fig. 4.40. Gamma-function: effect of parameters α, β, γ

$$\theta = \int_0^\infty f(I) \cdot I \cdot dI \qquad (4.159)$$

and by

$$\psi = \int_0^\infty f(I) \cdot I^2 \cdot dI . \qquad (4.160)$$

The first moment θ is the mean of the distribution function, and ψ is a measure for the variance σ. In the case of the gamma-function, θ and ψ can be expressed as

$$\theta = \alpha \cdot \beta + \gamma , \qquad (4.161)$$

$$\sigma^2 = \alpha \cdot \beta^2 , \qquad (4.162)$$

$$\psi = \theta^2 + \sigma^2 . \qquad (4.163)$$

The parameters α, β, and γ are chosen in order to achieve the best representation of the specific fuel of interest. The continuous thermodynamics can also be applied to single-component fuels by using extremely narrow distributions. Table 4.4 summarizes the distribution parameters for various fuels as given in the works of Lippert [77] and Pagel [100].

Besides a transport equation for the overall fuel mass fraction, two additional transport equations for the first and the second moment, describing the shape of the distribution function for the vaporized fuel, are necessary in order to determine the change of the vapor phase composition in the computational cell around the evaporating droplet. The relevant vapor phase transport equations have been derived by Tamin and Hallett [137]. For a single component k of the multi-component fuel the mass transport equation can be written as

$$\frac{\partial}{\partial t}(\rho Y_k) + \nabla \cdot (\rho \vec{u} Y_k) = \nabla \cdot (\rho D_{k,av} \nabla Y_k) + \frac{d}{dt}(\rho Y_k)^C + \frac{d}{dt}(\rho Y_k)^S , \qquad (4.164)$$

where Y_k is the mass fraction of the component k, and $D_{k,av}$ is the averaged binary diffusion coefficient. The resulting equations are derived by integrating the species transport equation over the complete distribution function in order to obtain

Table 4.4. Distribution parameters for multi-component and single-component fuels

	Diesel	Gasoline	n-Octane	n-Decane
α	18.5	5.7	100	100
β	10	15	0.1	0.1
γ	0	0	104.2	132.3
θ	185	85.5	114.2	142.3
σ	43	35.8	1	1

the transport equation for the species mass fraction, and by weighting the equation by I and by I^2 before integrating in order to get the transport equations for the first and the second moment [78, 137, 46]:

$$\frac{\partial}{\partial t}\left(\rho Y_f\right)+\nabla\cdot\left(\rho\bar{u}Y_f\right)=\nabla\cdot\left(\rho\bar{D}\nabla Y_f\right)+\left(\dot{\rho}Y_f\right)^S+\left(\dot{\rho}Y_f\right)^C, \tag{4.165}$$

$$\frac{\partial}{\partial t}\left(\rho Y_f\theta\right)+\nabla\cdot\left(\rho\bar{u}Y_f\theta\right)=\nabla\cdot\left(\rho\tilde{D}\nabla\left(Y_f\theta\right)\right)+\left(\dot{\rho}Y_f\theta\right)^S+\left(\dot{\rho}Y_f\theta\right)^C, \tag{4.166}$$

$$\frac{\partial}{\partial t}\left(\rho Y_f\psi\right)+\nabla\cdot\left(\rho u_i Y_f\psi\right)=\nabla\cdot\left(\rho\hat{D}\nabla\left(Y_f\psi\right)\right)+\left(\dot{\rho}Y_f\psi\right)^S+\left(\dot{\rho}Y_f\psi\right)^C. \tag{4.167}$$

The quantity Y_f is the overall fuel mass fraction in the gas, ρ is the mass-averaged density, and $\bar{D}, \tilde{D}$ and $\hat{D}$ are binary diffusion coefficients for the respective conserved scalars. It can be shown that $\bar{D}\approx\tilde{D}\approx\hat{D}$. The first term on the right hand side of Eqs. 4.165–4.167 is due to mass diffusion, while the second and the third ones are source terms due to spray evaporation and combustion.

The source term due to spray evaporation is determined by the evaporation of the multi-component fuel droplets. If the droplet interior is assumed to be well mixed, which is usually done in most of the 3D CFD calculations, the molar flux over the droplet surface is given by

$$\dot{n} = -\frac{1}{A_{drop}}\frac{d\left(c_l V_{drop}\right)}{dt}, \tag{4.168}$$

where $A_{drop} = 4\pi R^2$, $V_{drop} = (4/3)\pi R^3$, and $c_l = \rho_l/\theta_l$ is the liquid molar density in mol/cm^3. In contrast to the fuel vapor distribution function, the relevant parameters of the liquid fuel distribution inside the droplet are denoted by the subscript l. It should also be mentioned that all liquid-phase composition calculations are performed on a molar basis, since the vapor-liquid equilibrium is solved in a molar frame of reference. Using the expressions for drop surface area and volume in Eq. 4.168, the time-dependent change of droplet radius becomes

$$\frac{dR}{dt} = \frac{R}{3c_l\theta_l}\left(c_l\frac{d\theta_l}{dt}-\frac{d\rho_l}{dt}\right)-\frac{\dot{n}}{c_l}. \tag{4.169}$$

The mole flux $\dot{n}$, which is needed to determine the mean and the second moment of the droplet composition, can be calculated equating the fluxes at the liquid/gas interface. In the case of a single component k of the liquid phase, Eq. 4.168 yields

$$\dot{n}_k = -\frac{1}{A_{drop}}\frac{d\left(x_k c_l V_{drop}\right)}{dt} = x_k\dot{n}-c_l\frac{R}{3}\frac{dx_k}{dt}, \tag{4.170}$$

where x_k is the mole fraction of component k in the liquid phase. Because a well-mixed droplet interior is assumed, it is also the mole fraction inside the whole droplet. The mole flux $y_{f,R}$ in the gas phase at the liquid-vapor interface (index R) is given by assuming equal molar diffusion from droplet to gas,

$$\dot{n}_k = \dot{n} \cdot y_{k,R} - c \cdot D_{k,av} \left. \frac{dy_k}{dr} \right|_R . \tag{4.171}$$

Equating Eq. 4.170 and Eq. 4.171 gives

$$\dot{n}\left(x_k - y_{k,R}\right) = -c \cdot D_{k,av} \frac{dy_k}{dr} + c_l \frac{R}{3} \frac{dx_k}{dt} . \tag{4.172}$$

The resulting equations for the liquid composition are derived by integrating over the complete distribution function and by several further transformations [137]:

$$\dot{n}\left(1 - y_{f,R}\right) = -c \cdot \bar{D} \frac{dy_f}{dr} , \tag{4.173}$$

$$\frac{d\theta_l}{dt} = \frac{3}{c_l R} \left[\dot{n}\left(\theta_l - y_f \theta\right) + c \cdot \tilde{D} \frac{d}{dr}\left(y_f \theta\right) \right]_R , \tag{4.174}$$

$$\frac{d\psi_l}{dt} = \frac{3}{c_l R} \left[\dot{n}\left(\psi_l - y_f \psi\right) + c \cdot \hat{D} \frac{d}{dr}\left(y_f \psi\right) \right]_R . \tag{4.175}$$

The index R denotes the droplet surface. Next, equilibrium is assumed at the interface, resulting in

$$y_f(r) = 1 - \left(1 - y_{f,\infty}\right)\left(1 + B\right)^{-R/r} \tag{4.176}$$

and

$$\phi(r) = \phi_R - \frac{1}{B}\left(\phi_R - \phi_\infty\right)\left[\left(1 + B\right)\left(1 + B\right)^{-R/r} - 1\right], \tag{4.177}$$

where ϕ represents $y\theta$ and $y\psi$, respectively. The subscripts "R" and "∞" indicate the gas phase properties at the droplet surface and in the undisturbed surroundings, respectively. B is the Spalding transfer number,

$$B = \frac{y_{f,R} - y_{f,\infty}}{1 - y_{f,R}} . \tag{4.178}$$

Using Eqs. 4.176 and 4.177 in Eqs. 4.174 and 4.175 results in expressions that are only dependent on quantities calculated by assuming equilibrium conditions at the liquid-gas interface (R) and on those of the surrounding (∞). The final expressions describing the change of the mean and the second moment of the droplet composition are

$$\frac{d\theta_l}{dt} = \frac{3\dot{n}}{c_l R}\left[\left(\theta_l - y_{f,R} \cdot \theta_R\right) + \frac{1}{B}\left(\theta_\infty \cdot y_{f,\infty} - \theta_R \cdot y_{f,R}\right)\right], \tag{4.179}$$

$$\frac{d\psi_l}{dt} = \frac{3\dot{n}}{c_l R}\left[\left(\psi_l - y_{f,R} \cdot \psi_R\right) + \frac{1}{B}\left(\psi_\infty \cdot y_{f,\infty} - \psi_R \cdot y_{f,R}\right)\right]. \tag{4.180}$$

The time-dependent development of droplet temperature can be calculated by an energy balance in analogy to Eq. 4.139 as

$$\frac{dT_l}{dt} = \frac{3}{c_l \overline{c}_v R}\left[\dot{q}_{conv} - \dot{n}\overline{h}_{fg}\right], \tag{4.181}$$

where $\overline{h}_{fg}$ and $\overline{c}_v = c_v \cdot \theta$ are the molar heat of evaporation and the molar specific heat of the liquid. The convective heat flux and the molar flux may be approximated by

$$\dot{q}_{conv} = \frac{\lambda\left(T_\infty - T_l\right)}{2R}\frac{\ln\left(1+B\right)}{B} Nu \tag{4.182}$$

and

$$\dot{n} = \frac{c\overline{D}}{2R}\ln\left(1+B\right)Sh . \tag{4.183}$$

The appropriate Nusselt and Sherwood numbers including corrections for the influence of a relative velocity between droplet and gas are

$$Nu = 2.0 + 0.6\,Re^{1/2}\,Pr^{1/3} \text{ and} \tag{4.184}$$

$$Sh = 2.0 + 0.6\,Re^{1/2}\,Sc^{1/3}, \tag{4.185}$$

where Pr and Sc are the Prandtl and the Schmidt numbers, respectively.

As already mentioned, phase equilibrium is assumed to determine the vapor mole fraction in the gas phase at the droplet's surface, and Raoult's law is applied. In the case of a single component, this relation between the mole fraction of vaporized fuel y_k at the interface and the mole fraction x_k inside the liquid drop is

$$y_k = x_k\left(\frac{p_k}{p}\right), \tag{4.186}$$

where p is the static pressure in the gas phase and p_k is the partial pressure of the component k with molar mass I. In the case of a continuous distribution function, the mole fractions of the different components inside the liquid are described by a distribution function, which changes its shape during evaporation. Thus, the mole fractions at the surface also change, and in this case Raoult's law is derived by integrating over the complete distribution function:

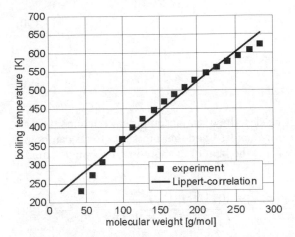

Fig. 4.41. Boiling temperature of alkanes as function of molecular weight

$$y_f = \int_0^\infty f_I(I) \left(\frac{p_k(I)}{p} \right) dI \; . \tag{4.187}$$

The value of p_k at the interface is given by the Clausius-Clapeyron equation,

$$p_k(I) = p_0 \exp\left[\frac{\overline{s}_{fg}}{\overline{R}} \left(1 - \frac{T_b(I)}{T_l} \right) \right] , \tag{4.188}$$

where $p_0 = 101.325$ kPa is a reference pressure, $T_b(I)$ is the corresponding boiling temperature of component I, $\overline{s}_{fg}$ is the molar entropy of evaporation in J/(mol·K), $\overline{R} = 8.314151$ J/(mol·K) is the molar gas constant, and T_l is the temperature of the liquid drop. Fig. 4.41 shows the boiling temperature of alkanes as a function of molar mass I. A linear correlation as proposed by Lippert [77] can be assumed,

$$T_b(I) = a_b + b_b I \; , \tag{4.189}$$

where a_b and b_b are constants obtained from regression of boiling point data. In the case of diesel fuel, Lippert [77] uses $a_b = 208.53$ K and $b_b = 1.5673$ K/mol.

Further quantities needed to calculate the multi-component fuel evaporation process using the continuous thermodynamics approach are the heat of evaporation, the heat conductivity of the gas mixture, the specific heat capacities of gas mixture and liquid, the binary diffusion coefficient between air and fuel vapor, and the critical temperature. It should be noted that for all correlations used to describe the boundary layer the 1/3rd rule of Hubbard et al. [54] is used in order to account for the strong gradients of the quantities of interest and to calculate representative reference values inside the boundary layer. For example, T_{ref} and $Y_{f,ref}$ are calculated as

$$T_{ref} = \frac{2}{3}T_R + \frac{1}{3}T_\infty, \tag{4.190}$$

and

$$Y_{f,ref} = \frac{2}{3}Y_{f,R} + \frac{1}{3}Y_{f,\infty}. \tag{4.191}$$

The heat of evaporation can be described via an empirical relation. It is assumed that ideal gas behavior can be applied and that pressure effects can be neglected. According to Reid et al. [113] the heat of evaporation for a single component is given as

$$\bar{h}_{fg} = \left(1.093 \cdot \bar{R} \cdot T_b\right)\left[\frac{\log\left(p_{crit}\right)-1.013}{0.930 - T_b / T_{crit}}\right]\left[\frac{T_{crit} - T}{T_{crit} - T_b}\right]^{0.38}, \tag{4.192}$$

where p_{crit} is the critical pressure and T_{crit} is the critical temperature. Tamin and Hallett [137] have modified this approach for multi-component fuel mixtures:

$$\bar{h}_{fg} = \left[a_h + b_h\left[\left(y_f\theta\right)_R - \frac{1}{B}\left[\left(y_f\theta\right)_\infty - \left(y_f\theta\right)_R\right]\right]\right]\cdot\left[\frac{\left(T_{crit} - T_l\right)^{0.38}}{6.959}\right], \tag{4.193}$$

where $a_h = 2.07 \cdot 10^{-7}$ and $b_h = 1.35 \cdot 10^{-5}$ for units of $\bar{h}_{fg}$ of J/kmol.

The thermal conductivity of a gaseous fuel component can be described as a function of temperature and molecular weight [137],

$$\lambda\left(I,T\right) = \left(a_{KC} + a_{KT}T\right) + \left(b_{KC} + b_{KT}T\right)\cdot I, \tag{4.194}$$

where $\lambda(I, T)$ is in W/(m·K), $a_{KC} = -2.37 \cdot 10^{-2}$, $a_{KT} = 1.09 \cdot 10^{-4}$, $b_{KC} = 3.47 \cdot 10^{-5}$, $b_{KT} = -1.91 \cdot 10^{-8}$.

A correlation for the specific heat capacity of the fuel vapor has been developed by Chou and Prausnitz [22]:

$$\bar{c}_p\left(I,T\right) = \bar{R}\left(a_{cp} + b_{cp} \cdot I\right),$$
$$a_{cp} = a_{c0} + a_{c1}T + a_{c2}T^2 + a_{c3}T^3, \tag{4.195}$$
$$b_{cp} = b_{c0} + b_{c1}T + b_{c2}T^2 + b_{c3}T^3,$$

where $c_p(I, T)$ is in J/(mol·K), $a_{c0} = 2.465$, $a_{c1} = -1.144 \cdot 10^{-2}$, $a_{c2} = 1,759 \cdot 10^{-5}$, $a_{c3} = -5.972 \cdot 10^{-9}$, $b_{c0} = 2.465$, $b_{c1} = -1-144 \cdot 10^{-2}$, $b_{c2} = 1.759 \cdot 10^{-5}$, and $b_{c3} = -5.972 \cdot 10^{-9}$ [77].

Tamin and Hallett [137] have given a correlation for the binary diffusion coefficient between alkanes and air. This correlation assumes a linear relation between molar mass and diffusion coefficient. However, for low and very large molar masses, this function underpredicts the diffusivity, see Fig. 4.42. Fuller (see [113]) uses a non-linear approach,

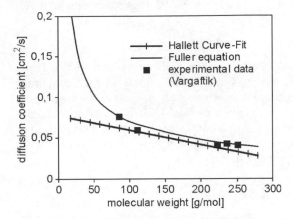

Fig. 4.42. Binary diffusion coefficients for alkanes and air [77]

$$D_{af} = \frac{0.00143 \cdot T^{1.75}}{p \cdot \sqrt{M_{af}} \left[\left(\sum v \right)_a^{1/3} + \left(\sum v \right)_f^{1/3} \right]^2} ,$$

$$M_{af} = 2 \left(\frac{1}{MW_a} + \frac{1}{\theta_f} \right)^{-1} ,$$

(4.196)

where T, p, and MW_a are the temperature, the pressure and the molecular weight of the surrounding gas, θ_f is the mean of the fuel vapor distribution, and Σv is the sum of the atomic diffusion volumes of air and fuel [113]. This method shows a much better representation of data from Vargaftik, Fig. 4.42, and it also includes the reciprocal dependency on pressure as predicted by the kinetic theory of gases [77]. D_{af} is used as binary diffusion coefficient in Eqs. 4.165–4.167.

A linear relation for the critical temperature as function of molecular weight is given by Tamin and Hallett [137],

$$T_{crit}(I) = a_c + b_c I , \tag{4.197}$$

where $a_c = 440.8$ and $b_c = 1.21$. Lippert [77] modified this approach introducing the critical volume V_{crit},

$$T_{crit} = \frac{\int V_{crit}(I) T_{crit}(I) f(I) dI}{\int V_{crit}(I) f(I) dI} . \tag{4.198}$$

The critical volume can be described by a linear function with parameters from Li et al. [75]:

$$V_{crit}(I) = a_v + b_v I , \tag{4.199}$$

where $a_v = 15.903$ cm^3/mol and $b_v = 1.21$ cm^3/mol^2. The combination of Eq. 4.198) and Eq. 4.199 results in [77]

$$T_{crit} = a_c a_v + a_v b_c \theta + a_c b_v \theta + b_c b_v \psi .$$
(4.200)

Typical results obtained with the multi-component evaporation model by Pagel [100] are shown in the following figures. First, the temperature rise inside a droplet (diesel fuel) during evaporation is shown in Fig. 4.43. The corresponding change of the distribution function is given in Fig. 4.44. In contrast to single-component fuels that first heat up and keep a constant temperature until the end of evaporation, Fig. 4.43 shows the typical strong temperature rise at the end of evaporation due to the multi-component fuel modeling.

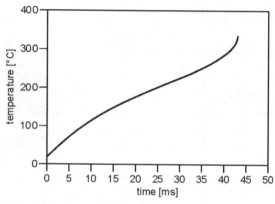

Fig. 4.43. Change of liquid temperature during evaporation of a multi-component diesel droplet (initial values: $D = 100$ μm, $u = 0$ m/s, $T_{drop} = 300$ K, $T_{gas} = 973$ K, $p_{gas} = 0.1$ MPa) [100]

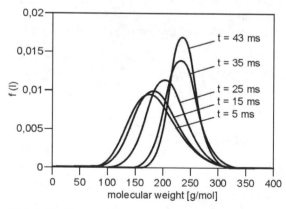

Fig. 4.44. Change of liquid composition during evaporation of a multi-component diesel droplet (initial values: $D = 100$ μm, $u = 0$ m/s, $T_{drop} = 300$ K, $T_{gas} = 973$ K, $p_{gas} = 0.1$ MPa) [100]

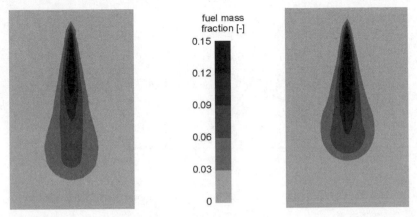

Fig. 4.45. Vapor phase fuel mass fractions for evaporating n-$C_{14}H_{30}$ (left) and continuous multi-component fuel (right) sprays in nitrogen after 2 ms, m_{fuel} = 40 mg, T_{fuel} = 300 K, p_{gas} = 5.7 MPa, T_{gas} = 800 K, p_{inj} = 110 MPa [100]

The application of Raoult's law for every component of the distribution guarantees that the diffusive mass transport of the components with low molecular weight and high volatility is larger than that of the heavier components. Hence, most of the evaporated fuel mass consists at first of the lightweight components, and only a few heavier components leave the liquid drop. Thus, the distribution function is shifted to larger molecular weights, and its variance decreases, Fig. 4.44. The value $f(I)$ of the distribution function increases because the curves are normalized and the area below is always 1. The droplet temperature is controlled by the balance of energy that is transferred from the hot gas to the drop and of energy that leaves the drop due to the evaporation of liquid mass. As long as the mean molecular weight of the evaporated mass is low, the liquid temperature can be kept down. As evaporation proceeds, the mean molecular weight of the evaporated mass increases. Due to the lower volatility of the heavier components (Raoult's law), the droplet temperature increases until the energy loss due to an enhanced evaporation at higher temperature compensates the energy input from the hot gas again. If the droplet composition does not change again, a steady-state as described in the case of a single-component fuel, Fig. 4.38, could be reached.
At the very end of evaporation, only a few components with increasing molecular weight and strongly decreasing volatility are left. Now the heat flux to the droplet results in a strong increase of droplet temperature.

In Fig. 4.45, the evaporation of a complete spray under engine-like conditions is shown in the case of multi-component diesel fuel and tetradecane (n-$c_{14}H_{30}$), which is usually used in standard single-component evaporation models in order to represent the behavior of diesel fuel. The figure shows the overall fuel mass fraction in the gas phase two milliseconds after the start of injection. Although the mean properties of the multi-component fuel and tetradecane are very similar,

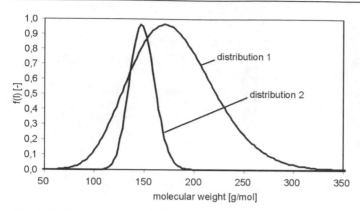

Fig. 4.46. Representation of different molecule classes by different distribution functions

significant differences concerning the overall mixture formation process are visible. Due to the stronger initial evaporation of the more volatile components near the nozzle, evaporation is enhanced, droplet sizes reduce faster, and the specific aerodynamic forces are increased resulting in a smaller penetration and a broader fuel vapor distribution. The simulations of Pagel [100] have also revealed that due to the strong evaporation at the edges of the spray the droplets at these positions have a higher molecular weight than those in the center of the spray. Although initial evaporation is increased in the case of multi-component diesel, ignition processes start later compared to a tetradecane spray. At first, the lightweight components with high volatility but low ignitability evaporate, and the molecules with longer carbon chains that produce radicals much more easily are the last ones to enter the vapor phase. In the case of tetradecane, its long molecules can earlier start to initiate ignition.

It should be mentioned that it is not always possible to represent the relevant fuel properties of multi-component fuels by using only one distribution function. Although the only use of the distribution of alkanes shows accurate results in the case of diesel, this simple approach is no more sufficient if for example tailored HCCI-fuels consisting of two or more groups of completely different molecule classes are considered. In this case, at least one more distribution function must be implemented describing the behavior of the second molecule group, Fig. 4.46. Most recently, a combination of two distribution functions has been investigated by Fischer et al. [37]. This way of modeling multi-component fuels promises a much more accurate representation of the evaporation behavior and is especially needed in the case of HCCI engine simulations, where ignition delays are long and the low-temperature ignition processes strongly depend on the temporal composition of the fuel vapor phase, see also Sect. 6.4.

In the case of conventional diesel engines, the Shell-model (see Sect. 4.9.1) is normally used in 3D CFD codes today to describe the ignition process. Numerical investigations have shown that the cetane number of the fuel has an important effect on the ignition behavior. The larger the cetane number, the larger the ignitability. Thus, the effect of a variable cetane number due to the change of fuel vapor composition in the gas during the evaporation of a multi-component fuel has to be taken into account. Heywood [51] has given a correlation describing the activation energy for auto-ignition as function of the cetane number (CN):

$$E_A = \frac{618840}{CN + 25}.$$ (4.201)

This approach was implemented by Ayoub [9] in the standard Shell-model. The expression for the activation energy of the chain propagation reaction

$$\dot{R} \xrightarrow{f_4 k_p} \dot{R} + Q,$$ (4.202)

(Sect. 4.9.1), which has been shown to be a crucial path for the imtermediate ignition species to transform into the branching species [68], was modified:

$$E^*_{f4} = E_{f4} \frac{65}{CN + 25}.$$ (4.203)

In the case of diesel fuel with a cetane number of 40, the activation energy remains unchanged. In order to apply this approach to multi-component fuels, further equations describing the relation between cetane number and mean molecular weight of the distribution function are necessary. Rose and Cooper [120] have measured the cetane number of pure hydrocarbons, Fig. 4.47. The relation between cetane number and mean molecular weight of the fuel vapor can approximated by the function

$$CN_{paraffins} = -4.2438 \cdot 10^{-6} \theta^3 + 1.7080 \cdot 10^{-3} \theta^2 + 0.14675 \theta + 29.295.$$ (4.204)

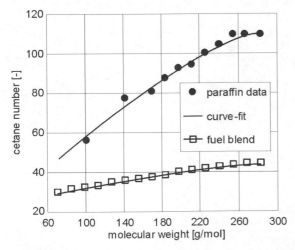

Fig. 4.57. Cetane number as function of molecular weight of paraffins (alkanes), data from [120]

However, diesel fuel does not only consist of n-alkanes but also of other components like iso-alkanes and aromats. Thus, the cetane number of Eq. 4.204 must be modified by another correlation given by Glavincevski et al. [42]:

$$CN = 30 + 0.221182 \cdot CN_{paraffins} - 47.18, \quad 48 \leq CN_{paraffin} \leq 110 . \quad (4.205)$$

This approach has been successfully used by Lippert [77] and Pagel [100] in order to simulate the ignition behavior of a multi-component fuel using the Shell-model.

4.5.3 Flash-Boiling

When a liquid, initially in a subcooled state, is rapidly depressurized to a pressure sufficiently below the saturated vapor pressure, it can no longer exist in the liquid state, and a rapid boiling process called flash-boiling is initiated. A portion of the fuel then evaporates instantaneously and cools the rest of the liquid down. This sudden evaporation results in a significant increase of spray volume and a faster spray break-up. In the case of high-pressure diesel injection, the phenomenon of flash-boiling can only be achieved if the fuel is sufficiently preheated before injection. In the case of gasoline injection, flash-boiling is much easier to obtain due to the lower boiling curve. Especially if gasoline is injected in the intake manifold, where the static pressure can fall below the saturated vapor pressure of some hydrocarbon fuel components. Such a condition will result in unintended flash-boiling. This causes significant changes in the fuel spray distribution and the fuel-air mixing.

Fig. 4.48 shows the conventional and the flash-boiling fuel injection in a pressure-enthalpy diagram. Subcooled liquid exists to the left of the liquid saturation line, and superheated vapor exists to the right of the vapor saturation line.

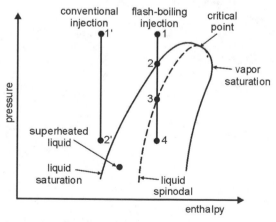

Fig. 4.48. Comparison of conventional injection and flash-boiling injection [99]

Superheated liquid can exist for a significant period of time without phase transition in a metastable condition between the liquid saturation line and the liquid spinodal, while to the right of the liquid spinodal there is no metastable state and liquid and vapor must coexist. During injection, the highly pressurized fuel leaves the nozzle through the injection hole, in which the liquid is strongly accelerated and the pressure decreases. In the case of conventional injection (line 1'- 2'), the fuel temperature, and thus the enthalpy, is too low to cross the liquid saturation line during pressure decrease. In the case of flash-boiling injection, the increased fuel temperature results in a higher fuel enthalpy, and the fuel undergoes a pressure reduction from point 1 to point 4 while passing through the nozzle hole. Between point 2 and point 3, vapor bubbles are formed and begin to grow. If there were be enough time, an equilibrium vapor fraction would be achieved. As point 3 is approached, the nucleation rates become large, and beyond point 3 the transition from vapor to liquid becomes explosively rapid.

The flash-boiling process consists of three stages: nucleation, vapor bubble growth, and atomization [99]. A nucleus is a vapor bubble in a metastable equilibrium with the surrounding liquid. The size r of the nucleus depends on a force balance between the surface tension force tending to reduce the bubble radius, and an opposing force due to the difference of saturated vapor pressure $p_{sat}(T)$ inside the bubble and the static pressure p in the surrounding liquid,

$$r_0 = \frac{2\sigma}{p_{sat}(T) - p}, \qquad (4.206)$$

where σ is the surface tension of the liquid in contact with its vapor. The larger the degree of superheat at a constant pressure of the liquid, the larger the saturated vapor pressure, and the smaller the bubble radius in equilibrium. A decrease of surrounding pressure p will result in bubble growth. Nuclei can form in crevices and other imperfections at the nozzle hole wall (e.g. roughness), on solid particles in the flow, etc. Small air bubbles entrained in the flow can also provide nucleation sites. At high degrees of superheating, nucleation even occurs at random locations throughout the liquid in the absence of particles or air bubbles.

During bubble growth, fuel evaporates at the bubble wall, and the vapor is added to the bubble volume. The theory of vapor bubble growth is based on the same basic equation as the theory of bubble collapse in the case of cavitation. Usually the Rayleigh-Plesset equation (e.g. [108, 48]) or some advanced forms of this equation are used in order to describe the bubble dynamics. In the case of vapor bubble growth due to flash-boiling, the latent heat of vaporization, which is transferred from the liquid to the bubble surface, must be included (e.g. [99]).

Atomization is the final stage of the flash-boiling injection process. Three possible mechanisms of atomization are discussed in the literature: bubble coalescence, inertial shattering, and-micro explosions of droplets. In the case of bubble coalescence, it is assumed that bubbles grow until they touch each other, resulting in a transition from bubbles in a liquid matrix to liquid drops in a vapor matrix. The release of surface tension energy may then result in further atomization.

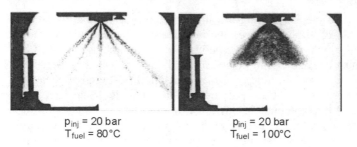

$p_{inj} = 20$ bar $p_{inj} = 20$ bar
$T_{fuel} = 80°C$ $T_{fuel} = 100°C$

Fig. 4.49. Comparison of flashing and non-flashing full-cone sprays [98]

In the case of inertial shattering, the rapid bubble growth outside the nozzle causes a momentum in radial direction that finally results in disintegration. In addition to these effects, bubble growth inside primary droplets that are still superheated can produce micro-explosions and result in further atomization.

There are three potential benefits of flash-boiling injection: enhanced atomization, increased initial spray cone angle for faster fuel-air mixing, and reduced spray penetration. These advantages can be attractive in direct injection diesel and stratified-charge gasoline engines as well as in HCCI combustion, in which fuel-air mixing rates and spray penetration must be carefully matched to the combustion chamber geometry and to the gas temperature and pressure in order to avoid wall wetting. Further on, a better atomization and fuel-air mixing due to flash-boiling might also enhance cold-starting processes.

Fig. 4.49 shows a comparison between a non-flashing gasoline spray from a multi-hole injector (full-cone spray plumes) and the flashing condition, which is achieved by an increase of initial fuel temperature from 80°C to 100°C (vapor pressure exceeds ambient pressure). Both sprays are injected into atmospheric air. As can be seen, the spray pattern changes completely: the instantaneous evaporation results in an increase of air-fuel mixing (the single spray plumes are no more visible) and in reduced penetration.

In contrast to Fig. 4.49, Fig. 4.50 shows a comparison between a non-flashing hollow-cone iso-octane spray and the corresponding flashing spray. The fuel has been injected into a pressure chamber ($T_{chamber} = 323$ K, $p_{chamber} = 50$ kPa, $p_{rail} = 8$ MPa). Again, the spray structure changes significantly if flash-boiling occurs: the hollow-cone spray collapses into a full-cone structure, but this causes a decrease of cone angle and an increase of spray penetration due to the compact spray structure. The droplets at the tip of the spray induce air motion and reduce the relative velocity between air and the following drops, and since the droplets are surrounded by saturated fuel vapor, evaporation is inhibited and drop sizes reduce slower [126].

Numerous experimental studies of flash-boiling injection have been performed to examine the phenomenon in detail. Gerrish and Ayer [41] observed an increase of spray cone angle when diesel fuel was preheated prior to injection. Kim et al. [62] performed flash-boiling studies with alcohol, in both an engine cylinder and a test chamber. The measurements of drop sizes, spray cone angle and penetration confirmed that flash-boiling provides the benefits listed above. Similar results

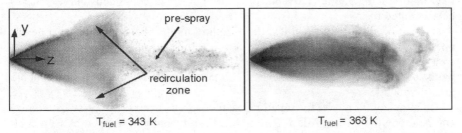

$T_{fuel} = 343$ K $T_{fuel} = 363$ K

Fig. 4.50. Comparison of flashing and non-flashing hollow-cone sprays [126]

were obtained by more fundamental experiments in atmospheric pressure chambers (e.g. [146, 99]).

The experimental investigations have revealed that there are two main categories of flash-boiling sprays, dependent on the degree of superheat: external flash-boiling and internal flash-boiling (e.g. [99, 103]). In the case of external flashing, evaporation and rapid bubble growth occur outside the nozzle in the spray. The rapid bubble growth shatters the liquid jet and results in an increased spray cone angle and a reduced penetration. In the case of internal flashing, rapid bubble growth occurs already inside the nozzle hole, resulting in an under-expanded compressible two-phase flow that expands immediately upon leaving the nozzle.

External flashing is difficult to obtain because a smooth nozzle entry geometry and/or low injection velocities must be used in order to avoid regions of locally low pressure inside the injection holes and to suppress internal flashing. Further on, careful matching of injection pressure and fuel temperature with combustion chamber pressure is necessary during the time of injection, because the degree of superheat and thus the change from the external to the internal flashing regime is very sensible to the chamber pressure. Hence, the relevant flashing regime is the internal flashing mode. However, internal flashing results in a reduction of effective cross-sectional area inside the injection hole and thus reduces the mass flow through the nozzle. At high degrees of superheat, the nozzle hole can become vapor-locked, and the mass flow reduces drastically [115]. In engine applications this has to be avoided at all costs.

Hence, the optimum degree of superheat is difficult to control, and this is one of the reasons why flash-boiling is not used today in series production engines. However, flash-boiling might become an important effect for future DISI and HCCI engines because of the low gas pressures inside the combustion chamber during early injection.

The development of flash-boiling models is challenging for several reasons. In the case of internal flashing, the nozzle hole flow must be closely linked to the primary spray formation process. Thus, some kind of nozzle hole flow modeling must be included. Non-equilibrium effects have to be included by sub-models describing the bubble dynamics, the nucleation rate in a metastable zone as well as the inception locations. Further on, the multi-component nature of the fuel (no distinct boiling temperature and vapor pressure, but rather continuous curves) has to be accounted for.

In spite of all these difficulties, flash-boiling models have already been developed and implemented in CFD-codes. Some authors have extended conventional spray models and included the effect of flash-boiling by changing the starting and boundary conditions. Others included already detailed nucleation and bubble growth models.

An atomization model for superheated fuel sprays from pressure-swirl atomizers including the effect of flash-boiling has been recently developed by Zuo et al. [150]. The model is based on the linearized instability sheet atomization model (LISA) of Senecal et al. [129] and Schmidt et al. [125], Sect. 4.1.6. It is assumed that under superheat conditions a hollow-cone spray sheet is still formed from the pressure-swirl atomizer, and the sheet flash-boiling is controlled by the rate of heat that can be conducted inside the sheet with an effective thermal conductivity. Hydrodynamic instability, cavitation and bubble growth finally break the sheet up to form drops. Models for the subsequent drop vaporization account for heat transfer under flash-boiling and sub-boiling conditions.

Further models considering the effect of flash-boiling on spray atomization, including detailed nucleation and bubble growth models, are published, for example, by Fujimoto et al. [38], Kawano [61] and Zeng and Lee [148].

4.5.4 Wall Film Evaporation

In Sect. 4.8, it is shown that spray wall impingement and the formation of liquid films may have an important effect on spray atomization and mixture formation. In the case of film formation on a hot wall, which can happen, for example, in a small-bore direct injection diesel engine if the spray impinges on the hot piston surface, its evaporation strongly influences the mixture formation process in the near wall region and must be included in CFD models.

The description of wall film evaporation is based on the wall film energy equation

$$\frac{DE}{Dt} = \sum \dot{Q} , \qquad (4.207)$$

where the left hand side represents the material derivate of the energy E ($E = \rho_l c_{v,l} \overline{T_l} dx dy dz$, $\overline{T_l}$: mean film temperature, $c_{v,l}$: liquid specific heat), and the right hand side is the sum of all external energy fluxes changing the energy inside a film cell like energy fluxes due to conduction, impinging droplets or splashing.

An early model for the simulation of wall film evolution and heat transfer from the wall to the film or to the impinging drops was developed by Eckhause and Reitz [31] and is described in Sect. 4.8.3.

A more detailed model of film evaporation has been developed by O'Rourke and Amsden [94]. The temperature profile in the film normal to the surface is approximated to be piecewise linear, Fig. 4.51, varying from the wall temperature

T_{wall} to $\overline{T}_l$ in the lower half of the film, and from $\overline{T}_l$ to a gas surface temperature T_s in the upper half of the film. The film energy equation is

$$\rho_l c_{v,l} dx dy dz \left[\frac{\partial \overline{T}_l}{\partial t} + \left(\vec{v}_{film} \cdot \nabla_s \right) \overline{T}_l \right]$$

$$= \lambda_l dx dy \left[\frac{T_s - \overline{T}_l}{\delta / 2} - \frac{\overline{T}_l - T_{wall}}{\delta / 2} \right] + \dot{Q}_{imp} - \dot{Q}_{splash} \quad , \tag{4.208}$$

where the coordinate system is shown in Fig. 4.52. The liquid specific heat $c_{v,l}$ and the liquid heat conductivity λ_l are temperature-dependent. $\dot{Q}_{imp} = \dot{m}_{imp} \cdot e_l(T_{drop})$ is the energy flux due to impingement ($e_l(T_{drop})$: specific internal energy of the impinging droplet mass), and $\dot{Q}_{splash} = \dot{m}_{splash} e_l(\overline{T}_l)$ is the energy flux due to splashing. The left hand side of Eq. 4.208 is the material derivate and consists of the time derivate and the convective term due to film movement. The first term on the right hand side expresses the effect of heat conduction ($\dot{Q}_{cond} = -\lambda_l dx dy (\Delta T / \Delta z)$) is the sum of the heat transfer between gas and film (upper half of the film) and between film and wall (lower half of the film), Fig. 4.51.

Dividing Eq. 4.208 by the wall surface area $A_{wall} = dx dy$ and remembering that $dz = \delta$ and $\dot{Q} / A = \dot{q}$ yields

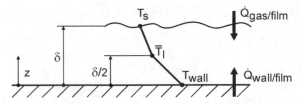

Fig. 4.51. Piecewise linear film temperature profile

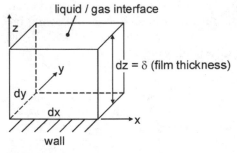

Fig. 4.52. Liquid film element and coordinate system

$$\rho_l c_{v,l} \delta \left[\frac{\partial \overline{T}_l}{\partial t} + \left(\vec{v}_{film} \cdot \nabla_s \right) \overline{T}_l \right] = \lambda_l \left[\frac{T_s - \overline{T}_l}{\delta / 2} - \frac{\overline{T}_l - T_{wall}}{\delta / 2} \right] + \dot{q}_{imp} - \dot{q}_{splash} \cdot \quad (4.209)$$

In order to calculate the mean film temperature, the interface conservation condition is used, which relates the gas-side heat transport $\dot{q}$, the energy used to vaporize the fuel mass $\dot{m}_{vap}$, and the liquid-side heat transport due to conduction:

$$\dot{q} = \frac{\dot{m}_{vap}}{A_{wall}} h_{fg} \left(T_s \right) + \lambda_l \frac{T_s - \overline{T}_l}{\delta / 2} . \quad (4.210)$$

In Eq. 4.210, h_{fg} is the latent heat of evaporation.

The expressions for $\dot{q}$ and for the vaporized fuel mass are still unknown. They are determined using wall functions. The gas-side heat transport depends on the velocity and temperature profiles in the boundary layer above the liquid film. Because vaporization results in additional velocity components normal to the wall, the structure of the turbulent boundary layer is altered and the standard wall functions must be modified in order to account for this effect. O'Rourke and Amsden [94] have derived modified wall functions, which account for an inhibition of mass, momentum, and energy transport in comparison to the standard non-evaporative situation, and which reduce to the standard wall functions in the absence of evaporation.

In the non-evaporative situation, the standard velocity profile is given by

$$u^+ = \begin{cases} y^+ & y^+ < y_c^+ \\ \dfrac{1}{\kappa} \ln y^+ + C & y^+ > y_c^+ , \end{cases} \quad (4.211)$$

where $\kappa = 0.433$, $u^+ = u/u_\tau$, u is the relative velocity between gas and liquid tangential to the surface, $u_\tau = \left(\tau_w / \rho_g \right)^{0.5}$, $y^+ = y \cdot u_\tau / v_{lam}$, $y_c^+ = 11.5$ is the transition between the laminar and the fully turbulent region, and v_{lam} is the laminar kinematic viscosity of the gas. Note that in the case of wall functions, the coordinate normal to the wall is traditionally called y. For this reason y is used in the following instead of z, although this coordinate is equal to the coordinate z in Fig. 4.52. Using $\tau_w = \rho_g C_\mu^{0.5} k$ (k: turbulent kinetic energy, $C_\mu = 0.09$) yields

$$y^+ = \frac{y C_\mu^{1/4} k^{1/2}}{v_{lam}} . \quad (4.212)$$

In the standard case, the shear stress can now be expressed as

$$\frac{\tau_w}{\rho_g u C_\mu^{1/4} k^{1/2}} = \begin{cases} \left(y^+\right)^{-1} & y^+ < y_c^+ \\ \left(\frac{1}{\kappa}\ln y^+ + C\right)^{-1} & y^+ > y_c^+ . \end{cases} \tag{4.213}$$

Using the analogy between the velocity boundary layer and the thermal boundary layer, the corresponding standard wall function for the boundary layer heat flux is

$$\frac{\dot{q}}{\rho_g c_{p,g} C_\mu^{1/4} k^{1/2} \left(T - T_s\right)} = \begin{cases} \left(y^+ Pr_{lam}\right)^{-1} & y^+ < y_c^+ \\ \left(y_c^+ Pr_{lam} + \frac{Pr_{turb}}{\kappa}\ln\left(y^+ / y_c^+\right)\right)^{-1} & y^+ > y_c^+ , \end{cases} \tag{4.214}$$

where Pr_{lam} and Pr_{turb} are the laminar and turbulent Prandtl numbers.

These standard wall functions have been modified by O'Rourke and Amsden [94] in order to account for the effect of vaporization. The authors use a dimensionless vaporization rate

$$M^* = \frac{\dot{m}_{vap} / A_{wall}}{\rho_g C_\mu^{1/4} k^{1/2}}, \tag{4.215}$$

where the mass vaporization rate $\dot{m}_{vap}$ is given by

$$\frac{\dot{m}_{vap}}{A_{wall}} = H_Y \ln\left(\frac{1 - Y_v}{1 - Y_{eq}}\right), \tag{4.216}$$

and H_Y is described in similarity to the heat transfer (Eq. 4.214) by replacing the Prandtl numbers by the Schmidt numbers:

$$\frac{H_Y}{\rho_g C_\mu^{1/4} k^{1/2}} = \begin{cases} \left(y^+ Sc_{lam}\right)^{-1} & y^+ < y_c^+ \\ \left(y_c^+ Sc_{lam} + \frac{Sc_{turb}}{\kappa}\ln\left(y^+ / y_c^+\right)\right)^{-1} & y^+ > y_c^+ . \end{cases} \tag{4.217}$$

In Eq. 4.216, Y_v is the fuel vapor mass fraction at y^+, and Y_{eq} is the equilibrium vapor mass fraction at film surface temperature. Finally, the modified wall functions for boundary layer shear stress and heat flux given by O'Rourke and Amsden [94] are

$$\frac{\tau_w}{\rho_g u C_\mu^{1/4} k^{1/2}} = \begin{cases} \dfrac{M^*}{e^{M^* y^+} - 1} & y^+ < y_c^+ \\[4mm] \dfrac{M^*}{e^{CM^*}\left(y^+\right)^{M^*/\kappa} - 1} & y^+ > y_c^+ \end{cases} \tag{4.218}$$

and

$$\frac{\dot{q}}{\rho_g c_{p,g} C_\mu^{1/4} k^{1/2} \left(T - T_s\right)} = \begin{cases} \dfrac{M^*}{e^{y^+ M^* Pr_{lam}} - 1} & y^+ < y_c^+ \\[4mm] \dfrac{M^*}{e^{y_c^+ M^* Pr_{lam}}\left(y^+ / y_c^+\right)^{\frac{M^* Pr_{turb}}{\kappa}} - 1} & y^+ > y_c^+ \end{cases} \tag{4.219}$$

A further detailed model for the simulation of wall film evaporation has been developed by Stanton and Rutland [133]. The main difference to the model of O'Rourke and Amsden [94] results from the use of different modified wall functions. In the model of Stanton and Rutland a film roughness due to evaporation is used in order to include the effect of evaporation on the boundary layer conditions. The film roughness modifies the constant C in the wall function for the boundary layer shear stress. The effect on the thermal boundary layer is expressed via an empirical correlation including the modification of heat transfer by roughened surfaces. The effect on mass transfer is obtained again by replacing the Prandtl numbers by the Schmidt numbers.

Because it is very difficult to get appropriate data to verify a wall film heat transfer and evaporation model, the film vaporization models have only been validated in combination with a spray wall impingement and a film movement model [94, 133]. In both cases, the combined models show reasonable results, and the calculated film thicknesses agree well with experimental data, see also Sect. 4.8.3.

4.6 Turbulent Dispersion

The relative velocity between the gas and droplets does not only result in droplet deceleration, deformation, and break-up, but also results in an additional dispersion or diffusion due to turbulent velocity fluctuations, and thus in a quicker and more homogeneous mixing of air and fuel than in case of laminar flow. The gas velocity can be considered as the sum of a time-averaged velocity $\bar{\vec{u}}$ and a fluctuating velocity $\vec{u}'$,

$$\vec{u} = \bar{\vec{u}} + \vec{u}', \tag{4.220}$$

where the fluctuating velocity is dependent on the turbulent kinetic energy and is caused by the turbulent eddies in the flow field. The turbulent kinetic energy k is predicted by a turbulence model, usually the k-ε model. If isotropic turbulence is assumed, the velocity fluctuations can be related to k as follows:

$$k = \frac{1}{2}\left(u_x'^2 + u_y'^2 + u_z'^2\right) = \frac{3}{2}u'^2 . \tag{4.221}$$

The additional drop motion due to the velocity fluctuations is superimposed to the average drop motion. The turbulent dispersion has been described by Faeth [35, 34] and Gosman and Ioannides [44], for example. As droplets pass through the flow, they are assumed to interact with the individual eddies, see Fig. 4.53.

Dependent on the size, mass, and velocity, the result of a droplet-eddy interaction can be quite different. Small drops can be completely captured by the eddy and follow the gas motion. Intermediate drop sizes only follow the large-scale structures, and very large drops are not influenced by the turbulent eddies. In order to determine the drop motion, the time t_{int} of interaction between drop and eddy must be specified. It is assumed that it is the smaller of either the eddy lifetime t_e or the transit time t_t that is needed to pass through the eddy. The eddy lifetime is given by the dissipation time scale,

$$t_e = \frac{l_e}{u'} = \frac{C_\mu^{3/4}}{\sqrt{2/3}} \frac{k}{\varepsilon} , \tag{4.222}$$

where the characteristic eddy size l_e is assumed to be the dissipation length scale

$$l_e = C_\mu^{3/4} \frac{k^{3/2}}{\varepsilon} , \tag{4.223}$$

and $C_\mu = 0.09$ (k-ε model).

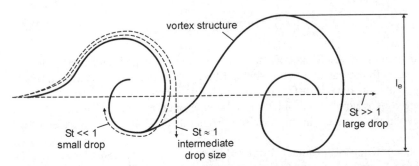

Fig. 4.53. Possible drop trajectories in a turbulent flow field [27]

The transit time

$$t_t = -\tau \cdot \ln\left[1 - \frac{l_e}{\tau\left|\vec{u}_g - \vec{u}_d\right|}\right]$$

(4.224)

to pass through an eddy of size l_e is found by solving the linearized droplet momentum equation in uniform flow. In Eq. 4.224, $\vec{u}_g$ is the gas velocity and $\vec{u}_d$ is the droplet velocity. Both velocities refer to the conditions at the start of the interaction of the drop with an eddy. The quantity τ is the aerodynamic response time and is a measure of the responsiveness of a droplet to a change in gas velocity. It is defined as

$$\frac{d\vec{u}_d}{dt} = \frac{\vec{u}_g - \vec{u}_d}{\tau}.$$

(4.225)

Using the droplet's equation of motion,

$$\frac{\pi}{6}\rho_l d_{drop}^3 \frac{d\vec{u}_d}{dt} = \frac{1}{2}\rho_g C_D \frac{\pi}{4} d_{drop}^2 \left|\vec{u}_g - \vec{u}_d\right|^2,$$

(4.226)

together with the Stoke's drag law, $C_D = 24/Re$, the following expression,

$$\tau = \frac{\rho_l d_{drop}^2}{18\mu_g},$$

(4.227)

can be derived, where μ_g is the dynamic viscosity of the gas. If $l_e \geq \tau\left|\vec{u}_g - \vec{u}_d\right|$, Eq. 4.224 has no solution. This can be interpreted as the eddy having captured a particle so that the interaction time becomes t_e.

Crowe et al. [27] have observed that small droplets follow the eddies better than the larger ones with their smaller drag-inertia ratio, see Fig. 4.53. The authors proposed a time scaling ratio, the Stokes number St, in order to assess the importance of an eddy structure on droplet dissipation:

$$St = \frac{\tau}{t_e}.$$

(4.228)

For small values of the Stokes number the droplets will maintain near velocity equilibrium with the gas and will disperse like the gas. For large Stokes numbers, the eddy structures have insufficient time to influence the particle motion. For intermediate Stokes numbers, droplets may be entrapped in the vortices and centrifuged beyond the vortices. In this case the droplet dispersion can even exceed that of the gas.

The phenomenon of turbulent droplet dispersion has to be considered in CFD calculations. It can be included by calculating the change in droplet motion during the interaction time t_{int} as a function of the resulting gas velocity $\vec{u}_g$ present at the beginning of the interaction period. Usually, a random number generator is used to represent the turbulence. It is assumed that velocity fluctuations are isotropic and have statistically independent Gaussian probability density functions in each coordinate direction. Hence, the actual value of the fluctuating velocity is usually sampled from a Gaussian distribution,

$$G(u') = \frac{1}{\sqrt{2\pi}\sqrt{2k/\varepsilon}} \exp\left(\frac{|u'|^2}{4k/3}\right),$$

(4.229)

with a variance equal to the turbulence intensity,

$$\sigma = |u'| = \sqrt{2k/3}.$$

(4.230)

Thus, the fluctuating velocities are piecewise linear functions with constant values during the interaction time t_{int}, that are sampled from Eq. 4.229. More details concerning the numerical implementation can be found in Amsden et al. [5] for example.

The interaction of the liquid droplets with the gas turbulence modulates the gas turbulence itself. A portion of the turbulent kinetic energy is used to disperse the droplets, and this reduces the rate of production of turbulent kinetic energy in the flow field. Experiments of Modarress et al. [88] have confirmed this effect. It is usually accounted for by including an additional source term $\dot{W}^s < 0$ in the k- and the ε- conservation equation of the k-ε turbulence model. For incompressible turbulence in the absence of gradients the equations become

$$\rho\frac{\partial k}{dt} = \dot{W}^s$$

(4.231)

and

$$\rho\frac{\partial \varepsilon}{dt} = C_s\frac{\varepsilon}{k}\dot{W}^s,$$

(4.232)

where $C_s = 1.5$ as suggested by Amsden et al. [5].

4.7 Collision and Coalescence

4.7.1 Droplet Collision Regimes

Droplet collision is an important effect in the dense spray region near the injection nozzle, where the number of droplets per unit volume is large and the probability of collisions is high. Droplet collisions in sprays are caused by differences in their velocities. These differences can arise from differences in the injection velocity (e.g. multiple injections), from different deceleration and velocity trajectories of droplets due to drag forces, from break-up, turbulence, wall impingement, etc. The result of a collision event depends on the impact energy, the ratio of droplet sizes, and ambient conditions like gas density, gas viscosity, and the fuel-air ratio of the gas surrounding the droplets during impact. Collision can result in a combination of droplet masses (coalescence), in pure reflection, or in break-up. Thus, after collision, the droplet velocity trajectories as well as their sizes and numbers are usually changed. This again effects mass, momentum, and energy transfer processes during mixture formation. For example, smaller droplets decelerate and evaporate faster, while an increase of drop radius results in an increase of penetration. Furthermore, it is known that especially in the case of low ambient temperatures (non-vaporizing sprays), spray drop size is the outcome of a competition between drop break-up and drop coalescence phenomena.

There are four important dimensionless parameters governing the collision phenomenon, the Reynolds number Re_{coll}, the Weber number We_{coll}, the drop diameter ratio Δ, and the impact parameter B:

$$Re_{coll} = \frac{\rho_l d_1 u_{rel}}{\mu_l}, \quad We_{coll} = \frac{\rho_l d_2 u_{rel}^2}{\sigma}, \quad \Delta = \frac{d_2}{d_1}, \quad B = \frac{2X}{d_1 + d_2}. \tag{4.233}$$

In Eq. 4.233, ρ_l is the liquid drop density, μ_l the viscosity, σ the surface tension, d_1 and u_1 are the diameter and velocity of the larger drop, and d_2 and u_2 are those of the smaller drop. An important parameter governing the collision outcome is the relative velocity u_{rel} of the drops. If the collision angle α between the trajectories of both drops is known, Fig. 4.54, the relative velocity is given as

$$u_{rel} = \sqrt{u_1^2 + u_2^2 - 2u_1 u_2 \cos\alpha}. \tag{4.234}$$

The impact parameter X is defined as the projection of the distance between the droplet centers in the direction normal to that of u_{rel}, Fig. 4.54. The quantity B is the dimensionless impact parameter, the value of which varies from one to zero. In the case of head-on collision ($B = 0$), the relative velocity vector coincides with the center-to-center line. If $B > 0$ the collision is off-axis, and $B = 1$ is called a tangential or grazing collision.

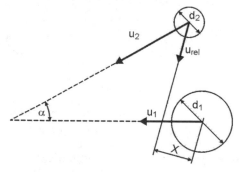

Fig. 4.54. Schematic illustration of geometric collision parameters

The possible outcomes of a collision event can be divided into five regimes: bouncing, coalescence, reflexive separation, stretching separation, and shattering. Fig. 4.55 shows a schematic diagram of the droplet collision regimes as function of the collision Weber number and impact parameter. Fig. 4.56 shows the corresponding mechanisms.

If two drops, which are surrounded by a gas, approach each other, a gas layer has to be squeezed out of the gap between both drops. In the case of bouncing collision (region II), the contact of the drop surfaces is prevented by this gas film, and the drops deform and bounce apart.

Coalescence refers to collisions in which the two drops permanently combine and form a single drop. On the one hand, this can happen at low Weber numbers, where the impact energy is small and where there is enough time to squeeze the air film out of the gap (slow coalescence, region I). On the other hand, coalescence is also possible at higher Weber numbers, where the normal velocity is high enough to eliminate the air layer (region III). Experiments have shown that in the case of increased ambient pressures the slow coalescence regime becomes undetectable [109] because the high pressure makes it difficult for the drops to push away the ambient gas without losing their kinetic energy. Hence, in engine applications, the slow coalescence regime can be neglected.

Separation collision occurs when the drops combine temporarily and then separate into a string of two or more drops. Reflexive separation (region IV) is found to occur for near head-on collisions. As shown in Fig. 4.56, the droplets combine and are flattened in the normal direction. Surface tension then makes the disk contract radially inward. This reflexive action generates a long cylinder, and if the Weber number is large enough, this cylinder will break up again. Otherwise, the cylinder will just oscillate until a spherical drop is formed (coalescence collision).

Stretching separation (region V) occurs for large impact parameter collisions. In this case, only a portion of each drop will come in direct contact with the other one, resulting in a small region of interaction. The remaining portions of the drops continue to flow in their original direction and stretch the region of interaction.

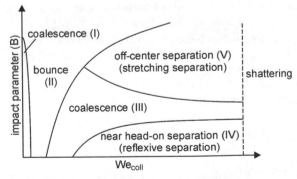

Fig. 4.55. Schematic of droplet collision regimes

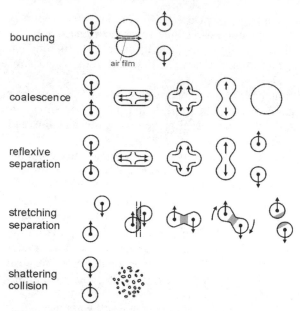

Fig. 4.56. Mechanisms of droplet collision

The outcome of collision depends on the competition between surface energy, which is striving to keep the drops together, and the kinetic energy, which is striving to separate the drops again. Assuming a constant Weber number, an increase of the impact parameter reduces the region of interaction and increases the stretching forces. The transition from coalescence to stretching separation occurs if a critical value of the impact parameter is reached, Fig. 4.55.

Shattering collision occurs at high relative velocities where the surface tension forces are only of secondary importance and the phenomenon is inertia-dominated. The droplets disintegrate into a cluster of many small droplets.

In addition to the effects described so far, the numbers and sizes of the new drops resulting from a collision event strongly depend on the diameter ratio Δ of the parent drops. Thus, the exact extensions of the collision regions in Fig. 4.55 are also a function of Δ. Furthermore it has been shown that an increase of gas density promotes bouncing, while a gas atmosphere with a high content of evaporated fuel promotes coalescence. Detailed experimental investigations on coalescence and separating collisions under different boundary conditions are given in Qian and Law [109] and Ashgriz and Poo [8], for example.

4.7.2 Collision Modeling

The collision models used in CFD codes today usually do not take all the different collision phenomena into account. One of the reasons for the use of more simple models in engine simulations is that a direct evaluation of a collision model by comparison with experimental data is not possible. The relevant spray parameters like droplet sizes and velocity components for example are always a result of many phenomena like break-up, evaporation, and collision, and it is impossible to quantify the single effects from measurements. The standard collision model used in most spray simulations today is the model of O'Rourke [97, 93].

In the O'Rourke's model [97, 93], only two main outcomes of droplet collision are regarded: permanent coalescence (region III in Fig. 4.55) and stretching separation (region V). Following the approach of Brazier-Smith et al. [17], an energy balance is used in order to predict whether the two drops, which coalesce at the moment of collision, separate again to re-form the original drops or combine to form a larger drop. If the rotational energy of the coalesced drops,

$$E_{rot} = \frac{L^2}{2J}, \qquad (4.235)$$

is larger than the extra surface energy,

$$\Delta E_{surf} = 4\pi\sigma\left(r_1^2 + r_2^2 - r_{eff}^2\right), \qquad (4.236)$$

stretching separation is assumed to happen. Otherwise permanent coalescence will occur. In Eqs. 4.235 and 4.236

$$L = \frac{m_1 m_2}{m_1 + m_2} u_{rel} X \qquad (4.237)$$

is the angular momentum about the center of mass of the coalesced drops,

$$J = \frac{2\left(m_1 + m_2\right)}{5} r_{eff}^2 \qquad (4.238)$$

is the corresponding moment of inertia,

$$r_{eff} = \left(r_1^3 + r_2^3\right)^{1/3} \qquad (4.239)$$

is the effective radius, and X is the off-center distance, see Fig. 4.54. Equating Eq. 4.235 and Eq. 4.236 gives the critical impact parameter B_{crit}, which represents the transition condition between regions III and V in Fig. 4.55:

$$B_{crit} = \frac{2X_{crit}}{d_1 + d_2} = \sqrt{\frac{24}{5We_{coll}} \frac{\left(\Delta^3 + 1\right)^{11/3}}{\Delta^5\left(\Delta + 1\right)^2}\left[\Delta^2 + 1 - \left(\Delta^3 + 1\right)^{2/3}\right]}. \qquad (4.240)$$

Thus, the transition between coalescence and stretching separation is a function of the collision Weber number and the drop diameter ratio. In Fig. 4.57, the general behavior of Eq. 4.240 is shown. For a fixed value of We_{coll} (relative velocity and diameter of the small droplet are constant) the critical impact parameter increases as the diameter ratio approaches zero (small diameter constant, large diameter increases), and the coalescence region extends. This behavior is reasonable, because a larger drop can more easily absorb the kinetic energy of a small drop.

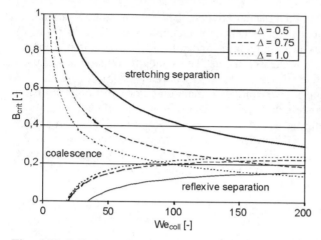

Fig. 4.57. Influence of droplet size ratio on the coalescence/stretching separation (Eq. 4.240) and on the coalescence/reflexive separation transition criterion (Eq. 4.245)

If the two droplets permanently coalescence, the velocity $\vec{u}_{new}$ and the temperature T_{new} of the combined drop are calculated as

$$\vec{u}_{new} = \frac{m_1\vec{u}_1 + m_2\vec{u}_2}{m_1 + m_2}, \tag{4.241}$$

$$T_{new} = \frac{m_1 T_1 + m_2 T_2}{m_1 + m_2}. \tag{4.242}$$

If stretching separation occurs, both droplets are re-formed again, and it is assumed that the temperature of the initial droplets is not altered. O'Rourke [93] derived the equations

$$\vec{u}_1^{new} = \frac{\left[m_1\vec{u}_1 + m_2\vec{u}_2 + m_2\left(\vec{u}_1 - \vec{u}_2\right)\right]\left[B - B_{crit}\right]}{\left[m_1 + m_2\right]\left[1 - B_{crit}\right]} \tag{4.243}$$

and

$$\vec{u}_2^{new} = \frac{\left[m_1\vec{u}_1 + m_2\vec{u}_2 + m_1\left(\vec{u}_2 - \vec{u}_1\right)\right]\left[B - B_{crit}\right]}{\left[m_1 + m_2\right]\left[1 - B_{crit}\right]} \tag{4.244}$$

in order to predict the velocities of both droplets after separation. Eqs. 4.243 and 4.244 include some simplifying assumptions about the fraction of energy that is dissipated during collision.

Figure 4.58 shows the effect of the collision model on the Sauter mean radius (SMR) of a full-cone high-pressure diesel spray under evaporating and non-evaporating conditions, which is injected in pressurized air. Table 4.5 summarizes the boundary conditions used in the simulation. All curves in Fig. 4.58 show ex-

tremely large values at the beginning of injection, because the initial drops have a
size equal to the nozzle hole diameter, and because the first break-up occurs after
a small time delay, the break-up time. In the case of low ambient gas temperatures, the effect of evaporation on the droplets' lifetimes is small, and the Sauter
mean radius is a result of a competition between coalescence and break-up. If coalescence effects are neglected, break-up will produce an enormous number of extremely small droplets, which results in unreasonably small SMR values. On the
other hand, if the collision model is used, there is even an increase in SMR over
time because of the rising number of droplets per unit volume that increase the
collision frequency and thus the probability of coalescence. Hence, the inclusion
of an appropriate collision model is very important in the case of non-evaporating
sprays, which are customarily used to validate break-up models and to investigate
the effect of nozzle geometry and injection strategy on the spray formation processes. In the case of high ambient gas temperatures evaporation has a dominant effect on the droplets' lifetimes. Especially the small droplets evaporate fast and the
chance to coalesce is reduced. Thus, compared to the non-evaporating case, there
are less droplets, the collision frequency is reduced, and the overall effect of the
collision model on the SMR is smaller.

Table 4.5. Boundary conditions used for the calculations shown in Fig. 4.58

	Evaporating spray	Non-evaporating spray
Ambient temperature [K]	800	298
Ambient pressure [MPa]	5.7	5.7
Fuel temperature [K]	298	298
Nozzle hole diameter [μm]	200	200
Injected mass [mg]	40,7	40.7
Injection duration [ms]	4	4
Break-up model	Blob + KH/RT	Blob + KH/RT

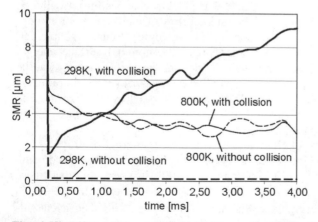

Fig. 4.58. Effect of the collision model on the SMR of an evaporating and a non-
evaporating diesel spray

In the model of O'Rourke [97, 93], only coalescence and stretching separation are considered. Reflexive separation, which is important for near head-on collisions, and shattering collisions are not included. Furthermore, the formation of satellite drops after stretching separation is ignored. For this reason, Tennison et al. [140] presented an enhanced version of the collision model, which also takes reflexive separation into account. The model extension is based on the theoretical work of Ashgriz and Poo [8], who derived the transition criterion between coalescence and reflexive separation that:

$$We_{coll} = -3\left[4\left(1+\Delta^2\right)-7\left(1+\Delta^3\right)^{2/3}\right]\cdot\frac{\Delta\left(1+\Delta^3\right)^2}{\Delta^6\eta_1+\eta_2}, \tag{4.245}$$

where

$$\eta_1 = 2\left(1-\varsigma\right)^2\left(1-\varsigma^2\right)^{0.5}-1, \tag{4.246}$$

$$\eta_2 = 2\left(\Delta-\varsigma\right)^2\left(\Delta^2-\varsigma^2\right)^{0.5}-\Delta^3, \tag{4.247}$$

$$\varsigma = 0.5B_{crit}\left(1+\Delta\right). \tag{4.248}$$

Reflexive separation occurs if the collision Weber number is above the value given by Eq. 4.245. Eq. 4.245 also includes the effect that reflexive separation is more likely to occur if the diameter ratio Δ is close to one, see Fig. 4.57, which is in agreement with experimental observations [8]. The consideration of the reflexive separation regime in diesel sprays was found to give a slight reduction of the overall Sauter mean diameter [17].

In Eqs. 4.246 and 4.247, η_1 and η_2 are the fractions of the drops' kinetic energy that participates in the reflexive separation process. The expressions are derived using a balance between kinetic energy and surface energy. It is assumed that the two combined drops form a flattened disc that quickly changes into a cylinder, which stretches out under the force of the internal flow of the fluid moving in opposite directions, Fig. 4.56. It would then seem to be logical to use a simple balance between this effective reflexive energy and the nominal surface energy. However, once the cylinder has stretched far enough, the surface energy can be reduced by forming two drops. Hence, separation can occur even if the reflexive energy is less than the surface energy. The criterion derived by Ashgriz and Poo [8] is that reflexive separation occurs if the reflexive energy is more than 75% of the nominal surface energy.

As suggested in the study of Hung and Martin [57], shattering of droplets during binary collisions can be important for $We_{coll} > 80$. A drop-shattering collision model is has been proposed by Georjon and Reitz [40]. It is assumed that first a large drop with radius $r_0 = (r_1^3 + r_2^3)^{1/3}$ and then a ligament (length: 2δ, radius: $r = (2r_0^2/(3\delta))^{0.5}$) are formed during the collision of two droplets with a sufficient collision Weber number, Fig. 4.59. The ligament elongates and capillary wave-induced disturbances grow (Rayleigh linear jet break-up theory). If the break-up

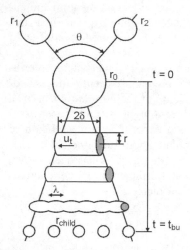

Fig. 4.59. Schematic representation of ligament formation and break-up [40]

time is shorter than the time taken by the two ends of the cylindrical ligament to retract again, it disintegrates into small droplets. The drop dynamics (elongation and retraction) are formulated based on the energy equation of a half-cylinder and yield a second-order non-linear differential equation, which is solved numerically in order to get the time-dependent diameter and length of the elongating ligament. Next, the Rayleigh linear jet break-up theory is applied [117], and this finally yields the wavelength λ of the disturbance, the break-up time of the ligament t_{bu}, and the radius r_{child} of the new child drops:

$$\lambda = 9.02 \cdot r \,, \tag{4.249}$$

$$t_{bu} = \frac{2\pi}{\Omega} = \frac{2\pi}{0.34} \sqrt{\frac{\rho_l r^3}{\sigma}} \,, \tag{4.250}$$

$$r_{child} = 1.89 \cdot r \,. \tag{4.251}$$

If the break-up time is reached before the cylinder contracts again, the ligament is assumed to break up into small droplets of size r_{child}. Otherwise, shattering collision does not occur. It is assumed that the shattering collision neither depends on the impact parameter B nor on the droplet size ratio Δ.

Georjon and Reitz [40] consider shattering collisions to be an extension of the stretching separation regime. For each pair of drops that have undergone stretching separation, it is tested whether a shattering collision is possible by solving the equation of motion of the ligament. If shattering is possible, the collision is calculated between N pairs of drops, where N is the minimum of droplets in the two colliding parcels. The remaining droplets do not change their properties and are put into a new parcel. The droplets taking part in the collision process disintegrate into smaller droplets and undergo velocity changes that reflect momentum and energy conservation. The model has only been validated against experimental data from non-evaporating sprays, and the results show a slight overprediction of child drop sizes and velocities.

In real engine sprays, collision phenomena are probably much more complicated than described by all of the above-mentioned models. A deeper understanding of spray physics in this regime is necessary. Some of the correlations used in the models come from experiments in entirely different regimes (rain drops, cloud physics) and need to be tested against fundamental experiments on collision of hydrocarbon droplets at high pressures. However, data about fundamental experiments under diesel engine conditions are scarce. Furthermore, it has been shown that coalescence of hydrocarbon droplets is promoted if the environment contains fuel vapor [109]. So far, no model accounts for this effect.

4.7.3 Implementation in CFD Codes

In general, two possible methods for the implementation of a collision model in CFD codes exist: the statistical and the deterministic approach. Using the deterministic approach, the exact positions and velocity vectors of all parcels are used in order to check the possibility of collision for all parcel pairs. Because this method is computationally extremely expensive, the statistical approach is usually preferred. Only collisions of parcels that lie in the same computational cell are considered. All pair combinations of parcels in a cell are checked according to the following procedure. It is assumed that the droplets of both parcels are homogeneously distributed inside the cell volume V_{cell}. Then, the number of collisions of one drop of parcel 1 containing the larger droplets (N_1 droplets of diameter d_1) with all the smaller droplets (parcel 2: N_2 droplets of diameter d_2) is predicted, and it is assumed that all the other large droplets of parcel 1 have identical behavior. The large droplet is called a collector droplet.

The probability that a collector collides with k droplets from parcel 2 follows a Poisson distribution [97],

$$P_k = \frac{\left(v_{12}\Delta t\right)^k}{k!}\exp\left(-v_{12}\Delta t\right),\tag{4.252}$$

where Δt is the time step of the computation and

$$v_{12} = \frac{N_2}{V_{cell}}\frac{\pi}{4}\left(d_1 + d_2\right)^2\left|\vec{u}_{rel}\right|E_{12}\tag{4.253}$$

is the collision frequency, which is modeled in analogy to the kinetic theory of gases and is the product of number density N_2/V_{cell}, collision cross section $\pi(d_1^2 + d_2^2)/4$, see Fig. 4.60, and the magnitude of the relative velocity $\vec{u}_{rel}$. The quantity E_{12} in Eq. 4.253 is the collision efficiency, which is shown to have a value of $E_{12} \approx$

1.0 in engine sprays [97]. In Eq. 4.252 the probability of no collision is $P_0 = \exp(-v_{12}\Delta t)$.

Next, the actual number n of collisions between one collector droplet of parcel 1 and the small droplets of parcel 2 must be specified. A random number ξ_1 is sampled from a uniform distribution in the interval [0, 1], and if $\xi_1 > P_0$ a collision is assumed to occur (otherwise no collision occurs and the next parcel pair is checked). If $\xi_1 > P_0$ the corresponding value of the integrated distribution function of P_k (normalized to lie also in the interval [0, 1]),

$$\int_{k=0}^{k=n} P_k \, dk = \xi_1, \tag{4.254}$$

is identified and solved for n. This procedure guarantees that the possibility of sampling a discrete number of collisions is given by Eq. 4.252. Further details are given in [5], for example.

Finally, the nature of collision must be specified. In the model of O'Rourke [97], it depends on the impact parameter B (interval [0, 1]). Again, a random number ξ_2 is sampled from a uniform distribution in the interval [0, 1], and the non-dimensional off-center distance B is determined from the relation

$$B^2 = \xi_2. \tag{4.255}$$

If $B > B_{crit}$ (B_{crit}: from Eq. 4.240 or by the following approximation from O'Rourke [97]: $B^2_{crit} = \min\{1.0, (2.4/We_{coll})(\Delta^3 - 2.4\Delta^2 + 2.7\Delta)\}$), the collision will result in stretching separation. Otherwise permanent coalescence will occur. If further collision regimes are included, they can be implemented as sub-regimes. For example, if reflexive separation is included, and if the O'Rourke model predicts permanent coalescence, then Eq. 4.245 is used in a second step to check whether reflexive separation occurs instead of coalescence.

If the outcome of the collision is permanent collision, one must check whether $n \cdot N_1 > N_2$, because then more collisions are predicted than droplets of parcel 2 are present (each collision erases a droplet of parcel 2). In this case, n is reduced to the maximum possible number (all N_2 droplets collide): $n = N_2 / N_1$. The mass of the new large droplet after n collisions is

$$m_1^{new} = m_1 + n \cdot m_2, \tag{4.256}$$

and the new velocity and temperature are (see also Eqs. 4.241 and 4.242)

$$\vec{u}_1^{new} = \frac{\vec{u}_1 m_1 + n \cdot \vec{u}_2 m_2}{m_1^{new}}, \tag{4.257}$$

$$T_1^{new} = \frac{T_1 m_1 + n \cdot T_2 m_2}{m_1^{new}}. \tag{4.258}$$

The remaining droplets of parcel 2 keep their properties, but their number is reduced to

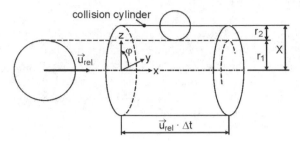

Fig. 4.60. Collision cylinder volume

$$N_2^{new} = N_2 - n \cdot N_1 . \qquad (4.259)$$

If the outcome of the collision is stretching separation, only one collision between a drop of parcel 1 with a drop of parcel 2 is considered. Temperature and droplet number of each parcel remain unchanged, and the new velocities are calculated according to Eqs. 4.243 and 4.244. If $N_1 \neq N_2$, the new velocity of the parcel containing more drops is calculated conserving momentum. In the case of $N_1 < N_2$ the new velocity of parcel 2 is

$$\vec{u}_{2,new} = \frac{N_1}{N_2} u_2^{new} + \left(1 - \frac{N_1}{N_2}\right) \vec{u}_2 . \qquad (4.260)$$

Eq. 4.255 can be derived as follows. It is assumed that the probability $P(X, \varphi)$ of collision is uniformly distributed over the collision cross section $A = \pi (r_1 + r_2)$, Fig. 4.60. The integrated probability function is

$$S(X) = \frac{1}{\pi (r_1 + r_2)^2} \int\limits_{r=0}^{r=X} (2\pi r) dr = \frac{X^2}{(r_1 + r_2)^2} , \qquad (4.261)$$

the values of which are between zero ($X = 0$) and one ($X = r_1 + r_2$). The value of $S(X) = \xi_2$ is sampled from a uniform distribution in the interval [0, 1], and finally $S(X)$ is solved for X using the inverse function

$$X = (r_1 + r_2)\sqrt{S(X)} = (r_1 + r_2)\sqrt{\xi_2} . \qquad (4.262)$$

Using $B = X/(r_1 + r_2)$, Eq. 4.262 can be transformed into Eq. 4.255.

4.8 Wall Impingement

Spray-wall impingement is an important process during mixture formation in direct injection small bore diesel engines as well as in direct injection and port injection gasoline engines. Usually, two main physical processes are involved. Wall-spray development and wall film evolution. Both processes may strongly influence combustion efficiency and the formation of pollutants.

In a small direct injection diesel engine, the liquid penetration is sometimes longer than the distance between the nozzle tip and the piston cavity wall, especially in engines with low swirl or during cold start. In this case, the spray-wall impingement may cause a significant increase of unburned hydrocarbon and soot emissions, especially if a wall film is formed. On the other hand, if no liquid wall film is generated, it promotes combustion under hot engine conditions, because spray heating and vaporization are intensified by drop shattering, and the large-scale gas vortex, which forms in the near-wall region, enhances gas entrainment.

Impingement in port injected engines causes difficulties in the transient control of the engine, because only a part of the injected fuel enters the combustion chamber during the corresponding cycle, and the rest is added to the wall film and slowly transported to the valve. At the valves, fuel separates from the film and enters the combustion chamber. However, this happens many cycles after the corresponding injection event and adds some uncontrollable amount of fuel to that of the actual injection. This effect is responsible for decreased engine response, increased fuel consumption and increased hydrocarbon emissions. In order to minimize the negative effect of a liquid film on the walls of the induction system, the fuel is sprayed directly on the back of the intake valves. In this case the interaction between valve and spray is an important source of atomization.

Hence, in diesel as well as gasoline engines, a detailed modeling of spray-wall impingement processes is necessary in order to predict their effects on engine performance and on the formation of pollutants.

4.8.1 Impingement Regimes

Figure 4.61 shows the various impingement regimes of a droplet-wall interaction. In the stick regime, a droplet with low kinetic energy adheres to the wall in nearly spherical form and continues to evaporate. In the case of spread, the droplet impacts with moderate velocity on a dry or wetted wall, spreads out and mixes with the wall film (wetted wall) or forms a wall film (dry wall). If rebound occurs, the droplet bounces off the wall (reflection) and does not break up. This regime is observed in the case of dry and hot walls, where the contact between drop and wall is prevented by a vapor cushion. Rebound also occurs in the case of a wet wall if the impact energy is low and an air film between drop and liquid film minimizes energy loss. In the boiling-induced break-up regime, the droplet disintegrates due to a rapid liquid boiling on a hot wall. The wall temperature must be near the Nakayama temperature T_N, at which a droplet reaches its maximum evaporation rate. In the case of break-up, the droplet deforms into a radial film on the hot surface, which breaks up due to thermo-induced instability. The splash regime occurs at very high impact energy. A crown is formed, jets develop on the periphery of the crown, become unstable and disintegrate into many droplets.

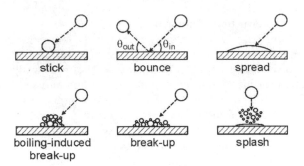

Fig. 4.61. Schematic illustration of different impact regimes [11]

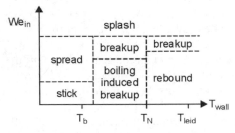

Fig. 4.62. Droplet impingement regimes and transition conditions for a dry wall [11]

There are a number of parameters characterizing the impingement regimes such as incident drop velocity, incidence angle, liquid properties such as viscosity, temperature, surface tension, wall properties like surface roughness and temperature, wall film thickness etc. Some of these parameters can be combined to yield dimensionless parameters. The two most important numbers are the Weber number

$$We = \frac{\rho_l v_n^2 d}{\sigma},$$
(4.263)

which represents the ratio of the droplet's kinetic energy (v_n: velocity component normal to the surface, ρ_l: liquid density, d: droplet diameter) and its surface energy, and the Laplace number,

$$La = \frac{\rho_l \sigma d}{\mu_l^2},$$
(4.264)

which measures the relative importance of surface tension and viscous forces acting on the liquid (μ_l: dynamic viscosity of liquid). The Laplace number is also represented by the Ohnesorge number $Z = La^{-1/2}$.

Another important parameter influencing the impingement process is the wall temperature. The characteristic temperatures

$$T_b < T_N < T_{leid} \tag{4.265}$$

are the liquid boiling temperature T_b, the Nakayama temperature T_N at which a droplet reaches its maximum evaporation rate, and the Leidenfrost temperature T_{leid} at which a thin layer of vapor forms between the surface and the drop and evaporation is minimized. Figure 4.62 gives an overview of droplet impingement regimes and transition conditions for a dry wall and fixed Laplace number and surface roughness [11]. In internal combustion engines the wall temperatures during injection are usually below the fuel boiling point [11]. This reduces the number of relevant impingement regimes in case of a dry wall to stick, spread and splash. In the case of a wet wall, Kolpakov et al. [66] revealed that with an increasing impact Weber number the regimes stick, rebound, spread, and splash are important.

4.8.2 Impingement Modeling

Naber and Reitz [90] developed one of the first impingement models. In their model, three regimes are considered: stick, reflection (rebound), and slide. In the slide regime a tangential motion along the surface like a jet with the same magnitude of velocity as before impact is predicted. In all regimes the size of the drops is not changed by the wall interaction. In the stick regime, droplets with low kinetic energy stick to the wall and continue to vaporize. In the case of reflection, drops rebound and the magnitude of their tangential and normal velocity components remains unchanged. However, the normal one changes its sign. This causes specular reflection and is in contrast to the experimental results of Wachters and Westerling [144], Fig. 4.63, in which the outgoing Weber number is generally smaller than the incident Weber number.

Wachters and Westerling [144] performed an experimental study of single drops falling on a hot surface in order to determine the relationship between the velocity components before and after impact. In contrast to the behavior of the tangential velocity component, the normal component is always significantly reduced. Gonzalez et al. [43] developed a numerical fit to their data, Fig. 4.63,

$$We_{out} = 0.678 \cdot We_{in} \cdot \exp\left(-0.04415 \cdot We_{in}\right). \tag{4.266}$$

For $We_{in} \leq 80$, the drops do not disintegrate during impact and bounce from the surface, while for $We_{in} > 80$ disintegration into small droplets on the surface occurs. In a later version of the model of Naber and Reitz [90], which was presented by Gonzalez et al. [43], a correction of the normal drop velocity (index "n") in the rebound regime is implemented,

$$v_{n,out} = -v_{n,in} \cdot \sqrt{\frac{We_{out}}{We_{in}}}, \tag{4.267}$$

where We_{out} is determined by Eq. 4.266.

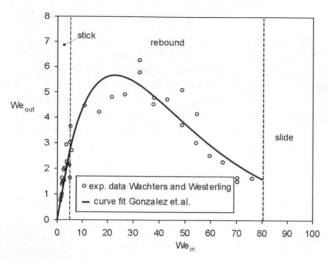

Fig. 4.63. Weber numbers of drops before and after impingement, data from [144]

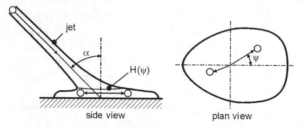

Fig. 4.64. Schematic diagram of the slide model (jet analogy model)

In the slide regime, a droplet striking the wall is given a velocity in the direction of the local tangent to the wall in the manner of a liquid jet. This model is an attempt to develop a simple version of a combined spread/wall film model in analogy to a liquid jet impinging on an inclined wall, Fig. 4.64. It is assumed that the sheet geometry is given by

$$H(\psi) = H_{\psi=\pi} \cdot \exp\left(\beta\left(1 - \psi/\pi\right)\right),$$ (4.268)

where the parameter β is determined from mass and momentum conservation as

$$\sin\alpha = \frac{\exp(\beta)+1}{\exp(\beta)-1} \cdot \frac{1}{1+(\beta/\pi)^2}.$$ (4.269)

In Eq. 4.269, α is the jet inclination angle (Fig. 4.64). The probability that an impinging drop slides along the surface at an angle ψ (Fig. 4.64) is assumed to be proportional to the sheet thickness $H(\psi)$, and ψ is obtained by integrating Eq. 4.268,

$$\psi = -\frac{\pi}{\beta}\ln\left(1 - p\left(1 - \exp(-\beta)\right)\right),$$
(4.270)

where p is a random number uniformly distributed in the interval [0, 1].

The most important limitation of the model of Naber and Reitz [90] is that the phenomenon of droplet shattering (splash), which occurs at high impact energy and is important in effecting wall spray dispersion and vaporization, is ignored. This is especially important if the distance between nozzle and wall becomes small. Furthermore, the formation of a liquid film adhering to the wall and its effect on the impingement behavior of drops are not included. Under normal engine conditions, the wall temperatures are below those used in the experiments of Wachters and Westerling [144], and, especially under cold-starting conditions, the transition criterion and the determination of momentum loss during impact are not applicable any more.

Bai and Gosman [11] have developed a more detailed impingement model considering also the splash regime. In the case of a dry wall, the stick and spread regimes are combined and called the adhesion regime, and the stick regime is neglected in the case of a wetted wall because of its typically very low impact energy.

For a dry wall, the transition criterion between adhesion and splash is given as

$$Adhesion \rightarrow Splash \qquad We_{crit} = A \cdot La^{-0.18},$$
(4.271)

where the coefficient A is depends on the surface roughness r_s, Table 4.6.

In the case of an already wetted wall the transition Weber numbers are:

$$Rebound \rightarrow Spread \qquad We_{crit} \approx 5,$$
(4.272)

$$Spread \rightarrow Splash \qquad We_{crit} = 1320 \cdot La^{-0.18}.$$
(4.273)

In Eq. 4.273 it is assumed that a liquid film can be approximated by a very rough dry wall. Next, the post-impingement characteristics have to be modeled. In the adhesion regime, the incoming droplets are assumed to coalesce to form a local liquid film. In the rebound regime, the droplet bounces off the wall and does not break up, but loses a small part of its kinetic energy by deforming the liquid film. Bai and Gosman [11] use a relationship developed by Matsumoto and Saito [85] for a solid particle bouncing on a solid wall in order to determine the rebound velocity components,

$$v_{t,out} = \frac{5}{7}v_{t,in},$$
(4.274

and

Table 4.6. Coefficient A as function of surface roughness r_s [11]

r_s [μm]	0.05	0.14	0.84	3.10	12.00
A [/]	5264	4534	2634	2056	1322

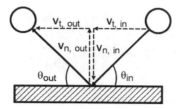

Fig. 4.65. Velocity components and respective angles for an impinging drop

$$v_{n,out} = -e \cdot v_{n,in} , \tag{4.275}$$

where $v_{t,in}$ and $v_{n,in}$ are the tangential and normal velocity components of the incoming drop, and $v_{t,out}$ and $v_{n,out}$ are the components of the outgoing drop, Fig. 4.65. The quantity e is the restitution coefficient, which is assumed to follow the relation

$$e = 0.993 - 1.76 \cdot \theta_{in} + 1.56 \cdot \theta_{in}^2 - 0.49 \cdot \theta_{in}^3 . \tag{4.276}$$

In Eq. 4.276, θ_{in} (measured in radians) is the incident angle of the incoming drop, see Fig. 4.65.

In the splash regime, several more post-impingement quantities must be determined. These are the ratio of splashed mass and total incoming mass m_s/m_{in}, the sizes, velocities, and ejection angles of the new droplets. The mass ratio is modeled by the following relation:

$$\frac{m_s}{m_{in}} = \begin{cases} 0.2 + 0.6 \cdot \alpha, & dry\ wall \\ 0.2 + 0.9 \cdot \alpha, & wetted\ wall \end{cases}, \tag{4.277}$$

where α is randomly chosen in the interval [0, 1]. Eq. 4.277 corresponds with experimental investigations. In the case of a dry wall there is always some part of liquid lost at the wall, and for the wet wall, the mass ratio is allowed to become larger than 1, because the splashing drops may entrain liquid from the film.

In order to predict the sizes of the new splashed droplets, it is assumed that each incoming droplet parcel produces two new parcels with equal mass $m_s/2$, but different drop sizes and velocities. The number of new droplets $N = N_1 + N_2$ is determined using a correlation based on experimental results,

$$N = 5 \cdot \left(\frac{We}{We_{crit}} - 1 \right) . \tag{4.278}$$

N_1 is randomly chosen in the interval [1, N] and $N_2 = N - N_1$. The diameters d_1 and d_2 of the two new droplet classes are given by mass conservation:

$$N_1 d_1^3 = \frac{1}{2} \frac{m_s}{m_{in}} d_{in}^3 N_{in} , \tag{4.279}$$

$$N_2 d_2^3 = \frac{1}{2} \frac{m_s}{m_{in}} d_{in}^3 N_{in} .$$
(4.280)

The velocity components $\vec{v}_{1,out} = (v_{1,n}, v_{1,t})$ and $\vec{v}_{2,out} = (v_{2,n}, v_{2,t})$ of the two new droplet classes are derived from an energy balance. It is assumed that the splashing energy $E_{k,s}$

$$\frac{1}{4} m_s \left(|\vec{v}_1|^2 + |\vec{v}_2|^2 \right) + \pi \sigma \left(N_1 d_1^2 + N_2 d_2^2 \right) = E_{k,s} = E_k - E_{k,crit} ,$$
(4.281)

which consists of kinetic and surface energy, is the difference between the kinetic energy E_k of the incoming parcel and the critical kinetic energy $E_{k,crit}$, below which no splashing occurs. $E_{k,crit}$ is believed to be lost in deformation, dissipation, and in producing liquid film energy. $E_{k,crit}$ is calculated from the critical Weber number:

$$E_{k,crit} = \frac{We_{crit}}{12} \pi \sigma d_{in}^2 .$$
(4.282)

From experimental data on the size-velocity correlation of the new droplets after splashing, the relation

$$\frac{|\vec{v}_1|}{|\vec{v}_2|} \approx \ln \left(\frac{d_1}{d_{in}} \right) / \ln \left(\frac{d_2}{d_{in}} \right)$$
(4.283)

is deduced. This relation expresses the observation, that the larger the size of the new droplets, the smaller the magnitude of their velocity. Finally, application of the tangential momentum conservation law yields

$$\frac{m_s}{2} |\vec{v}_1| \cos (\theta_1) + \frac{m_s}{2} |\vec{v}_2| \cos (\theta_2) = c_f m_{in} |\vec{v}_{in}| \cos (\theta_{in}) .$$
(4.284)

The wall friction coefficient c_f is assumed to be in the range of 0.6 to 0.8, and θ_1 and θ_2 are the ejection angles of the two new parcels. θ_1 is randomly chosen inside an assumed ejection cone (approximately $10° < \theta_1 < 160°$), and θ_2 is determined from Eq. 4.284. Together with

$$|\vec{v}_1|^2 = v_{1,n}^2 + v_{1,t}^2 \quad \text{and} \quad |\vec{v}_2|^2 = v_{2,n}^2 + v_{2,t}^2$$
(4.285)

the above set of equations can be solved in order to calculate the normal and tangential velocity components of the two new parcels.

The model of Bai and Gosman [11] has been successfully validated against experimental data. A typical spray pattern (side view, half spray) simulated with this model and showing all the characteristic features of a wall spray is given in Fig. 4.66. The computational parcels are shown as black points.

Fig. 4.67 shows comparisons between predicted and measured adhered fuel ratio on the wall, wall spray radius, and wall spray height. Comparisons of local droplet diameters are not given. The experiments were performed by Saito et al. [123], and the boundary conditions are given in Table 4.7.

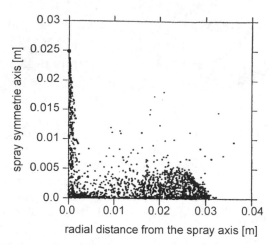

Fig. 4.66. Simulation of an impinging spray [11], t = 2 ms after SOI, atmospheric conditions

Table 4.7. Boundary conditions for the experiments of Saito et al. [123]

Case No	N4	N5	N6
Wall distance [mm]	25	25	25
Gas pressure [MPa]	0.2	0.2	0.2
Gas temperature [K]	293	293	293
Nozzle Diameter [mm]	0.25	0.25	0.25
Injection duration [ms]	2.85	2.1	1.425
Injection angle [deg.]	90	90	90
Injected fuel [mm³/pulse]	35	35	35
Fuel	diesel	diesel	diesel
Injection pressure [MPa]	30	55	120

Measurement and simulation agree well. The cusp in the curves of Fig. 4.67a corresponds to the end of injection. After the end of injection, the adhered fuel ratio, which is defined as total mass deposited on the wall divided by the total mass injected at that time, increases, because the deposited mass continues to increase for a short time while the injected mass remains constant. The predictions of wall spray radius (Fig. 4.67b) and wall spray height (Fig. 4.67c) capture the trend that an increase of injection pressure results in larger values of wall spray radius and height. The droplets in the wall spray get larger tangential velocities due to the increased gas velocities along the wall, and larger normal velocities because of a more intense splashing.

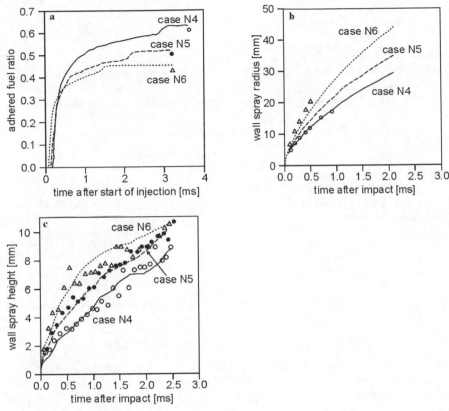

Fig. 4.67. a Measured and predicted adhered fuel ratio, **b** wall spray radius, **c** wall spray height, [11]

Other detailed wall-impingement models have been developed by O'Rourke and Amsden [95], Lee and Ryou [72], and Stanton and Rutland [132]. Differences between these models and that of Bai and Gosman [11] appear mainly in the splash regime.

In the model of O'Rourke and Amsden [95], the film momentum equation is used in combination with experimental results from Mundo et al. [89] in order to derive the spread-splash transition criterion

$$E^2 = \frac{\rho_l v_{n,in}^2 d_{in}}{\sigma} \cdot \left[\min\left(\frac{h_0}{d_{in}}, 1\right) + \frac{\delta_{bl}}{d_{in}} \right]^{-1} > (57.7)^2 = E_{crit}^2, \qquad (4.286)$$

where

$$E = \frac{v_{n,in}}{\dfrac{1}{d_{in}}\sqrt{\dfrac{\sigma h_0}{\rho_l}}} \qquad (4.287)$$

is the splash Mach number, h_0 is the initial film thickness before impact, and $\delta_{bl} = d_{in}/Re)^{1/2}$ is a boundary layer thickness. The ratio of splashed mass and incident mass is modeled on the basis of the experimental data of Yarin and Weiss [147],

$$\frac{m_s}{m} = \begin{cases} 1.8 \cdot 10^{-4}\left(E^2 - E_{crit}^2\right) & for\ E_{crit}^2 < E^2 < 7500 \\ 0.75 & for\ E^2 > 7500 \end{cases}. \qquad (4.288)$$

The prediction of drop radius r is based on the experimental results of Yarin and Weiss [147] and Mundo et al. [89], and is sampled from a truncated Nukiyama-Tanasawa distribution

$$f(r) = \frac{4}{\sqrt{\pi}}\frac{r^2}{r_{max}^3}\exp\left[-\left(\frac{r}{r_{max}}\right)^2\right], \quad r_{max} = r_{in}\cdot\max\left(\frac{E_{crit}^2}{E^2},\frac{6.4}{We},0.06\right). \quad (4.289)$$

In order to predict the velocity components of the new droplets, O'Rourke and Amsden [95] approximated the velocity distributions reported in [89] and [147]:

$$\vec{v} = v'_{n,out}\cdot\vec{n} + \left(0.12\cdot v_{n,in} + v'_{t,out}\right)\cdot\left(\cos\psi\vec{e}_t + \sin\psi\vec{e}_p\right) + 0.8v_{t,in}\vec{e}_t, \qquad (4.290)$$

where $\vec{n}$ is the unit vector normal to the wall surface, $\vec{e}_t$ is the unit vector tangential to the surface in the plane of $\vec{n}$ and the incident drop, $\vec{e}_p = \vec{n}\times\vec{e}_t$, $v'_{n,out}$ and $v'_{t,out}$ are the normal and tangential velocity fluctuations, and Ψ is the angle that the fluctuating velocity makes with the vector $\vec{e}_t$ in the plane of the wall, see Fig. 4.64. The quantities $v'_{n,out}$, $v'_{t,out}$, and Ψ are random variables chosen from the following distributions:

$$f\left(v'_{n,out}\right) = \frac{4}{\sqrt{\pi}}\frac{\left(v'_{n,out}\right)^2}{\left(0.2v_{n,in}\right)^3}\exp\left[-\left(\frac{v'_{n,out}}{0.2v_{n,in}}\right)^2\right] \quad \text{(Nukiyama-Tanasawa),} \quad (4.291)$$

$$f\left(v'_{t,out}\right) = \frac{1}{\sqrt{2\pi\left(0.1\cdot v_{n,in}\right)^2}}\exp\left[-\left(\frac{v'_{t,out}}{2\left(0.1\cdot v_{n,in}\right)^2}\right)^2\right] \quad \text{(Gaussian dist.),} \quad (4.292)$$

and the angle Ψ is chosen as suggested in the model of Naber and Reitz.

A further wall impingement model has been developed by Lee and Ryou [72]. The main difference to the model of Bai and Gosman [11] is a modified calculation of the dissipated energy due to the impact of a drop on the wetted surface, and the subsequent calculation of the total velocity from the energy conservation law.

Furthermore, the spread-splash transition criterion is determined by using an empirical correlation based on the experimental results of Mundo et al. [89], and only one instead of two new parcels is created in the splash regime. The size and number of the splashed new droplets are determined from the conservation of mass and from experimental data.

4.8.3 Wall Film Modeling

A detailed modeling of wall film evolution includes the simulation of all effects increasing or decreasing the liquid mass on the surface of a wall cell like deposition of droplets, evaporation and film movement, see Fig. 4.68. Film movement can be due to the gravity force or due to exchange of momentum between gas and film or between incident droplets and film. Different classes of wall film models exist. One method is to set up and solve the governing equations for wall films. The other method is to simplify the treatment and to use empirical correlations for the relevant effects.

An early model for the simulation of wall film evolution and heat transfer from the wall to the film or to the impinging drops was developed by Eckhause and Reitz [31]. This model includes the impingement model of Naber and Reitz [90], Sect. 4.8.2. The model distinguishes between a flooded (a wall film is present) and a non-flooded case (no wall film), see Fig. 4.69. According to the model of Naber and Reitz [90], only droplets with $We_{in} > 80$ (slide) contribute to the formation of a liquid film. The film thickness δ is calculated separately for each surface cell by taking the total mass of fuel on the surface of a wall cell and dividing by the fuel density ρ_l and the wall cell area A_{cell}. The transition criterion between the non-flooded and the flooded case is given as $\delta = 2$ μm. The total heat transferred to a film or a drop is

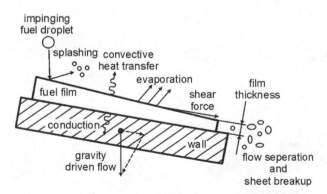

Fig. 4.68. The major physical phenomena governing film flow [133]

$$Q = A \cdot \alpha \cdot (T_s - T) \cdot t_{res} + Q_g , \tag{4.293}$$

where A is the contact area, α is the convective heat transfer coefficient ($Nu = (\alpha \, \delta)/\lambda$, Nu: Nusselt number, δ: film thickness or drop diameter, λ: thermal conductivity), T is the liquid drop or film temperature, T_s is the surface temperature, t_{res} is the residence time of the liquid on the surface, and Q_g is added in case of wall film formation and is the heat transferred from the gas to the film.

Droplets may either impinge on a dry wall (non-flooded case) or on a liquid wall film (flooded case). Furthermore, droplets may rebound or slide in both cases.

In the case of rebound, the residence time of a drop is assumed to be the first-order vibration period of an oscillating drop [144],

$$t_{res} = \frac{\pi}{4} \sqrt{\frac{\rho_l d_{in}^3}{\sigma}} , \tag{4.294}$$

and the contact area between the deformed drop and the wall or the film is predicted based on the spreading equation of Akao et al. [3]:

$$A = \frac{\pi}{4} D_{max}^2 , \quad D_{max} = 0.613 \cdot d_{in} \cdot We_{in}^{0.39} . \tag{4.295}$$

The surface temperature T_s is the film temperature or the wall temperature, depending on whether a wall film is present or not. The Nusselt number is taken to be $Nu = 2.0$ for individual droplets.

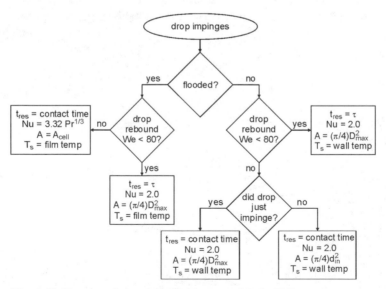

Fig. 4.69. Heat transfer model of Eckhause and Reitz [31]

For the case of slide, Eckhause and Reitz [31] distinguish between the flooded and the non-flooded regime. In the non-flooded regime the contact area is calculated according to Eq. 4.295 during the time step of first contact, and then the area is $A = (\pi/4)\, d_{in}^2$ during the time the droplet slides along the wall. Again, $Nu = 2.0$ is used and the corresponding surface temperature is the wall temperature. In the flooded regime the droplet becomes part of the wall film. The film temperature is the mass average temperature of all droplets belonging to the film, and the contact area A is equal to the surface area A_{cell} of the respective wall cell. The gas-phase heat transfer is calculated based on the modified law of the wall (Reitz [116]), while the liquid-phase heat transfer is based on the heat transfer correlation $Nu = 3.32 \cdot Pr^{1/3}$ for the boundary layer flow.

The temperature change of a film element or a drop is given by

$$\Delta T = \frac{Q}{m \cdot c_{v,l}}, \qquad (4.296)$$

where $c_{v,l}$ is its specific heat capacity and m is the liquid mass of the fuel on a wall cell (flooded case) or the one of a single drop (non-flooded case and rebound).

A much more detailed wall film model, which simulates thin fuel film flow on solid surfaces of arbitrary configuration, is presented in Stanton and Rutland [132, 133]. The major mechanical and physical processes like mass, momentum, and pressure contributions to the film due to spray impingement and splashing effects as well as the effects of shear forces, piston acceleration, and gravity are included. The following assumptions are used in the formulation of the model: First, the liquid film is treated as a thin film and the momentum and continuity equations are integrated across the film thickness $\bar{x}_3$, see Fig. 4.70. Thus, a two dimensional film is regarded that flows over a three dimensional surface. Fig. 4.70 shows the local coordinate system used for the description of a wall film cell.

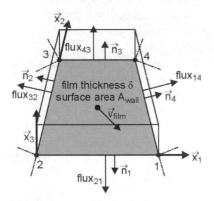

Fig. 4.70. Wall film cell and local coordinate system of the film model

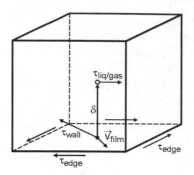

Fig. 4.71. Momentum cell

The second assumption is that the mass flux, the tangential momentum, and the dynamic pressure, which are added to the liquid film due to impinging drops, are averaged over the cell area. Further on, it is assumed that the velocity profile in the cross-film direction, which is needed when integrating across the film thickness, is either laminar or turbulent.

After integrating in the cross-film direction, the continuity equation yields

$$\rho_l A_{wall} \frac{\partial \delta}{\partial t} = -\sum_{i=1}^{N_{side}} (\vec{v}_{film} \cdot \vec{n})_i \delta_i l_i \rho_l A_{wall} + \dot{m}_{imp} - \dot{m}_{evap} , \qquad (4.297)$$

and after some rearrangement it results in

$$\frac{\partial \delta}{\partial t} + \frac{1}{A_{wall}} \sum_{i=1}^{N_{side}} (\vec{v}_{film} \cdot \vec{n})_i \delta_i l_i = \frac{\dot{m}_{imp}}{\rho_l A_{wall}} - \frac{\dot{m}_{evap}}{\rho_l A_{wall}} , \qquad (4.298)$$

where A_{wall} is the wall cell area, l_i and δ_i are the length of side i and the film thickness of side i, $\vec{v}_{film}$ is the mass-averaged film velocity relative to the wall, and $\dot{m}_{imp}$ is the difference of incoming mass flux due to impinging drops and outgoing mass flux due to splashing. The quantity $\dot{m}_{evap}$ is the rate of fuel vaporization and must be predicted by a film vaporization model, Sect. 4.5.4.

The momentum equation is derived in the same manner. Fig. 4.71 shows a typical momentum cell containing gas (upper part) and the liquid film at the bottom. The momentum equation yields

$$\rho_l A_{wall} \frac{\partial (\delta \cdot \vec{v}_{film})}{\partial t} + \sum_{i=1}^{N_{side}} \vec{v}_{film} (\vec{v}_{film} \vec{n})_i \rho_l \delta_i l_i \phi_i$$

$$= -\sum_{i=1}^{N_{side}} (p \, \vec{n}) \delta_i l_i + \rho_l A_{wall} \delta \vec{g} + M_{tang} + \sum_{i=1}^{N_{edge}} (\vec{\tau} \, A_i) + \rho_l A_{wall} \delta \vec{a} , \qquad (4.299)$$

and after some rearrangement the following form can be obtained:

$$\frac{\partial \left(\delta \cdot \vec{v}_{film} \right)}{\partial t} + \frac{1}{A_{wall}} \sum_{i=1}^{N_{side}} \vec{v}_{film} \left(\vec{v}_{film} \vec{n} \right)_i \delta_i l_i \phi_i$$

$$= -\frac{\sum_{i=1}^{N_{side}} \left(p\, \vec{n} \right) \delta_i l_i}{\rho_l A_{wall}} + \delta \vec{g} + \frac{M_{tang}}{\rho_l A_{wall}} + \frac{\sum_{i=1}^{N_{edge}} \left(\vec{\tau}_i\, A_i \right)}{\rho_l A_{wall}} + \delta \vec{a} \ . \tag{4.300}$$

The left-hand side of Eq. 4.300 is the material derivate of the film momentum. The convective momentum term (second term) is an approximation of the integration of the non-linear convective term in the cross-film direction (due to the velocity profile), where

$$\phi = \frac{1}{1 - \frac{\delta_t}{\delta}} - \frac{\Theta_t}{\delta} \cdot \frac{1}{\left(1 - \frac{\delta_t}{\delta} \right)^2} \tag{4.301}$$

is used in order to compensate the effect of using the mass-averaged film velocity instead of a momentum-averaged velocity derived from the exact velocity profile. In order to calculate the displacement thickness δ_t and the momentum thickness Θ_t, a velocity profile has to be specified in the cross-film direction.

The right-hand side of Eq. 4.300 is the sum of all relevant forces. The first term describes the pressure force. The pressure

$$p = p_{amb} + p_{imp} \tag{4.302}$$

does not vary in the cross-film direction and includes the ambient pressure p_{amb} and the dynamic pressure p_{imp} due to drop impingement and splashing effects, which can be calculated using the momentum equation for one-dimensional flows,

$$p_{imp} = \rho_l \sum_{k=1}^{N_{drop}} v_{n,k}^2 \frac{A_k}{A_{wall}} + \rho_l \sum_{j=1}^{N_{splash}} v_{n,j}^2 \frac{A_j}{A_{wall}} \ . \tag{4.303}$$

In Eq. 4.303, $v_{n,k}$ is the velocity component of the incident drop normal to the surface, and $v_{n,j}$ is the normal velocity of the droplets resulting from splashing. N_{drop} and N_{splash} are the total numbers of incident and splashed droplets, and A_k and A_j are the corresponding droplet areas.

The second term on the right hand side describes the gravity effect, which is important for wall films on vertical surfaces. The third term is the change of tangential momentum (index τ) due to spray impingement and splashing and is given as

$$M_{tang} = \sum_{i=1}^{N_{drop}} \left(m_i \vec{v}_{\tau,i} \right) - \sum_{j=1}^{N_{splash}} \left(m_j \vec{v}_{\tau,j} \right) . \tag{4.304}$$

In Eq. 4.304, m_i and m_j are the masses of the incident drops and the droplets resulting from splashing.

The effect of viscous shear forces is given by the fourth term on the right-hand side,

$$\sum_{i=1}^{N_{edge}} \left(\vec{\tau}_i A_i \right) = \sum_{j=1}^{N_{side}} \left(\vec{\tau}_{edge} \delta l \right)_j + \vec{\tau}_{wall} A_{wall} + \vec{\tau}_{liq/air} A_{wall} , \qquad (4.305)$$

where $\vec{\tau}_{edge}$ is the viscous shear along the edges of the film cell, $\vec{\tau}_{wall}$ is the wall shear, and $\vec{\tau}_{liq/air}$ is the shear at the liquid-gas interface.

The last term on the right-hand side describes the effect of an acceleration of the solid surface, which is only relevant for fuel films on piston or valves.

The film model has been combined with a film evaporation model and has been successfully validated against experimental data under diesel engine conditions [133]. The relevant engine parameters and operating conditions are summarized in Table 4.8. As an example, Fig. 4.72 shows a comparison of measured and simulated film thickness as a function of crank angle at different positions on the piston

Table 4.8. Engine parameters and operating conditions [133]

Bore [mm]	150
Stroke [mm]	225
Compression ratio [/]	14
Wall temperature [°C]	400
Swirl ratio [/]	0.0
Fuel injected per hole [mm^3/cycle]	25.17
Orifice diameter [mm]	0.35
Start of injection [deg. CA]	-19
Injection duration [deg. CA]	34

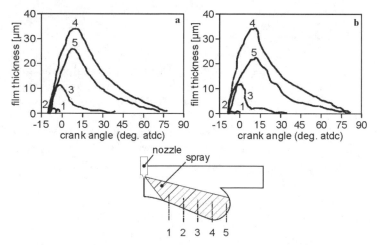

Fig. 4.72. Comparison of film thickness. **a** simulation, **b** measurement, data from [133]

bowl surface for the full load condition. Despite some small discrepancies between measurement and experiment, a good overall qualitative and quantitative agreement is obtained. A high level of agreement is also attained in the case of reduced loads and different injection timings, emphasizing the suitability of the model for the simulation of film development and film evaporation.

4.9 Ignition

4.9.1 Auto-Ignition

The auto-ignition of hydrocarbons in diesel engines is a chain-branching process including the four reaction classes of chain initiation, chain propagation, chain branching, and chain termination. After the start of injection, ignition occurs after a certain induction time, the ignition delay. During this time delay, fuel evaporates until a first region of ignitable mixture with an air-fuel ratio of $0.5 < \lambda < 0.7$ is formed. Furthermore, the chemical reactions in this region have to produce enough fuel radicals in order to start the combustion process. The chain initiation produces these first radicals from stable fuel molecules. This reaction proceeds slowly, because stable molecules are involved in the process. Then, if a certain radical concentration is reached, the chain propagation and the chain branching reactions form additional radicals. The chain propagation reactions change the nature of the radicals but not their number. Some of the chain propagation reactions produce radicals, which then take part in the chain branching reactions that increase the number of radicals and result in a considerable acceleration of the reactions, leading finally to explosion. The ignition delay is strongly temperature-dependent, a rise of temperature decreases this time.

The multi-stage ignition process can be divided into three temperature regions, the low temperature reactions (cool flame regime), the intermediate temperature region, and the high temperature oxidation. Because details concerning the relevant reactions in the three temperature regimes are given in Sect. 6.4.2, only a brief description of the relevant mechanisms will be given in the following. The cool flame regime typically occurs at gas temperatures between 600 and 800 K. Reactions proceed slowly with only a small temperature rise. However, for increasing temperatures, the production of radicals by the cool flame reactions is reduced because the reverse reactions become faster (degenerate chain branching). Hence, this intermediate temperature region is characterized by the so-called negative temperature coefficient (NTC), which represents an increased ignition delay for increased temperatures. As soon as the temperature, which is increased by the heat release of the cool flame reactions and the further compression of the cylinder charge, is reached, the high temperature chain branching reactions ($T > 1000$ K) lead to explosion.

Auto-ignition takes place on time scales that are relatively long compared to the relevant hydrodynamic time scales [134]. Hence, the influence of convective and diffusive species transport must be taken into account. This is usually done by solving mass conservation equations for a certain radical indicator species, also including source terms for the chemical reactions. These source terms are expressed by Arrhenius-type reaction rates. Ignition timing is predicted by defining a certain threshold of the indicator species that has to be reached. Then, the ignition model is switched off and the calculation continues with a combustion model. Details concerning appropriate combustion models are given in [134], for example.

The most widely used auto-ignition model is the so-called Shell model, which was developed by Halstead et al. [47]. The name of the model stems from the affiliation of the authors. The model was originally developed to predict knock in spark ignition engines and was later adjusted to predict auto-ignition in diesel engines (e.g. Kong et al. [67], Sazhina et al. [124]). Because it is not possible to model all of the several hundreds of relevant reactions during ignition, the model is based on a class chemistry concept and includes only eight reaction steps between five species. It represents a virtual mechanism between generic species and is formulated to reflect the multistage ignition behavior of hydrocarbon air mixtures including a degenerate chain branching mechanism. The eight reaction steps are given as

$$RH + O_2 \xrightarrow{k_q} 2R^* \qquad \text{(chain initiation)} \qquad (4.306)$$

$$R^* \xrightarrow{k_p} R^* + P + heat \qquad \text{(chain propagation)} \qquad (4.307)$$

$$R^* \xrightarrow{f_1 k_p} R^* + B \qquad \text{(chain propagation forming B)} \qquad (4.308)$$

$$R^* \xrightarrow{f_4 k_p} R^* + Q \qquad \text{(chain propagation forming Q)} \qquad (4.309)$$

$$R^* + Q \xrightarrow{f_2 k_p} R^* + B \qquad \text{(chain propagation forming B)} \qquad (4.310)$$

$$B \xrightarrow{k_b} 2R^* \qquad \text{(degenerate branching)} \qquad (4.311)$$

$$R^* \xrightarrow{f_3 k_p} termination \qquad \text{linear termination} \qquad (4.312)$$

$$2R^* \xrightarrow{k_t} termination \qquad \text{quadratic termination.} \qquad (4.313)$$

In Eqs. 4.306–4.313, R^* represents the radical, RH is the fuel, Q is an unstable intermediate agent, B is the branching agent, and P represents oxidized products. The concentrations of the different species can be calculated solving the differential equations for their change rates, which are given as

$$\frac{d[R^*]}{dt} = 2k_q[RH][O_2] + 2k_b[B] - f_3 k_p[R^*] - k_t[R^*]^2 , \qquad (4.314)$$

$$\frac{d[B]}{dt} = f_1 k_p[R^*] + f_2 k_p[R^*][Q] - k_b[B], \qquad (4.315)$$

$$\frac{d[Q]}{dt} = f_4 k_p [R^*] - f_2 k_p [R^*][Q], \tag{4.316}$$

$$\frac{d[O_2]}{dt} = -p k_p [R^*], \tag{4.317}$$

$$\frac{d[RH]}{dt} = \frac{[O_2] - [O_2]_{(t=0)}}{p \cdot m} + [RH]_{(t=0)} . \tag{4.318}$$

The quantity m depends on the number of hydrogen atoms in the original fuel molecule C_nH_{2m}, $p = (n(2-\gamma) + m)/2m$, and $\gamma \approx 0,67$ is the CO/CO_2 ratio. The rate coefficients

$$f_1 = A_{f1} \exp\left(-E_{f1} / RT\right) [O_2]^{x1} [RH]^{y1}, \tag{4.319}$$

$$f_2 = A_{f2} \exp\left(-E_{f2} / RT\right), \tag{4.320}$$

$$f_3 = A_{f3} \exp\left(-E_{f3} / RT\right) [O_2]^{x3} [RH]^{y3}, \tag{4.321}$$

$$f_4 = A_{f4} \exp\left(-E_{f4} / RT\right) [O_2]^{x4} [RH]^{y4}, \tag{4.322}$$

$$k_i = A_i \exp\left(-E_i / RT\right), \quad i = 1, 2, 3, 4, q, b, t, \tag{4.323}$$

are of Arrhenius-type, and

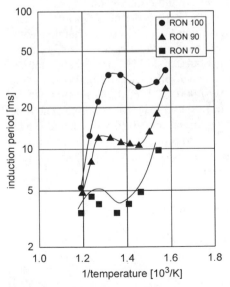

Fig. 4.73. Comparison of simulated and measured ignition delays [47], RON: Research Octane Number

$$k_p = \left[\frac{1}{k_1[O_2]} + \frac{1}{k_2} + \frac{1}{k_3[RH]} \right]^{-1}. \tag{4.324}$$

The model includes 26 parameters that have to be adjusted for each fuel of interest. More details are given in Kong et al. [67], for example.

The Shell model is the most-used model in simulations of diesel engine auto-ignition today. It is capable of describing the negative temperature coefficient effect as well as the influence of temperature, pressure, and fuel-air mixture on ignition delay. Fig. 4.73 shows a comparison between simulated and experimentally obtained ignition delays as a function of temperature for three primary reference fuels of different octane quality.

However, if auto-ignition processes with ignition delays much longer than those of conventional diesel engines are regarded (e.g. HCCI processes), the Shell model is no longer capable of predicting the auto-ignition process with sufficient accuracy. In these cases, more complex chemical models that include a much more detailed description of the low temperature reactions have to be used, see also Sect. 6.4.2.

4.9.2 Spark-Ignition

Gasoline-air mixtures in spark-ignition engines do not auto-ignite due to compression. In these engines, the start of combustion is initiated by the energy of an electrical spark. The spark forms a high-temperature zone between the two electrodes of the spark plug, in which radicals are produced and combustion begins. In this zone, enough heat must be released in order to heat up the neighboring mixture and to initiate a flame front that can propagate into the combustion chamber without any further energy input. During the time the electrical spark exists, a highly ionized plasma with temperatures of about 60000 K [134] forms a channel between the electrodes and then expands to become a spherical ignition kernel, while the temperature decreases to approximately 4000 K, Fig. 4.74. Further details about the kernel growth process are given in [83] and [51], for example. The time span needed for this ignition kernel growth is in the micro- to millisecond range. The flame front starts from this ignition kernel and propagates into the combustion chamber. At first, it propagates with laminar flame speed, and as the flame surface increases, turbulence increases the flame speed.

Fig. 4.74. Ignition kernel growth

In CFD models, a phenomenological sub-model usually describes the growth of the ignition kernel. If the kernel has reached a certain size, the calculation proceeds with a combustion model for turbulent flames.

The modeling of ignition kernel growth and flame speed is usually based on the assumption that the temperature inside the kernel is uniform and equal to the adiabatic flame temperature of the associated fuel-air mixture. Because the time needed to form a spherical ignition kernel is in the same range as the computational time step Δt_0 (about 1 μs), an isothermal ignition kernel of adiabatic flame temperature is assumed to exist at the end of the first numerical time step of ignition. The radius $r_{k,0}$ of this kernel is determined from an energy balance: the sum of the electrical energy W_{sp} of the spark discharge and the chemical energy released due to the combustion of the mixture inside the kernel results in a rise of the temperature inside the kernel from the unburned temperature T_u to the adiabatic temperature T_{ad}:

$$\frac{4}{3}\pi r_{k,0}^3 \rho_k c_p \left(T_{ad} - T_u\right) = \eta_{sp}\dot{W}_{sp}\Delta t_0 + \frac{4}{3}\pi r_{k,0}^3 \rho_k LHV_{mix}. \tag{4.325}$$

In Eq. 4.325, LHV is the lower heating value of the air-fuel mixture per gram mixture, and η_{sp} is the energy transfer efficiency, which can be assumed to be $\eta_{sp} \approx 1.0$ in this first phase of kernel growth. From Eq. 4.325 the radius

$$r_{k,0} = \left[\frac{3\eta_{sp}\dot{W}_{sp}\Delta t_0}{4\pi\rho_k \left(c_p \left(T_{ad} - T_u\right) - LHV_{mix}\right)}\right]^{1/3} \tag{4.326}$$

of the kernel after the first time step can be derived. During the following time steps the kernel size increases. Because all of the energy of the mixture inside the kernel with radius r_k is already released, only the surplus volume, which is the product of the kernel surface and the increase of kernel radius, can be released. Thus, the energy balance gives

$$4\pi r_k^2 \frac{dr_k}{dt} \rho_u c_p \left(T_{ad} - T_u\right) = \eta_{sp}\dot{W}_{sp} + 4\pi r_k^2 \frac{dr_k}{dt}\rho_u LHV_{mix}. \tag{4.327}$$

The energy transfer in this later phase of kernel growth is less efficient than in the first phase, the value of η_{sp} beeing about 30–50% [134]. The quantity dr_k/dt is called plasma velocity s_{pl}. Eq. 4.327 yields

$$s_{pl} = \frac{\eta_{sp}\dot{W}_{sp}}{4\pi r_k^2 \rho_u \left(c_p \left(T_{ad} - T_u\right) - LHV_{mix}\right)}. \tag{4.328}$$

Because s_{pl} is inversely proportional to r_k^2, it is reduced with increasing kernel radius. However, as the kernel growths, the effect of turbulence on the effective flame velocity increases and leads to an increasing velocity of the flame front propagating into the combustion chamber, see Fig. 4.75.

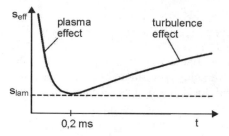

Fig. 4.75. Characteristic behavior of the effective flame speed after ignition [134]

Fan et al. [36] published the discrete particle ignition kernel (DPIK) model, which is based on the ideas of ignition kernel modeling as presented above. In this model, the initial flame kernel is represented by Lagrangian marker particles, allowing to reduce the influence of grid size effects on the ignition process. The initial ignition kernel is assumed to be spherical. The initial diameter $d_{k,0}$ is assumed to be in the range of the gap size between the two electrodes, usually 1 mm. The radial kernel growth rate is calculated as a function of the laminar flame speed s_{lam} and the turbulent kinetic energy k of the flow field,

$$u_k = \frac{dr_k}{dt} = \frac{T_{ad}}{T_u} s_{lam} + \sqrt{\frac{2k}{3}}.$$
(4.329)

The parameter T_{ad}/T_u accounts for the effect of thermal expansion. T_{ad} is the adiabatic flame temperature, and T_u is the local unburned gas temperature, which is determined assuming adiabatic compression from the conditions inside the kernel before (index 1) and after the start of ignition (index 2):

$$\frac{T_u}{T_{k,1}} = \left(\frac{p_{k,2}}{p_{k,1}}\right)^{(\kappa-1)/\kappa}.$$
(4.330)

The laminar flame speed s_{lam} is given by a relation of Metghalchi and Keck [87], which is

$$s_{lam} = (1.0 - 2.1R) \cdot \left[26.32 - 84.72(\varPhi - 1.13)^2\right] \cdot$$
$$\left[\frac{T_u}{298\ K}\right]^{2.18-0.8(\varPhi-1)} \cdot \left[\frac{p_{k,2}}{101.3\ kPa}\right]^{-0.16+0.22(\varPhi-1)}$$
(4.331)

In the case of gasoline [135]. In Eq. 4.331 $\varPhi$ is the equivalence ratio in the spark region, and R is the residual mass fraction. Hence, the diameter of the ignition kernel can be calculated as

$$d_k = 2u_k\left(t - t_{ign}\right) + d_{k,0}.$$
(4.332)

A one-step reaction is chosen during this early stage of combustion,

$$C_8H_{18} + 12.5O_2 \Rightarrow 8CO_2 + 9H_2O \,, \tag{4.333}$$

and the burn rate is calculated by

$$\frac{d\rho_i}{dt} = -C_w \min\left\{ \frac{\rho_f}{MW_f C_{sto,f}}, \frac{\rho_{O_2}}{MW_{O_2} C_{sto,O_2}} \right\} \cdot \sum \cdot s_l \cdot MW_i C_{sto,i} \,. \tag{4.334}$$

In Eq. 4.334 MW_i is the molecular weight of species i, $C_{sto,i}$ are the stoichiometric coefficients, and C_w is a constant used to account for the wrinkling effect of the kernel surface. For example, Fan et al. [36] use a value of $C_w = 80$ while Stiesch et al. [135] use $C_w = 20$. The quantity

$$\sum = \frac{N_{p,cell}}{N_{p,total}} \frac{\pi d_k^2}{V_{cell}} \tag{4.335}$$

is the flame surface density within a particular cell, where $N_{p,total}$ is the total number of marker particles, and $N_{p,cell}$ is the number of particles in the specific cell having a volume V_{cell}.

The combustion is initiated by the ignition model, and if the kernel size reaches the order of the integral length scale of turbulence,

$$d_k \geq C_k l_t \,, \tag{4.336}$$

the model switches to an appropriate turbulent combustion model. In Eq. 4.336 $l_t = 0.16\, k^{1.5}/\varepsilon$ is the turbulence length scale, and $C_k = 3.5$.

References

[1]   Agmon N, Alhassid Y, Levine RD (1979) An Algorithm for Finding the Distribution of Maximal Entropy. J Comp Phys, vol 30, no 2, pp 250–258

[2]   Ahmadi M, Sellens RW (1993) A Simplified Maximum-Entropy-Based Drop Size Distribution. Atomisation and Sprays, vol 3, pp 291–310

[3]   Akao F, Araki K, Mori S, Moriyama A (1980) Deformation Behaviours of a Liquid Droplet Impinging onto a Hot Metal Surface. Trans I S I J, vol 20, pp 737–743

[4]   Allocca L, Corcione, FE, Fusco A, Papetti F, Succi S (1994) Modelling of Diesel Spray Dynamics and Comparison with Experiments. SAE paper 941895

[5]   Amsden AA, O'Rourke PJ, Butler TD (1989) KIVA-II – A Computer Program for Chemically Reactive Flows with Sprays. Los Alamos National Laboratories, LA-11560-MS

[6]   Amsden AA, Ramshaw JD, O'Rourke PJ, Dukowicz JK (1985) KIVA.. A Computer Program for Two- and Three-Dimensional Fluid Flows with Chemical Reactions and Fuel Sprays. Los Alamos National Labs, Rept LA-10245-MS

[7]   Arcoumanis C, Gavaises M, French B (1997) Effect of Fuel Injection Process on the Structure of Diesel Sprays. SAE paper 970799

[8]   Ashgriz N, Poo JY (1990) Coalescence and Separation in Binary Collisions of Liquid Drops, J. Fluid Mech., vol. 221, pp. 183–204

[9]   Ayoub NA (1995) Modeling Multicomponent Fuel Sprays in Engines with Application to Diesel Cold-Starting. Ph D Thesis, University of Wisconsin-Madison

[10]   Babinsky E, Sojka PE (2002) Modeling Drop Size Distributions. Progress in Energy and Combustion Science 28. pp 303–329

[11]   Bai C, Gosman AD (1995) Development of Methodology for Spray Impingement Simulation. SAE paper 950283

[12]   Baumgarten C, Merker GP (2004) Modelling Primary Break-Up in High-Pressure Diesel Injection. Motortechnische Zeitschrift (MTZ worldwide) 4/2004, vol 65, pp 21–24

[13]   Baumgarten C. (2003) Modellierung des Kavitationseinflusses auf den primären Strahlzerfall bei der Hochdruck Dieseleinspritzung. Dissertation, Fortschritt-Berichte VDI, Reihe 12, Nr 543

[14]   Baumgarten C, Stegemann J, Merker GP (2002) A New Model for Cavitation Induced Primary Break-up of Diesel Sprays. ILASS-Europe 2002

[15]   Bellman R, Pennington RH (1954) Effects of Surface Tension and Viscosity on Taylor Instability. Quarterly of Applied Mathematics, 12, pp 151–162

[16]   Borman GL, Johnson JH (1962) Unsteady Vaporization Histories and Trajectories of Fuel Drops Injected into Swirling Air. SAE paper 598 C

[17]   Brazier-Smith PR, Jennings SG, Latham J (1972) The Interaction of Falling Water drops: Coalescence. Proc R Soc Lond A 326, pp 393–408

[18]   Brennen EC (1995) Cavitation and Bubble Dynamics. Oxford University Press

[19]   Chan M, Das S, Reitz RD (1997) Modeling Multiple Injection and EGR Effects on Diesel Engine Emissions. SAE paper 972864

[20]   Chang SK (1991) Hydrodynamics in Liquid Jet Sprays. Ph D Thesis, University of Wisconsin-Madison

[21]   Chehroudi B, Chen SH, Bracco FV, Onuma Y (1985) On the Intact Core of Full-Cone Sprays. SAE paper 850126

[22]   Chou GF, Prausnitz JM (1986) Adiabatic Flash Calculations for Continuous or Semicontinuous Mixtures Using an Equation of State. Fluid Phase Equilibria, vol 30, pp 75–82

[23]   Chryssakis CA, Assanis DN, Lee JK, Nishida K (2003) Fuel Spray Simulation of High-Pressure Swirl-Injector for DISI Engines and Comparison with Laser Diagnostic Measurements. SAE paper 2003-01-0007

[24]   Cotterman RL, Bender R, Prausnitz JM (1985) Phase Equilibria for Mixtures Containing very many Components – Development and Application of Continuous Thermodynamics for Chemical Process Design. Ind Eng Chem Process Des Dev, vol 24, pp 194–203

[25]   Cousin J, Desjonqueres PH (2003) A New Approach for the Maximum Entropy Formalism on Sprays. ICLASS 2003

[26]   Cousin J, Yoon SJ, Dumouchel C (1996) Coupling of Classical Linear Theory and Maximum Entropy Formalism for Prediction of Drop Size Distributions in Sprays: Application of Pressure-Swirl Atomizers. Atomization and Sprays, vol 6, pp 601–622

[27]   Crowe CT, Chung JN, Troutt TR (1988) Particle Mixing in Free Shear Flows. Prog Energy Combust Sci, vol 14, pp 171–194

[28]   Dobre M, Bolle L (1998) Theoretical Prediction of Ultrasonic Spray Characteristics Using the Maximum Entropy Formalism. ILASS-Europe'98

[29] Dombrowski N, Hooper PC (1962) The Effect of Ambient Density on Drop Forma-
      tion in Sprays. Cem Eng Sci, vol 17, p 291
[30] Dombrowski N, Johns WR (1963) The Aerodynamic Instability and Disintegration
      of Viscous Liquid Sheets. Chem Eng Sci, vol 18, p 203
[31] Eckhause JE, Reitz RD (1995) Modelling Heat Transfer to Impinging Fuel Sprays in
      Direct Injection Engines. Atomization and Sprays, vol 5, pp 213–242
[32] El Wakil MM, Uyehara OA, Myers PS (1954) A Theoretical Investigation of the
      Heating-Up Period of Injected Fuel Droplets Vaporizing in Air. NACA TN 3179
[33] El Wakil MM, Priem RJ, Brikowski HJ, Myers PS, Uyehara OA (1956) Experimen-
      tal and Calculated Temperature and Mass Histories of Vaporizing Fuel Drops.
      NACA TN 3490A
[34] Faeth GM (1983) Evaporation and Combustion of Sprays. Prog Energy Combust
      Sci, vol 9, pp 1–76
[35] Faeth GM (1987) Mixing, Transport and Combustion in Sprays. Prog Energy Com-
      bust Sci, vol 13, pp 293-345
[36] Fan L, Li F, Han Z, Reitz RD (1999) Modeling Fuel Preparation and Stratified
      Combustion in a Gasoline Direct Injection Engine. SAE paper 1999-01-0175
[37] Fischer M (2005) Ph.D. Thesis, Institute of Technical Combustion, University of
      Hannover, Germany
[38] Fujimoto H, Nishikori T, Hojyo Y, Tsukamoto T, Senda J (1994) Modeling of At-
      omization and Vaporization Process in Flash Boiling Spray. ICLASS-94, Rouen,
      France, pp 680–687, paper VI-13
[39] Gavaises M, Arcoumanis C (2001) Modelling of Sprays from High-Pressure Swirl
      Atomizers. Int J Engine Research, vol 2, no 2, pp 95–117
[40] Georjon TL, Reitz RD (1999) A Drop Shattering Collision Model for Multidimen-
      sional Spray Computations. Atomization and Sprays, vol 9, pp 231–254
[41] Gerrish HC, Ayer BE (1933) Influence of Fuel-Oil Temperature on the Combustion
      in a Prechamber Compression-Ignition Engine. NACA Technical Note 565
[42] Glavincevski B, Gülder OL, Gardner L (1984) Cetane Number Estimation of Diesel
      Fuels from Carbon Type Structural Composition. SAE paper 841341
[43] Gonzalez MA, Borman GL, Reitz RD (1991) A Study of Diesel Cold Starting Using
      Both Cycle Analysis and Multidimensional Calculations. SAE paper 910180
[44] Gosman AD, Ioannides E (1981) Aspects of Computer Simulation of Liquid-Fueled
      Combustors. AIAA paper no 81-0323
[45] Hagerty WW, Shea JF (1995) A Study of the Stability of Plane Fluid Sheets. J Appl
      Mech, vol 22, pp 509–514
[46] Hallett WLH (1997) A Simple Quasi-Steady Droplet Evaporation Model Using Con-
      tinuous Thermodynamics. The Combustion Institute, Canadian Section, Spring
      Technical Meeting
[47] Halstead MP, Kirsch LJ, Quinn CP (1977) The Autoignition of Hydrocarbon Fuels at
      High Temperatures and Pressures – Fitting of a Mathematical Model. Combustion
      and Flame, vol 30, pp 45–60
[48] Hammitt FG (1980) Cavitation and Multiphase Flow Phenomena. McGraw-Hill
      International Book Company
[49] Han Z, Parrish S, Farrell PV, Reitz RD (1997) Modeling Atomization Processes of
      Pressure-Swirl Hollow-Cone Fuel Sprays. Atomization and Sprays, vol 7, no 6, pp
      663–684

[50]   Herring C (1941) Theory of the Pulsations of the Gas Bubble Produced by an Underwater Explosion. Columbia University NDRC Rep C-4-sr10-010

[51]   Heywood JB (1988) Internal Combustion Engine Fundamentals. McGraw-Hill, New York

[52]   Hinze OJ (1975) Turbulence. McGraw-Hill, 2nd ed, New York

[53]   Hiroyasu H, Arai M (1990) Structures of Fuel Sprays in Diesel Engines. SAE-paper 900475

[54]   Hsieh KC, Shuen JS, Yang V (1991) Droplet Vaporization in High-Pressure Environments, I: Near Critical Conditions. Combust Sci Tech, vol 76, pp 111–132

[55]   Huh KY, Lee EJ, Koo JY (1998) Diesel Spray Atomization Model Considering Nozzle Exit Turbulence Conditions. Atomization and Sprays, vol 8, pp 453–469

[56]   Huh KY, Gosman AD (1991) A Phenomenological Model of Diesel Spray Atomization. Proc Int Conf on Multiphase Flows '91, Tsukuba, Japan

[57]   Hung CC, Martin JK (1997) Collisional Behavior of Hydrocarbon Droplets. ILASS Americas Conference

[58]   Hwang SS, Liu Z, Reitz RD (1996) Breakup Mechanisms and Drag Coefficients of High-Speed Vaporizing Liquid Drops. Atomization and Sprays, vol 6, pp 353–376

[59]   Ibrahim EA, Yang HQ, Przekwas AJ (1993) Modeling of Spray Droplets Deformation and Breakup. AIAA J Propulsion and Power, vol 9, no 4, pp 652–654

[60]   Jin JD, Borman GL (1985) A Model for Multicomponent Droplet Vaporization at High Ambient Pressures. SAE paper 850264

[61]   Kawano D(2004) Numerical Study on Flash-Boiling Spray of Multicomponent Fuel. FISITA 2004 World Automotive Congress, 23–27 May 2004, Barcelona, Spain

[62]   Kim YK, Iwai N, Suto H, Tsuruga T (1980) Improvement of Alcohol Engine Performance by Flash Boiling Injection. JSAE Review, no 2, pp 81–86

[63]   Klingsporn M (1995) Modellierung der Mehrkomponenten-Verdunstung bei der dieselmotorischen Einspritzung. VDI-Verlag

[64]   Knapp RT, Daily JW, Hammitt FG (1970) Cavitation. McGraw-Hill International Book Company

[65]   Kneer R (1993) Grundlegende Untersuchungen zur Sprühstrahlausbreitung in hochbelasteten Brennräumen: Tropfenverdunstung und Sprühstrahlcharakterisierung. Dissertation, Universität Karlsruhe

[66]   Kolpakov AV, Romanov KV, Titova EI (1985) Calculation of the Rebound Condition for Colliding Drops of Sharply Different Sizes. Kolloidn Zn, vol 47, p 953

[67]   Kong SC, Han Z, Reitz RD (1995) The Development and Application of a Diesel Ignition and Combustion Model for Multidimensional Engine Simulation. SAE paper 950278

[68]   Kong SC, Reitz RD (1993) Multidimensional Modeling of Diesel Ignition and Combustion Using a Multistep Kinetics Model. ASME paper 93-ICE-22, also Journal of Engineering for Gas Turbines and Power, vol 115, pp 781–789

[69]   Krzeczkowski SA (1980) Measurement of Liquid Droplet Disintegration Mechanisms. Int J Multiphase Flow, vol 6, pp 227–239

[70]   Kuensberg Sarre C, Kong SC, Reitz RD (1999) Modeling the Effects of Injector Nozzle Geometry on Diesel Sprays. SAE-paper 1999-01-0912

[71]   Launder BE, Spalding DB (1974) The Numerical Computation of Turbulent Flows. Comput Meth Appl Mech Engng, vol 3, pp 269–289

[72]   Lee SH, Ryou HS (2001) Development of a New Model and Heat Transfer Analysis of Impinging Diesel Sprays on a Wall. Atomization and Sprays, vol 11, pp 85–105

[73]   Lefebvre HA (1989) Atomization and Sprays, Combustion: An International Series, by Hemisphere Corporation

[74]   Levy N, Amara S, Champoussin JC (1998) Simulation of a Diesel Jet Assumed Fully Atomised at the Nozzle Exit. SAE paper 981067

[75]   Li R, Boulos M (1993) Modeling of Unsteady Flow Around Accelerating Sphere at Moderate Reynolds Numbers. Canadian J Chem Eng, vol 71, pp 837–843

[76]   Lichtarowicz A, Duggins RK, Markland E (1965) Discharge Coefficients for Incompressible Non-Cavitating Flow Through Long Orifices. J of Mechanical Eng Science, vol 7, no 2, pp 210–219

[77]   Lippert AM (1999) Modeling of Multicomponent Fuels with Application to Sprays and Simulation of Diesel Engine Cold Start. Ph D Thesis, University of Wisconsin-Madison

[78]   Lippert AM, Reitz RD (1997) Modeling of Multicomponent Fuels Using Continuous Distributions with Application to Droplet Evaporation and Sprays. SAE paper 972882

[79]   Liu AB, Mather D, Reitz RD (1993) Modelling the Effects of Drop Drag and Breakup on Fuel Sprays. SAE paper 930072

[80]   Liu AB, Reitz RD (1992) Mechanisms of Air-Assisted Liquid Atomization. Atomization and Sprays, vol 3, pp 1–21

[81]   Liu Z, Obokata T, Reitz RD (1997) Modelling Drop Drag Effects on Fuel Spray Impingement in Direct Injection Diesel Engines. SAE paper 970879

[82]   Long WQ, Hosoya H, Mashimo T, Kobayashi K, Obokata T, Durst F, Xu TH (1994) Analytical Functions to Match Size Distributions in Diesel Sprays. COMODIA 94

[83]   Maly R, Vogel M (1978) Initiation and Propagation of Flames in Lean CH_4-Air Mixtures by the Three Modes of the Ignition Spark. 17th Symp (Int) Combust, pp 821–831, The Combustion Institute, Pittsburgh, PA

[84]   Martinelli L, Bracco FV, Reitz RD (1984) Comparisons of Computed and Measured Dense Spray Jets. Progress in Astronautics and Aeronautics, vol 95, pp 484–512

[85]   Matsumoto S, Saito S (1970) On the Mechanism of Suspension of Particles in Horizontal Conveying: Monte Carlo Simulation Based on the Irregular Bouncing Model. J Chem Engng Japan, vol 3, pp 83–92

[86]   Meingast U, Reichelt L, Renz U, Müller D Heine B (2000) Nozzle Exit Velocity Measurements at a Multi-Orifice CR-Nozzle. ILASS-Europe 2000

[87]   Metghalchi M, Keck J (1982) Burning Velocities of Mixtures of Air with Methanol, Isooctane, and Indolene at High Pressure and Temperature. Combustion and Flame, vol 48, pp 191–210

[88]   Modarress D, Wuerer J, Elghobashi S (1984) An Experimental Study of a Turbulent Round Two-Phase Jet. Chem Eng Commun, vol 28, pp 341–354

[89]   Mundo C, Sommerfeld M, Tropea C (1995) Droplet-Wall Collisions: Experimental Studies of the Deformation and Breakup Process. Int J Multiphase flow 21, pp 151–173

[90]   Naber JD, Reitz RD (1988) Modeling Engine Spray/Wall Impingement. SAE paper 880107

[91]   Nishimura A., Assanis DN (2000) A Model for Primary Diesel Fuel Atomization Based on Cavitation Bubble Collapse Energy. 8th Int Conf on Liquid Atomization and Spray Systems (ICLASS), pp 1249–1256

[92]   Nurick WH (1976) Orifice Cavitation and its Effect on Spray Mixing. Journal of Fluids Engineering, 98, pp. 681–687

[93]   O'Rourke PJ (1981) Collective Drop Effects on Vaporizing Sprays. Ph D Thesis, Department of Mechanical and Aerospace Engineering, Princeton University

[94]   O'Rourke PJ, Amsden AA (1996) A Particle Numerical Model for Wall Film Dynamics in Port-Injected Engines. SAE paper 961961

[95]   O'Rourke PJ, Amsden AA (2000) A Spray/Wall Interaction Submodel for the KIVA-3 Wall Film Model. SAE paper 2000-01-0271

[96]   O'Rourke PJ, Amsden AA (1987) The TAB Method for Numerical Calculation of Spray Droplet Breakup. SAE-paper 872089

[97]   O'Rourke PJ, Bracco FV (1980) Modelling of Drop Interactions in Thick Sprays and a Comparison with Experiments. Proc I Mech E, vol 9, pp 101–116

[98]   Otto F (2001) Fluid Mechanical Simulation of Combustion Processes. Class Notes, University of Hanover, Germany

[99]   Oza RD, Sinnamon JF (1983) An Experimental and Analytical Study of Flash-Boiling Fuel Injection. SAE paper 830590

[100] Pagel S (2003) Verdampfungs- und Selbstzündungsmodelle für Mehrkomponentengemische. VDI-Verlag, Reihe 12, Nr. 565

[101] Pagel S, Stiesch G, Merker GP (2002) Modeling the Evaporation of a Multicomponent Fuel. Proc 12th Int Heat Transfer Conf, pp 899–904, Grenoble, France

[102] Papoulis A (1984) Probability, Random Variables and Stochastic Processes. McGraw-Hill, New York, 2nd edition

[103] Park BS, Lee SY (1994) An Experimental Investigation of the Flash Atomization Mechanism. Atomization and Sprays, vol 4, pp 159–179

[104] Park SW, Kim HJ, Lee CS (2002) An Experimental and Numerical Study on Atomization Characteristics of Gasoline Injector for Direct Injection Engine. ILASS Americas, 15th Annual Conference on Liquid Atomization and Spray Systems, pp 47–51

[105] Park SW, Kim HJ, Lee CS (2003) Investigation of Atomization Characteristics and Prediction Accuracy of Hybrid Models for High-Speed Diesel Fuel Sprays. SAE paper 2003-01-1045

[106] Park SW, Lee CS (2003) Macroscopic Structure and Atomization Characteristics of High-Speed Diesel Spray. International Journal of Automotive Technology, vol 4, pp 157–164

[107] Patterson MA, Reitz RD (1998) Modelling the Effect of Fuel Spray Characteristics on Diesel Engine Combustion and Emission. SAE-Paper 980131

[108] Prosperetti A, Lezzi A (1986) Bubble Dynamics in a Compressible Liquid, Part 1. First-Order Theory. J Fluid Mech, vol 168, pp 457–478

[109] Qian J, Law K (1997) Regimes of Coalescence and Separation in Droplet Collision. J Fluid Mech, vol 331, pp 59–80

[110] Ra Y, Reitz, RD (2003) The Application of a Multicomponent Droplet Vaporization Model to Gasoline Direct Injection Engines. Int J Engine Res, vol 4, no 3, pp 193–218

[111] Ranger AA, Nicholls JA (1969) Aerodynamic Shattering of Liquid Drops. AIAA J vol 7, no 2, p 285

[112] Ranz WE, Marshall WR (1952) Internal Combustion Engine Modeling. Hemisphere Publishing, New York

[113] Reid RC, Prausnitz JM, Poling BE (1986) The Properties of Gases and Liquids. McGraw-Hill Book Company

[114] Reitz RD (1987) Modeling Atomization Processes in High-Pressure Vaporizing Sprays. Atomization and Spray Technology 3, pp 309–337

[115] Reitz RD (1990) A Photographic Study of Flash-Boiling Atomization. Aerosol Science and Technology, vol 12, pp 561–569

[116] Reitz RD (1991) Assessment of Wall Heat Transfer Models for Premixed-Charge Engine Combustion Computations. SAE paper 910267

[117] Reitz RD, Bracco FV (1986) Mechanisms of Breakup of Round Liquid Jets. In Encyclopedia of Fluid Mechanics, Gulf Pub, NJ, 3, pp 233–249

[118] Reitz RD, Diwakar R (1987) Structure of High-Pressure Fuel Sprays. SAE-paper 870598

[119] Ricart LM, Xin J, Bower GR, Reitz RD (1997) In-Cylinder Measurement and Modeling of Liquid Spray Penetration in a Heavy-Duty Diesel Engine. SAE-paper 971591

[120] Rose JW, Cooper JR (1977) Technical Data on Fuel. John Wiley and Sons, New York

[121] Rosseel E, Sierens R (1995) Modeling the Vaporization of a Gasoil Droplet in a Diesel Engine. 21st CIMAC

[122] Rotondi R, Bella G, Grimaldi G, Postrioti L (2001) Atomization of High-Pressure Diesel Spray: Experimental Validation of a New Breakup Model. SAE paper 2001-01-1070

[123] Saito A, Kawamura K, Watnabe S, Takahashi T, Naoyuki T (1993) Analysis of Impinging Spray Characteristics Under High-Pressure Fuel Injection (1st Report, Measurement of Impinging Spray Characteristics), Trans Jap Soc Mech Engrs B, vol 59, pp 3290–3295

[124] Sazhina EM, Sazhin SS, Heikal MR, Babushok VI, Johns RJR (2000) A Detailed Modelling of the Spray Ignition Process in Diesel Engines. Combust Sci and Tech, vol 160, pp 317–344

[125] Schmidt DP, Nouar I, Senecal PK, Rutland CJ, Martin JK, Reitz RD, Hoffman JA (1999) Pressure-Swirl Atomization in the Near Field. SAE paper 1999-01-0496

[126] Schmitz I, Ipp W, Leipertz A (2002) Flash Boiling Effects on the Development of Gasoline Direct-Injection Engine Sprays. SAE paper 2002-01-2661

[127] Schugger C, Meingast U, Renz U (2000) Time-Resolved Velocity Measurements in the Primary Breakup Zone of a High Pressure Diesel Injection Nozzle. ILASS-Europe 2000

[128] Sellens RW, Brzustowski TA (1985) A Prediction of the Drop Size Distribution in a Spray from First Principles. Atomisation and Spray Technology 1, pp 89–102

[129] Senecal PK, Schmidt DP, Nouar I, Rutland CJ, Reitz RD, Corradini ML (1999) Modeling High-Speed Viscous Liquid Sheet Atomization. Int Journal of Multiphase Flow 25, pp 1073–1097

[130] Shannon CE, Weaver W (1949) The Mathematical Theory of Communication. University of Illinois Press, Urbana

[131] Simmons HC (1977) The Correlation of Drop Size Distributions in Fuel Nozzle Sprays, Part II: The Drop Size/Number Distribution. ASME Trans, Journal of Engineering for Power, vol 99, pp 315–319

[132] Stanton DW, Rutland CJ (1996) Modelling Fuel Film Formation and Wall Interaction in Diesel Engines. SAE paper 960628

[133] Stanton DW, Rutland CJ (1998) Multi-Dimensional Modeling of Heat and Mass Transfer of Fuel Films Resulting from Impinging Sprays. SAE paper 980132

[134] Stiesch G (2003) Modelling Engine Spray and Combustion Processes. Springer-Verlag Berlin Heidelberg New York, ISBN 3-540-00682-6

[135] Stiesch G, Merker GP, Tan Z, Reitz RD (2001) Modeling the Effect of Split Injections on DISI Engine Performance. SAE paper 2001-01-0965

[136] Su TF, Patterson M, Reitz RD (1996) Experimental and Numerical Studies of High Pressure Multiple Injection Sprays. SAE-paper 960861

[137] Tamin J, Hallett WLH (1995) A Continuous Thermodynamics Model for Mulitcomponent Droplet Vaporization. Chem Eng Sci, vol 50, no 18, pp 2933–2942

[138] Tanner FX (1997) Liquid Jet Atomization and Droplet Breakup Modeling of Non-Evaporating Diesel Fuel Sprays. SAE paper 970050

[139] Taylor GI (1963) The Instability of Liquid Surfaces when Accelerated in a Direction Perpendicular to their Planes. In Batchelor GK (1963) The Scientific Papers of Sir Geoffery Ingram Taylor. vol 3, pp 532–536, Cambridge University Press

[140] Tennison PJ, Georjon TL, Farrell PV, Reitz RD (1998) An Experimental and Numerical Study of Sprays from a Common Rail Injection System for Use in an HSDI Diesel Engine. SAE paper 980810

[141] Trilling L (1952) The Collapse and Rebound of a Gas Bubble. J Appl Phys, 23, pp 14–17

[142] Turns SR (1996) An Introduction to Combustion – Concepts and Applications. McGraw-Hill Book Company, pp 80–87

[143] von Berg E, Alajbegovic A., Greif D, Poredos A, Tatschl R, Winkelhofer E, Ganippa LC (2002) Primary Break-Up Model for Diesel Jets Based on Locally Resolved Flow Field in the Injection Hole. ILASS-Europe 2002

[144] Wachters LHJ, Westerling NAJ (1966) The Heat Transfer from a Hot Wall to Impinging Water Drops in the Spheroidal State. Chemical Engineering Science, vol 21, pp 1047–1056

[145] Walther J, Schaller JK, Wirth R, Tropea C (2000) Characterization of Cavitating Flow Fields in Transparent Diesel Injection Nozzles using Fluorescent Particle Image Velocimetry (FPIV). ILASS-Europe 2000

[146] Wu KJ, Steinberger RL, Bracco FV (1981) On the Mechanism of Breakup of Highly Superheated Liquid Jets. The Combustion Institute Central States Section, 1981 Spring Meeting, Warren, Michigan, March 1981

[147] Yarin AL, Weiss DA (1995) Impact of Drops on Solid Surfaces: Self-Similar Capillary Waves, and Splashing as a New Type of Kinematic Discontinuity. J Fluid Mech 283, pp 141–173

[148] Zeng Y, Lee CFF (2001) An Atomization Model for Flash Boiling Sprays. Combust Sci and Tech, vol 169, pp 45–67

[149] Zhu GS, Reitz RD, Xin J, Takabayashi T (2000) Characteristics of Vaporizing Continuous Multi-Component Fuel Sprays in a Port Injection Gasoline Engine. SAE paper 2001-01-1231

[150] Zuo B, Gomes AM, Rutland CJ (2000) Modelling Superheated Fuel Sprays and Vaporization. Int J Engine Research, vol 1, no 4, pp 321–336

5 Grid Dependencies

5.1 General Problem

Since the use of Computational Fluid Dynamics for spray simulations, it is well known that the results are relatively sensitive to the resolution of the numerical grid. It is agreed that adequate resolution is required to reproduce the structure of sprays with sufficient accuracy, but there are often different opinions about what this resolution should be. In this chapter, the reasons for the sensitivity of spray sub-processes like droplet-gas and gas-droplet momentum transfer, evaporation, collision etc. to the grid size as well as possibilities to minimize these mesh dependencies are discussed.

First, only continuous fluids shall be regarded. Continuous fluids like the gas phase in a combustion chamber are described using the Eulerian approach (Sect. 3.1). It is known that the numerical solution of the Navier-Stokes equations becomes more and more accurate if the grid is refined. The reason for this effect originates in the way the differential equations are numerically treated: the differential equations are discretized, which means that the differential quotients are approximated by difference quotients. The differences are expressed using the values of the flow quantities at the nodes of the computational grid. Hence, the finer the grid, the more accurate the differentials can be approximated, and the better strong spatial gradients of flow quantities are resolved. In order to achieve accurate results, the grid resolution should at least be fine enough to resolve the physical scales of the problem [1]. In the case of a gas jet being injected into a gas atmosphere (both phases are continua and described by the Eulerian approach) this implies that near the nozzle the jet cross-sectional area has to be resolved by at least four grid cells. Abraham [1] has shown that if such a resolution is not used, the structure of the jet cannot be reproduced with adequate accuracy and is significantly influenced by the selection of ambient turbulent length and time scales. The prediction of mixing and entrainment rates as well as jet penetration may then be erroneous. The largest errors are caused by an inadequate mesh size near the nozzle, where the velocity and species density gradients are strong and can only be resolved by extremely fine meshes.

However, the maximum grid resolution that is practically feasible is limited by the available computer power and time. For this reason, grids smaller than the orifice size are impractical, and usually much coarser ones are used.

Johnson et al. [8] have proposed a simple and very effective method to reduce the effect of grid resolution on the computation of gas jets. As already shown in the work of Abraham [1], it was found that the calculation results are most sensitive to the mesh size near the nozzle and the treatment of turbulence. Based on the observation that on a coarse mesh unrealistic turbulent length scales and thus large diffusion are generated, Johnson et al. [8] modified the k-ε model in order to suppress unphysical effects by a limitation of the largest length scales in the region of the gas jet. Because the largest scale inside a jet is the penetration, the length scale is simply limited to a maximum value equal to the distance away from the orifice within a given cone along the spray axis. This method implies that the jet cone angle is known a priori. Although this method is somewhat arbitrary, it has turned out to improve the predictive quality of the computations significantly, and it was found to apply universally to different gaseous jets. The application of this length scale limiter can also be extended to the calculation of sprays in order to improve the prediction of the gas phase behavior inside the spray cone angle.

For sprays consisting of liquid droplets which penetrate into a gas atmosphere, the situation becomes much more complex. The most popular way to simulate sprays is the Eulerian-Lagrangian approach (Sect. 3.2). Due to the inadequate space resolution of the gas phase flow quantities (e.g. velocity, temperature and vapor concentration), the Eulerian field is not properly computed in the vicinity of the liquid phase, resulting in an over-estimation of diffusion and thus in inaccuracies in the modeling of fuel-air mixing. Although it would also be impractical, it is (in contrast to the pure Eulerian approach) not possible to reduce the grid size until the gas phase is properly resolved and mesh-independent results are obtained, because this would violate a basic requirement for the Lagrangian liquid phase description. This description is based on the assumption that the void fraction within a cell is close to one. Hence, the volume of liquid drops inside a cell must be small compared to the cell volume. If very fine meshes, which resolve the flow field of the gas phase near the orifice, are used, this assumption is no longer justified.

For this reason, the gas flow can never be resolved accurately near the nozzle, and the exact flow quantities at the position of a liquid drop, which are needed in order to predict mass, momentum, and energy transfer between gas and liquid, are unknown.

Usually it is assumed that the required values are those of the nearest node. Fig. 5.1 shows the two-dimensional example of a gas cell and its four nodes. The dashed lines indicate a kind of staggered grid, and the cells of this dotted grid represent the volumes governed by the different nodes. For example, the velocity at node 4 is valid inside the whole volume of the dotted grid cell containing node 4. Velocity gradients inside this volume are not resolved. The coarser the grid, the worse the spatial resolution of the gas velocity, and the worse the prediction of gas-droplet momentum exchange, which depends on the relative velocity at the drop location. Hence, if the nearest node approximation is applied, the prediction of gas-droplet interactions suffers from the low spatial resolution of the gas phase.

In just the same manner, the momentum, which is transferred from the liquid parcel to the gas (source term in the Navier-Stokes equations), is added to the nearest node. This momentum gain is then immediately and uniformly distributed over the complete cell (dashed grid) governed by the node. This way of modeling the liquid-gas interactions gives rise to two unphysical effects on the grid size.

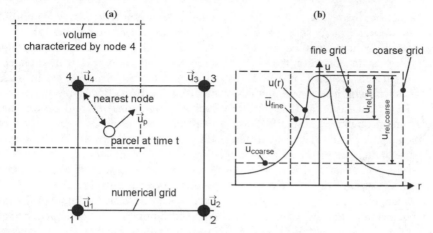

Fig. 5.1. Momentum exchange between liquid and gas phase

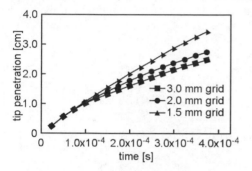

Fig. 5.2. Effect of grid resolution on spray penetration [5]

First, the immediate and uniform distribution of the momentum gain over the complete cell volume leads to an unphysically fast diffusion of momentum. The lower the grid resolution, the faster this diffusion. In reality, only the gas in the vicinity of the drop is be accelerated, and the momentum is transferred to a much smaller area of influence.

Second, if the cell volume is large, the change of gas velocity due to momentum exchange with liquid drops is small, and if the grid resolution is high, the gas velocity increases faster. Hence, the absolute value of the gas velocity after the interaction with a liquid drop depends on the cell size. Fig. 5.1b highlights the effect of this treatment. In the case of small grid sizes, the droplets being injected at early times transfer their momentum to small gas volumes and cause a fast increase of gas velocity. Due to the smaller relative velocities, the next droplets are less decelerated, resulting in an increase of spray penetration. For this reason, spray penetration increases if the grid is refined, and decreases if a coarser grid is used, Fig. 5.2.

Further on, the orientation of the grid relative to the spray axis may also influence the results [5]. As shown in Fig. 5.3, the number of nearest nodes (and thus the total gas volume) that are included in the direct exchange of mass, momentum, and energy is dependent on the angle between spray axis and grid. If the momentum transferred to the gas is distributed to a larger number of nodes, the increase of gas velocity in the spray region is slower. If there are less nodes, the gas velocity will increase faster, and the relative velocity between droplets and gas will decrease more quickly. These effects may result in different penetration lengths of identical sprays with orientation different only relative to the mesh. One possibility of reducing the effect of grid orientation is the use of spray adapted grids, where the grid arrangement is adjusted to the spray direction, such that the main spray direction is always more or less perpendicular to the cells, Fig. 5.4.

As mentioned already, the momentum exchange is only one example of a grid-dependent interaction between gas and dispersed liquid. The same dependencies are noticed in the case of droplet evaporation, where mass and heat are exchanged: the rate of mass loss of a drop is strongly influenced by the resolution of the temperature and vapor concentration fields inside the gas phase. The exchange of mass and heat with only the nearest node leads to an unrealistic transfer of the

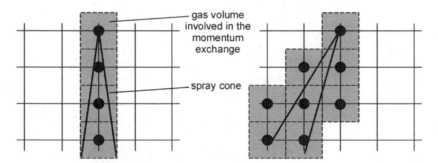

Fig. 5.3. Effect of grid orientation on spray penetration

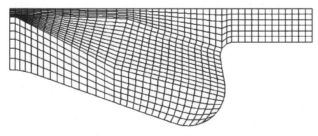

Fig. 5.4. Spray adapted grid for a DI diesel engine

source terms to the gas phase, and the distribution of the source terms over the complete cell volume results in an unphysically fast diffusion of fuel vapor, especially near the nozzle where the cells size is much larger than the jet diameter.

Altogether, three sources of mesh dependencies can be specified: (1) the calculation of representative gas phase flow quantities at the exact location of the drop, (2) the distribution of mass, momentum, and energy to the nodes of gas phase, and (3) the unrealistically fast diffusion due to immediate averaging of the source terms over the whole cell volume.

A fourth source of uncertainty is the problem of insufficient statistical convergence in the spray region close to the nozzle. Because it is impractical to follow each individual drop inside a spray, the combination of Monte-Carlo method and stochastic parcel technique (Sect. 3.2) is used in order to reduce the number of individual drops the behavior of which has to be directly calculated. However, the more parcels are used, the better the behavior of the dispersed liquid phase is resolved, and the better the statistical convergence. Due to break-up processes and droplet deceleration, the number density of parcels, and thus the statistical convergence, increases considerably with distance from the nozzle. Usually, computer power limits the maximum number of parcels to a value that does not allow for the realization of sufficient resolution of the liquid phase near the nozzle. However, in the remaining spray region, where most of the mixture formation processes take place, an adequate resolution of the liquid phase is achieved.

Now the reader might believe that the Eulerian-Lagrangian spay calculations are always extremely erroneous and misleading. This is not true. Strong grid dependencies appear only if the range of recommended grid resolution is left, i.e. if the computational mesh is too coarse or too fine. Usually, the model constants of each of the various sub-models in these codes have been adjusted in order to ensure maximum agreement between simulation (using the recommended grid resolution) and a large number of experimental results. However, if unusual grid spacing is used, or if extremely precise predictions are necessary, mesh dependency may become a serious source of uncertainty.

The problem of grid dependency will become more and more important in the near future. The more precise the modeling of the various physical and chemical sub-processes, the larger the relative uncertainty due to numerical errors. Altogether, this will result in a growing importance of research activities in this field. Successful approaches to reduce the influence of the numerical grid are discussed in the following sections.

Besides numerical errors due to the standard inter-phase coupling, the collision algorithm is a further source of grid dependency [14, 15]. The standard collision model in CFD-codes today is the one proposed by O'Rourke in 1981 [12], Sect. 4.7. This algorithm suffers from large spatial discretization errors because the cells of the gas phase mesh are used to calculate the collision probability. The collision probability depends on number density, and the strong spatial gradients in number density inside the dense spray, where most of the collision events occur, can only poorly be resolved by these cells. This topic is further discussed in Sect. 5.2.

5.2 Improved Inter-Phase Coupling

In order to improve inter-phase coupling, the calculation of representative gas phase flow quantities at the exact location of the drop (gas-drop interaction), the distribution of source terms to the nodes of the gas phase (drop-gas interaction), and the unrealistically fast diffusion due to immediate averaging of the source terms over the whole cell volume have to be addressed.

In order to replace the nearest node approach and to provide each parcel with accurate information about the gas phase boundary conditions at its exact location, Schmidt and Senecal [15] tested a simple interpolation scheme [11] that calculates a weighted average of the gas phase flow quantities from the nearby nodes, Fig. 5.5. For example, if the gas velocity $\vec{u}$ at the parcel location $\vec{x}_p$ is the quantity of interest, the velocities $\vec{u}_i$ at each of the eight corner nodes (location: $\vec{x}_i$) of the gas phase cell containing the parcel are used and simply weighted by their distance to the parcel location:

$$\vec{u} = \frac{\sum_{i=1}^{8} \vec{u}_i \left| \vec{x}_i - \vec{x}_p \right|^{-m}}{\sum_{i=1}^{8} \left| \vec{x}_i - \vec{x}_p \right|^{-m}} . \tag{5.1}$$

In the work of Schmidt and Senecal [15], the value of $m = 2.0$ was chosen. The droplet-gas momentum exchange can be modeled using the same scheme, but just the other way round. However, the authors did not obtain superior results with this simple averaging technique. Bella et al. [5] used a modified but similar approach. The amount of momentum exchange is also proportional to the inverse of the droplet-node distance, but the weighting function is given as

$$\vec{u} = \frac{\sum_{i=1}^{n} (\sum_{j \neq i=1}^{n} \left| \vec{x}_j - \vec{x}_p \right|^m) \cdot \vec{u}_i}{\sum_{i=1}^{n} (\sum_{j \neq i=1}^{n} \left| \vec{x}_j - \vec{x}_p \right|^m)} , \tag{5.2}$$

where $m = 1.0$. In contrast to the approach of Schmidt and Senecal [15], not only the vertexes of the cell containing the parcel but also the surrounding cells ($n = 27$ in a three-dimensional simulation) were included. The results showed that the variation of tip penetration due to grid effects could be highly reduced, but not completely eliminated, Fig. 5.6. The remaining influence of grid resolution may be explained by the fact that the total volume of the 27 cells included in the weighting process is still dependent on mesh size.

Based on the fact that a polar coordinate system is much more suited to resolve a spray which is usually also polar, Schmidt and Senecal [15] developed a tech-

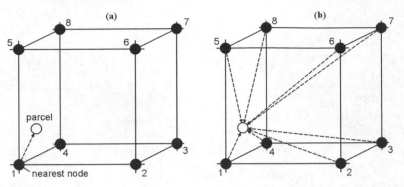

Fig. 5.5. Gas-droplet momentum exchange: **a** standard method, **b** enhanced approach

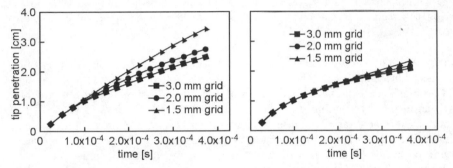

Fig. 5.6. Reduction of grid dependency due to an advanced modeling of momentum exchange between gas and liquid [5]

nique which is specially suited to predict accurate boundary conditions for the gas-droplet coupling in the radial direction (perpendicular to the spray axis). This algorithm establishes an additional polar coordinate system (aligned with the injection), transforms the velocity components at each of the eight neighbor nodes to the corresponding polar components, calculates a weighted average, and transforms the weighted average back to the original coordinate system. This technique is especially suitable to avoid ninety-degree artifacts in Cartesian meshes that may appear due to the poor azimuthal resolution, Fig. 5.7. However, it does not contribute much to the reduction of grid dependency in axial spray direction.

Beard et al. [3] also attempted to avoid grid dependency, and they developed a sub-grid scale model, which is specially suited to reduce the over-prediction of exchange rates between gas and liquid due to the effect of numerical diffusion. The numerical diffusion of vapor and momentum is limited by retaining both quantities along the parcel trajectory as long as the mesh is insufficient to resolve the steep gradients. Then, mass and momentum are gradually released and transferred to the nodes. Following the idea of Gonzalez et al. [7], the mass that leaves a drop due to evaporation is not immediately released but is collected in a gaseous

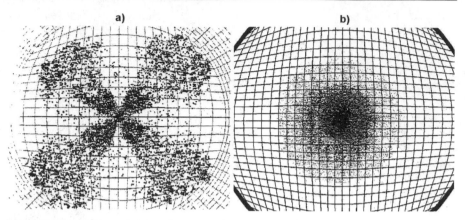

Fig. 5.7. Elimination of ninety-degree artifacts in Cartesian meshes. Both pictures show a hollow-cone spray directed towards the viewer [15]: **a** standard approach, **b** enhanced method

sphere with radius proportional to $(D{\cdot}t)^{0.5}$ around the drop (D: vapor diffusivity in the gas, t: droplet lifetime), Fig. 5.8. This gaseous sphere directly follows the liquid drop as long as it exists. It contains only vapor, and its temperature is equal to the one of the surrounding gas. The gas sphere expands because of vapor diffusion, and its mass varies because of evaporation and transfer of mass to the gas phase. The vapor concentration inside the gaseous sphere is used as a boundary condition for the prediction of further evaporation, resulting in accurate and mesh-independent local vapor concentrations around the drop. Remember that in the standard case the vapor is usually distributed over the whole cell volume, which varies with grid resolution. The exchange of mass with the surrounding gas phase starts if the gaseous particle is seen by the grid, which is the case if the radius r of the sphere is larger than the mean distance r_m of the sphere center from the eight cell nodes. Then, vapor mass is released until $r = r_m$ again, Fig. 5.8.

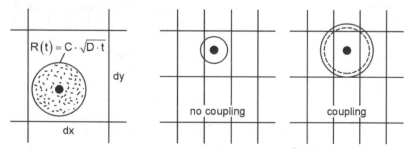

Fig. 5.8. Limitation of numerical diffusion: sub-grid scale vapor and momentum diffusion model [7, 3]

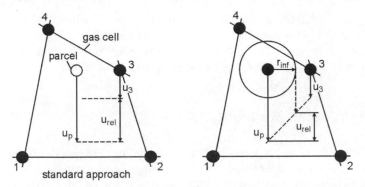

Fig. 5.9. Enhanced gas-drop momentum exchange: calculation of a representative gas velocity [3]

A similar treatment is applied to the drop-gas momentum transfer. Again, a sphere of influence with increasing radius is used, and this sphere releases momentum to the gas cell as soon as its radius reaches a critical size. However, the momentum is only transferred to the nearest node and not distributed to all neighbor nodes using a weight function. The gas phase velocity at the drop location, which is needed as boundary condition for the calculation of relative velocity and drag force, is obtained by a simple linear interpolation using the drop velocity (located at the drop center) and the gas velocity at the nearest node, Fig. 5.9. The representative gas velocity is the one corresponding to the radius of influence. Altogether, the approach of Beard et al. [3] results in an improved prediction of liquid and vapor penetration and in a reduction of mesh dependency. However, the influence of grid resolution cannot be completely suppressed. The critical value of the radius of influence still depends on grid resolution. Further on, the application of a weighting technique for the inter-phase mass, momentum, and energy transport could certainly lead to a further reduction of grid dependency.

A special case of grid dependency, which cannot be eliminated by an advanced treatment of inter-phase coupling, occurs directly at the walls, where velocity and thermal boundary layers cannot be resolved by the gas phase grid and have to be described by so-called wall functions, Sect. 3.1.4.3. The use of these wall functions can only produce reliable results if the dimensionless distance y^+ of the first grid points from the wall is smaller than 200 (optimum values: $y^+ = 30...120$). Usually the grid cells at the wall are too coarse, and the range of applicability of the logarithmic velocity and temperature profiles is exceeded. A simple approach would be to refine the mesh near the walls until the desired grid spacing is obtained. However, due to the strongly changing flow quantities, the boundary layer thickness is highly transient, and the sizes of all wall cells must be adapted dynamically to the required values. This approach has been realized by Lettmann et al. [10]. At the beginning of each computational time step, the normal distance of the first grid points from the wall is readjusted to $y^+ = 80$, and then the grid points next to the wall cells are moved in accordance with a grading factor in order to reduce gradually the cell size when approaching the wall. This treatment minimizes numerical convergence problems. Although this method requires considerable re-adjustment of the complete grid after each computational time step, the approach produces promising results and offers the opportunity to eliminate the grid de-pendency of momentum and heat transfer from the gas phase to the walls.

5.3 Improved Collision Modeling

The standard collision algorithm in today's CFD codes is the one proposed by O'Rourke [12], Sect. 4.7. This algorithm suffers from large spatial discretization errors, because the cells of the gas phase mesh are used to calculate the collision probability. The collision probability depends on number density, and the strong spatial gradients in number density inside the dense spray, where most of the colli-sion events happen, can only poorly be resolved by the gas phase mesh. Hence, the collision calculation is grossly under-resolved, resulting in an under-prediction of collision incidence by the algorithm. On the other hand, the use of the Poisson distribution for multiple collisions greatly over-estimates the number of collisions between drops [15].

Schmidt et al. [14, 15] developed a new collision algorithm, the so-called No Time Counter (NTC) collision model, which includes a more advanced descrip-tion of collision events and is completely mesh independent. The NTC algorithm, which is a popular method of calculating intermolecular collisions, was extended by the authors for the use in sprays. In practical use, this algorithm is much faster than the standard collision algorithm. However, the most important improvement is the use of a separate collision grid, which is not used for the gas phase, and which does not have to conform to boundaries. This cylindrical mesh is created automatically around the spray axis and is extended far enough to include the complete spray. The mesh is fine enough to guarantee optimum spatial resolution

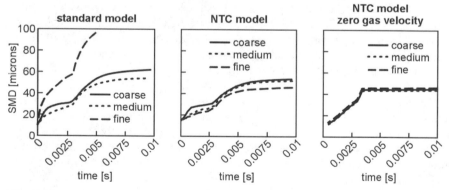

Fig. 5.10. Mesh dependency of standard and NTC collision model [15], diesel-type spray, Cartesian mesh

in order to provide perfect grid independence. As an example, Fig. 5.10 shows a comparison between average drop sizes for a diesel-type spray using the standard collision model of O'Rourke [12] and the NTC model of Schmidt et al. [15]. The calculations were performed on a Cartesian mesh. The strong grid dependency of the standard model is clearly visible. The use of the NTC algorithm results in a significant reduction of mesh dependency. The remaining mesh sensitivity is caused by the dependence of the coupling between spray and gas phase momentum on grid resolution: the droplet velocities, which determine the outcome of a collision event, are influenced by the gas phase mesh resolution. If the gas velocity is constrained to zero, the results are perfectly grid independent, Fig. 5.10.

5.4 Eulerian-Eulerian Approaches

As already discussed in Sect. 5.1, the Eulerian-Lagrangian description is based on the assumption that the void fraction inside all gas cells is close to one. For this reason it is not possible to increase grid resolution until the strong gradients of the flow quantities can be accurately resolved near the nozzle, even if the required computer power would be available. Appropriate sub-grid scale models have to be used. Furthermore, statistical convergence in the dense spray near the nozzle is usually not given.

In order to overcome these numerical problems in a more direct approach, some authors have developed alternative methods, in which the liquid phase in the dense spray region is modeled with a Eulerian approach instead of using the Lagrangian description.

This method was first introduced by Wan and Peters [17] and further optimized for the application to engine sprays by Krüger [9]. In this so-called Interactive Cross-sectionally Averaged Spray (ICAS) method, the transport equations for the multi-dimensional gas flow in the cylinder are solved by the CFD-code (Eulerian description), but the multi-dimensional Eulerian transport equations for both gas and liquid phases in the dense spray region are integrated over the cross-sectional area of the spray cone [13], resulting in a one-dimensional system of equations for the cross-sectionally averaged variables. This system is solved independently of the CFD-code. The integration of the multi-dimensional Eulerian transport equations of the two-phase flow is based on simple presumed profiles for velocity, mixture fraction and other relevant variables. The assumption made for the derivation of the integrated equations is that the cone angle and the length of the dense spray are constant and do not depend on injection velocity. The solution of these one-dimensional equations gives the source terms along the spray axis. In order to couple the dense spray region with the gas phase, the source terms have to be redistributed again in radial direction over the whole spray cone using presumed distribution functions.

Because the one-dimensional treatment of the spray has the disadvantage that the effect of multi-dimensional flow effects like swirl or tumble can hardly be included, the ICAS-method is only applied to the dense spray region where the spray itself dominates the mixture formation process. As soon as the dispersion of the liquid phase is high enough, the treatment of the liquid phase switches over to the well-known Lagrangian description, Fig. 5.11.

Wan and Peters [17] have proved that grid dependency in the dense spray region near the nozzle, which is by far the most important source of uncertainty due to grid effects, can be completely circumvented by employing the ICAS-model. However, this approach also has its shortcomings. The inclusion of relevant primary break-up mechanisms is more complex than in the Eulerian-Eulerian approach, and the spray cone angle must be known a priori and thus cannot be predicted. The most severe limitation however is due to the fact that the cross-sectionally averaging of the Eulerian transport equations as well as the subsequent redistribution of the one-dimensional source terms in the dense spray region depend on presumed distribution functions. These distributions should be a consequence of the primary break-up process and not be defined a priori.

Abraham and Magi [2] developed a similar approach called VLS (Virtual Liquid Source). In this model the liquid phase is treated as a liquid core, from which mass, momentum, and energy are released to the gas phase cells adjacent to the liquid core. Further Eulerian-Eulerian approaches for the description of the dense spray region have been published by Berg et al. [16], Blokkeel et al. [6], and Beck and Watkins [4] for example.

Altogether, it can be concluded that the most important advantage of the Eulerian-Eulerian over the Eulerian-Lagrangian approach is that the problem of statistical convergence can be completely eliminated. Because it is quite difficult to take into account the numerous phenomena occurring in engine sprays, the Eulerian approaches often do not contain as much physics concerning the relevant sub-processes as the Lagrangian ones. However, this way to describe engine sprays is just at the beginning of being explored.

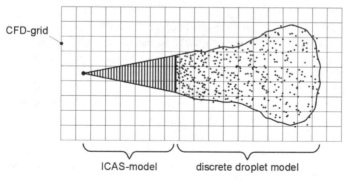

Fig. 5.11. Schematic illustration of the Interactive Cross-sectionally Averaged Spray (ICAS) method

References

[1] Abraham J (1997) What is Adequate Resolution in the Numerical Computations of Transient Jets? SAE paper 970051

[2] Abraham J, Magi V (1999) A Virtual Liquid Source (VLS) Model for Vaporizing Diesel Sprays. SAE paper 1999-01-0911

[3] Beard P, Duclos JM, Habchi C, Bruneaux G, Mokkadem K, Baritaud T (2000) Extension of Lagrangian-Eulerian Spray Modeling: Application to High Pressure Evaporating Diesel Sprays. SAE paper 2000-01-1893

[4] Beck JC, Watkins AP (2004) The Simulation of Fuel Sprays Using the Moments of the Drop Number Size Distribution. Int J Engine Res, vol 5, no 1, pp 1–21

[5] Bella G, Rotodini R, Corcione FE, Valentino G (1999) Experimental and Numerical Analysis of Diesel Spray. ICE

[6] Blokkeel G, Mura A, Demoulin FX, Borghi R (2003) A Continous Modeling Approach for Describing the Atomization Process from Inside the Injector to the Final Spray. ICLASS 2003

[7] Gonzalez MA, Lian ZW, Reitz RD (1992) Modeling Diesel Engine Spray Vaporization and Combustion. SAE paper 920579

[8] Johnson NL, Amsden AA, Naber JD, Reitz RD (1995) Three-Dimensional Computer Modeling of Hydrogen Injection and Combustion. High Performance Computing Symposium, The Society for Computer Simulation, Phoenix, Arizona, 9–13 April

[9] Krüger C (2001) Validierung eines 1D-Spraymodells zur Simulation der Gemischbildung in direkteinspritzenden Dieselmotoren. Ph.D. Thesis, University of Aachen, Germany

[10] Lettmann H, Eckert P, Baumgarten C, Merker GP (2004) Assessment of Three-dimensional In-Cylinder Heat Transfer Models in DI Diesel Engines. 4th European Thermal Sciences Conference, Birmingham

[11] Nordin N (2000) Complex Chemistry Modeling of Diesel Spray Combustion. Ph.D. Thesis, Chalmers University of Technology

[12] O'Rourke PJ (1981) Collective Drop Effects on Vaporizing Liquid Sprays. Ph.D. Thesis, Princeton University

[13] Peters N, Göttgens J (1991) Scaling of Buoyant Turbulent Jet Diffusion Flames. Combustion and Flame 85, pp 206–214

[14] Schmidt DP, Rutland CJ (2001) Reducing Grid Dependency in Droplet Collision Modeling. Proceedings of ASME: International Combustion Engine Division Chicago, Fall 2001

[15] Schmidt DP, Senecal PK (2002) Improving the Numerical Accuracy of Spray Simulations. SAE paper 2002-01-1113

[16] von Berg E, Edelbauer W, Alajbegovic A, Tatschl R (2003) Coupled Calculation of Cavitating Nozzle Flow, Primary Diesel Fuel Break-Up and Spray Formation with an Eulerian Multi-Fluid Model. ICLASS 2003

[17] Wan YP, Peters N (1997) Application of the Cross-Sectional Average Method to Calculations of the Dense Spray Region in a Diesel Engine. SAE paper 972866

6 Modern Concepts

6.1 Introduction

The internal combustion engine is by far the most important power train for all kind of vehicles, including on-road applications like passenger cars, motor-cycles, and trucks, and non-road applications like construction machines, agricultural vehicles, locomotives, and ships. Up to now there is no alternative to this kind of engine, and it is certain that the internal combustion engine will keep its leading position for at least the next three to five decades.

The only potential alternative, the fuel cell, is still in the initial stage of development. Due to the direct transformation of chemical energy into electrical energy it seems to be perfectly suited for electric power trains. The most important advantage is the low emission of pollutants. However, besides serious problems concerning lifetime, weight, and costs, this kind of energy source has to cope with the fuel problem, which is possibly the most important challenge. The fuel cell needs hydrogen, and because there is no satisfactory technique for the cost-effective mass production as well as the storage of hydrogen, it has to be produced on-board by a so-called reformer using ordinary fuels like gasoline and diesel, or special fuels like methane and natural gas. Especially the realization of a transient operation of such a reformer is a great challenge of its own. Furthermore, it is still in question if the combination of on-board reformer and fuel cell will ever be superior to the modern internal combustion engine regarding CO_2 emission and fuel consumption.

Altogether, the internal combustion engine will keep its leading position in the near future. However, it has to be continuously improved, and great efforts have to be made in order to increase efficiency and to fulfill future emission legislations on the one hand, while keeping the costs low on the other hand.

Besides the development of new exhaust gas after-treatment techniques as well as clean and tailored fuels, the reduction of engine raw emissions due to improved mixing formation and combustion concepts will be one of the key measures to keep the internal combustion engine up to date. Because mixing formation and combustion are mainly influenced by fuel injection, the requirements and potentialities of modern and future injection systems concerning their use for conventional as well as new combustion concepts shall be outlined in the following sections.

6.2 DI Diesel Engines

6.2.1 Conventional Diesel Combustion

Due to its high efficiency, the diesel engine has always been the favorite power train for heavy-duty applications, especially in trucks and all non-road applications. Regarding passenger car applications, the diesel engine suffered from disadvantages in noise, transient behavior (small low-end torque), and weight for a long time. During the last decade, the realization of high-pressure direct injection combined with modern turbo-charging techniques has completely revolutionized the diesel engine technology. High power output and high low-end torque combined with excellent fuel economy and a significant reduction of noise are the main reasons for the enormous rise in popularity of the passenger car diesel engine.

Regarding future demands, the modern diesel engine suffers from high soot (particulate matter, PM) and nitrogen oxide (NO_x) emissions, which have to be significantly reduced in order to fulfill emission legislation. Besides the use of exhaust gas after-treatment techniques like particulate filters and catalysts, one of the main measures will be to reduce the engine raw emissions. Significant improvements can be achieved using highly flexible high-pressure injection systems.

Before the potentials of modern and future injection systems and strategies are discussed in detail, the conventional diesel combustion as well as the formation of pollutants shall be briefly discussed.

In a direct injection diesel engine, the fuel is injected into the hot compressed air inside the combustion chamber. Due to the increasing gas temperature during compression, the combustion process is initiated by auto-ignition. The start of combustion is controlled by injection timing, and the desired engine torque is adjusted via the amount of fuel injected per cycle. Due to the short time available for mixture formation, the fuel-air mixture always consists of fuel-rich and lean regions, and thus it is strongly heterogeneous.

Fig. 6.1 shows the three phases of a conventional diesel combustion process, which will be discussed in detail in the following.

The first phase begins with the start of injection and ends after the so-called premixed combustion. The direct injection of fuel into the cylinder usually starts some degrees of crank angle before top dead center, the exact timing is depending on engine speed and load. The injection duration depends on the amount of fuel that has to be injected per cycle. As soon as the cold fuel jet starts to penetrate into the combustion chamber, it begins to mix with the hot compressed air. As penetration increases, more and more hot air is entrained, the droplets begin to vaporize, and a sheath of vaporized heated fuel-air mixture forms around the jet's periphery. When the temperatures inside this fuel-rich zone reach about 750 K, the first reactions resulting in a breakdown of high cetane fuel begin to occur. These reactions as well as the further entrainment of hot air and the additional compression of the cylinder charge increase the temperature and the rate of the reactions. The products of these basic reactions are largely C_2H_2, C_2H_4, and C_3H_3 fuel fragments as well as CO and H_2O [22]. Now temperature and reaction rate increase quickly, re-

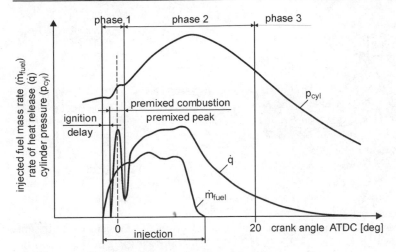

Fig. 6.1. Phases of the conventional diesel combustion process

sulting in a burning of the complete fuel-air mixture that has been formed during the ignition delay, which is the time between start of injection and ignition. This sudden combustion of well-prepared fuel-air mixture results in a strong and sudden increase of heat release and cylinder pressure, the so-called premixed peak. The strong pressure gradients ($dp/d\varphi$) cause considerable noise, and the high temperatures in the premixed zone are responsible for a first production of nitric oxides (NO_x). Due to the triple bond of N_2-molecules, temperatures of approx. 2000 K must be reached in order to split the molecules and to enable the formation of NO_x (Zeldovich mechanism, [47]). However, as soon as these temperatures are reached, the NO_x-production increases exponentially with temperature. The heat being released during the premixed combustion depends on the amount of fuel being injected and evaporated during the ignition delay.

The premixed combustion consumes all fuel-air mixture around the inner spray region. The temperatures of the inner spray region increases to about 1600–1700 K, and all available oxygen in this region is consumed by partial oxidation of the fuel fragments. From now on, air and partial burnt products, which diffuse from inside the spray to the outer regions, are burnt in a very thin reaction zone at the periphery of the spray, the so-called diffusion flame, Fig. 6.2. This kind of combustion, so-called diffusion burning, characterizes phases 2 and 3 in Fig. 6.1, and is limited by the mixing of partial burnt products and air, resulting in a slower reaction compared to the premixed one. During phase 2, the fuel drops being injected heat up due to the entrainment of hot air and combustion products into the jet near the nozzle. As the evaporated fuel further penetrates into the spray, it breaks down into small molecules, which are then subject to partial oxidation due to the lack of sufficient oxygen inside the hot spray cloud. The partial oxidation reactions and the heat transferred form the diffusion flame keep the temperature of the inner zone at values of about 1600–1700 K [22]. These high temperatures in combination with the low oxygen content are ideal conditions for the formation of

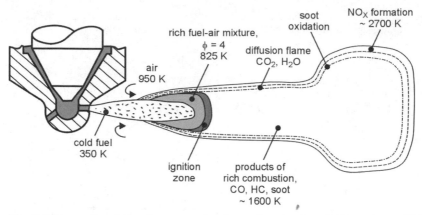

Fig. 6.2. Conventional diesel spray combustion and production of NO_x and soot (PM)

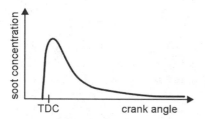

Fig. 6.3. Soot concentration as function of crank angle, schematic diagram

soot, and large amounts of soot are produced in the inner cloud. The products from inside this partial oxidation zone diffuse to the boundary and are consumed by the diffusion flame. While only 10–15% of the fuel energy is released in the partial burning zone [22], the diffusion flame releases the rest of the energy. The diffusion flame is fed with oxygen from the surrounding of the burning spray, and near stoichiometric equivalence ratios are reached at its outer zone. Due to the very high temperatures of about 2700 K, nearly all soot entering the diffusion flame is consumed. At the outer boundary of the hot diffusion flame, there is enough oxygen to produce large amounts of NO_x.

After the end of the injection process, the partial burning zone is no longer fed with fuel vapor, and it can now be consumed by the diffusion flame. The reduction of soot finally becomes stronger than the production, and the overall amount of soot inside the combustion chamber decreases. Fig. 6.3 shows the overall soot concentration inside a combustion chamber as a function of crank angle. Most of the large amount of soot produced at early crank angles is consumed again at later crank angles, and the remaining mass of soot, which is finally detected in the exhaust gases, is only a very small fraction of the initial one. At the end of injection, it is very important to achieve a complete closing of the injector, because unintended post-injections of low-speed large fuel drops, that remain near the nozzle and do not reach the region inside the diffusion flame, undergo partial oxidation and produce soot. This soot cannot be consumed by the diffusion flame any more and significantly increases the remaining mass of soot in the exhaust gases.

Fig. 6.1 shows that the diffusion burning can be divided into two sub-phases. In the mixing-controlled phase 2, which has already been described, the burning rate is only limited by mixing of fuel fragments and air. The velocity of the chemical reactions is much faster. In the subsequent phase 3 the final oxidation of the remaining unburned and partial oxidized fuel fragments as well as most of the soot particles takes place. However, due to the decrease of gas temperature during the expansion stroke and due to the strong decrease of oxygen, the chemical reactions become slower and are finally the limiting factor. Most of the diesel soot emissions are the result of quenching this final phase of oxidation. A sufficient reduction of soot is only achieved as long as the gas temperatures are not below 1600 K.

Now that the mechanisms of soot and NO_x formation have been discussed in detail, the problem of reducing both pollutants in the engine raw emissions, the so-called soot-NO_x trade-off, becomes obvious, Fig. 6.4. The problem is, that both components show opposite behavior: conditions that reduce the formation of nitric oxides increase the production of soot, and vice versa. In order to reduce the NO_x formation, the combustion should be influenced in such a way that local temperatures above 2000–2200 K are avoided or only reached for very short times. A possible conventional measure is to start the injection at later crank angles in order to shift the main combustion phase into the expansion stroke. This results in a significant reduction of maximum temperatures and NO_x formation. However, soot oxidation is less effective due to the lower temperatures, and fuel consumption increases due to a smaller thermal efficiency, caused by the late combustion. An early injection start results in an opposite effect: engine efficiency and NO_x production increase, while soot emissions are significantly reduced.

One of the most important influence factors is the fuel injection. The design of the injection nozzle (e.g. number, size and geometry of holes, spray direction,

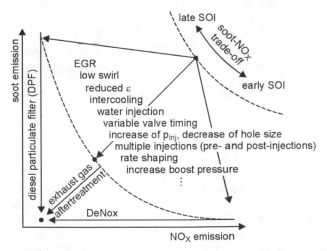

Fig. 6.4. Soot-NO_x trade-off: measures for reduction of soot and NO_x in engine emissions

VCO or sac-hole type) as well as injection pressure, start, and duration of injection, shape of injection rate, etc. must be carefully adjusted to the boundary conditions like combustion camber geometry, air motion, and pressure inside the cylinder. For example, in the case of a strong contribution of air motion (swirl) to the mixture formation process, less nozzle holes and lower injection pressures are necessary than in the case of low-swirl combustion concepts. The generation of strong swirl increases the pressure losses in the intake system and tends to increases fuel consumption. Further on, ignition delay and premixed peak are usually increased. If too many nozzle holes are used, the burning spray plumes may be displaced by the air motion in a way that fuel is injected in the burnt gases of the neighbor plume. This strongly increases soot formation. Today, low-swirl combustion concepts are often used, and the energy for mixture formation is more or less solely provided by the spray. For this reason, injection pressures and hole numbers are increased, and wide piston bowls, which allow the necessary spray penetration in order to include the complete cylinder charge in the combustion process, are in use.

Further measures to reduce pollutants in engine raw emissions are intercooling, cooled and non-cooled exhaust gas recirculation (EGR, see also Sect. 6.4), water injection (especially ship engines [36, 1, 99, 40, 90]) and the early closing of the intake valves (Miller cycle). In the case of early intake valve closing, the cylinder charge is cooled down by expansion, and the subsequent compression starts from a lower temperature level. In order to obtain equal cylinder charges compared to the conventional valve timing, the boost pressure must be increased. Because the Miller cycle is not favorable in all operating points of the engine, this measure usually requires variable valve timing.

An increase of charge pressure (reduction of soot due to increased amount of oxygen) combined with reduced compression ratios (prevention of increased peak temperatures and NO_x-formation) is also advantageous. However, highly efficient supercharging systems are necessary in order to provide the desired pressures at high mass flows.

In order to reduce both soot and NO_x emissions, a careful combination and adjustment of the different measures has to be applied. The most important influence factor however is the fuel injection. Highly flexible modern injection systems offer significant potentialities to reduce soot and NO_x raw emissions simultaneously. These potentials shall be discussed in the following sections.

6.2.2 Multiple Injection and Injection Rate Shaping

Modern injection systems offer the possibility to perform multiple injections (Sect. 2.2.1). Figure 6.5 shows an example of such a multiple injection strategy. The injection event consists of three parts, the pre-, main, and post-injection. The possibility of performing pre- and post-injections enables us to influence the basic heterogeneous diesel combustion and the formation of emissions in a very advantageous way: a simultaneous reduction of soot and NO_x is possible, and the trade-off curve in Fig. 6.4 can be moved closer to the origin.

Pre-Injection

As shown in the previous section, the amount of fuel injected during the ignition delay is responsible for the so-called premixed combustion. Because of the strong pressure gradients ($dp/d\varphi$) due to the rapid combustion of the premixed fuel-air mixture, this phase significantly increases engine noise. The more fuel injected during the ignition delay, the stronger the initial temperature rise in the combustion chamber, and the better the boundary conditions for NO_x formation. For these reasons, it is advantageous to shorten the ignition delay. This can be achieved by the injection of a small amount of fuel (pilot injection) shortly before the start of main injection. The pre-injected fuel mass evaporates quickly and ignition reactions start. Due to these reactions, which result in a pre-conditioning of the combustion chamber, the ignition delay of the main injection, and thus also the premixed peak, can be reduced significantly. Altogether, an optimized pre-injection is a very effective measure to reduce noise and to keep NO_x-emissions small during the first phase of combustion, Fig. 6.6.

Depending on the extent of NO_x-reduction, sometimes an earlier injection and combustion start is possible, resulting in an increase of thermal efficiency and a reduction of specific fuel consumption. It is important to adjust pre-injected mass, injection pressure, and dwell time between pre- and main injection carefully. Detailed investigations concerning such an optimization are reported in [68]. Usually short dwell times are advantageous. The higher the speed and load, the smaller the optimum fuel quantity. The pre-injection usually results in a slight increase of soot production, especially at part load (less premixed burning, lower initial temperatures, more diffusion burning). However, this effect can be completely compensated by a post-injection without affecting the fuel consumption, see below. Compared to a single pre-injection, the double pre-injection is an even more effective measure to reduce engine noise (e.g. [68, 33]), but injection timing, dwell times, and injected masses must be very accurately adjusted to each operating point in order to minimize an increase in emissions. Again, a post-injection is advantageous.

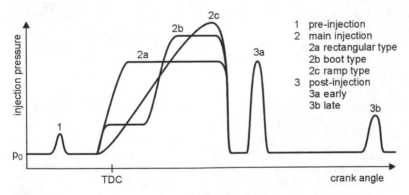

Fig. 6.5. Schematic diagram of a multiple injection

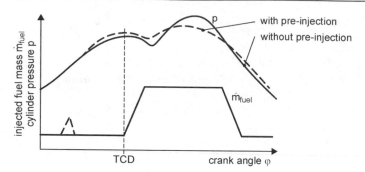

Fig. 6.6. Reduction of the premixed peak using a pre-injection, schematic diagram

Main Injection

The fuel mass injected during the main injection phase determines engine torque and power. Depending on the operating point of the engine, it may be advantageous to vary the injected mass as a function of time in order to reduce emissions, fuel consumption, and noise (e.g. [52, 21, 58, 59, 33, 14, 75]). This way of influencing the injection rate profile is called rate shaping. In Fig. 6.5, three possible injection rate profiles are shown. The ramp and the boot profiles reduce the amount of fuel mass injected at early times. The intention is to reduce the rate of heat release and to decrease the NO_x-formation at the beginning of combustion. Moderate initial injection pressures are favorable in order to limit the first disintegration and evaporation of the injected fuel. The combustion is slightly shifted to later crank angles, which usually results in a slight increase of fuel consumption due to a decrease of thermal efficiency. At the end of main injection high injection pressures are necessary in order to inject the complete fuel mass in the remaining time. Furthermore, high injection pressures are now advantageous, because they enhance mixture formation and support the reduction of soot at later crank angles. Altogether, a simultaneous reduction of soot and NO_x can be achieved, Fig. 6.7.

An increase of injection pressure over crank angle is typical for camshaft driven injection systems, Sect. 2.2.1. Injection rate shaping is especially advantageous in the case of low engine speed and high load [21], because then the time available for injection is long enough to realize an effective rate shaping. At high engine speeds an increasingly rectangular shape is often needed in order to inject the desired fuel mass in the available time. The rectangular profile is produced by common rail injection systems. This shape of injection rate usually results in lower fuel consumption and soot production, but in an increased formation of nitric oxides. However, in the case of high load and EGR, it has been shown that the effect of rate shaping is of minor importance [52]. In that case, the rectangular injection profile is recommended, and EGR-rate and -temperature must be adjusted instead. Further on, high initial and absolute injection pressures in combination with a fast needle opening and minimized cavitation inside the nozzle (k- and ks-nozzles, Sect. 2.2.1) increase spray penetration and lead to a better inclusion of the complete cylinder charge in the combustion process. This allows realizing higher

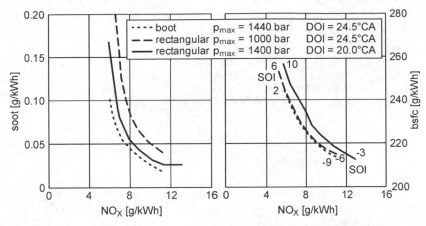

Fig. 6.7. Comparison of rectangular and boot-type injection, single cylinder engine, displacement: 1 liter, 1400 rpm, 100% load, no EGR [52]. SOI and DOI: start and duration of injection

specific power and decreasing the soot emissions, which are known to increase when applying EGR. In all cases, a very fast needle closing is advantageous in order to avoid the formation of soot due to a deteriorated mixture preparation.

Post-Injection

There are two kinds of post-injections, the late one and the early one. Late post-injection is applied in order to provide the exhaust gases with a well-defined quantity of evaporated but unburned fuel, which is then used in the catalytic reactions inside the exhaust gas after-treatment system. Injection pressure, amount of fuel, and injection timing must be carefully adjusted in order to prevent oil dilution. In the case of EGR, the late post-injection can alter the ignition delay, because some of the injected and unburned fuel is transported back into the cylinder.

The early post-injection directly follows the main injection and is a very effective measure to reduce soot emissions. Small amounts of fuel are injected with high pressure. Due to the small dwell time between main and post-injection, the injection is cut, and air is entrained in the near nozzle region and in the partial oxidation zone of the main combustion. It is very important to achieve a fast and complete closing of the injector needle in order to realize this additional entrainment of air and to avoid an unintended post-injection of low-speed large fuel drops which would significantly increase the emissions of soot and unburned hydrocarbons. The high-pressure post-injection contains enough kinetic energy to achieve an effective mixture formation and to entrain further oxygen into the partial oxidation zone. The combustion of this mixture increases the temperature during the late combustion phase and thus enhances the reduction of soot Fig. 6.8. The heat released by the post-injection is not sufficient to increase the NO_x emissions. Altogether, this kind of post-injection can result in a significant reduction of

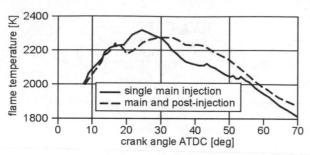

Fig. 6.8. Increase of temperature due to an early post-injection [12]

soot emissions (up to 70% reduction), while the fuel consumption is not affected (e.g. [21, 52]). The more soot produced by the main combustion, the more effective the post-injection.

6.2.3 Piezo Injectors

It has been shown in the previous sections that modern and future injection systems must be capable of performing multiple injections with variable injection timing and dwell times between two subsequent injections as well as variable injection pressures and extremely small pilot masses in order to reduce both, engine-out raw emissions and fuel consumption. The more flexible the injection system, the more accurate the optimization of the injection at each single operating point of the engine map can be. In addition to this, an extremely high reproducibility of injection timing and injected masses is necessary, especially in the case of short injection pulses. High needle speeds and a precise and complete needle closing are necessary in order to improve mixture formation and to avoid a deteriorated mixture formation and unintended post-injections at the end of each injection. Most of these demands can be fulfilled if piezo actuators instead of the slower and less precise solenoid valves are used in order to control the motion of the needle. The basic approach of the piezo technique is to use the elongation of a piezo quartz, which is proportional to the voltage impressed between two opposing surfaces, as the actuator. Piezo quartzes are extremely stiff, and the elongation is also extremely precise and fast. If the voltage is kept constant, a constant elongation can be realized. Because the elongation is proportional to the voltage, any elongation between zero and a maximum value can be precisely adjusted. Usually, a number of piezo quartzes are mechanically arranged in line (so-called piezo stack) in order to increase the resulting amplitude. In general, two main concepts are possible, the indirect and the direct needle control.

Fig. 6.9 shows the piezo-inline-injector from Bosch, where an indirect needle control is realized. The assembly consists of a piezo stack, a hydraulic coupling device, a control valve, and the nozzle module. The hydraulic coupling device amplifies the elongation of the piezo stack and transfers it to the control valve. In order to start the injection event, the needle of the control valve moves downwards and the pressure on top of the injection needle deceases because more fuel leaves

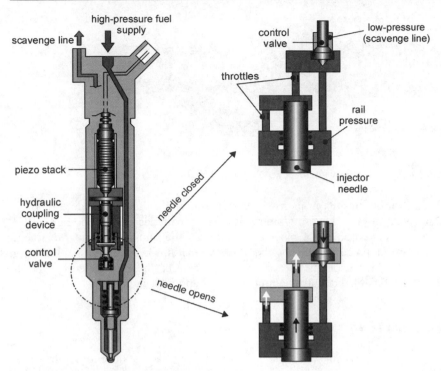

Fig. 6.9. Piezo in-line injector from Bosch [70]

the volume above this needle (larger cross-sectional area of the throttle) than enters from the high-pressure side (smaller throttle). At the end of injection the control valve closes and the increasing pressure above the injection needle pushes it downward again. Altogether, the stiff and close arrangement of the piezo actuator and needle, combined with small moving masses, allows the realization of an extremely reproducible, precise, and fast needle movement. Up to five injections per cycle with variable timings and extremely small injected fuel quantities can be performed [70].

The direct needle control offers even more flexibility, because well-defined partial needle lifts can be easily realized. Such a concept is shown in Fig. 6.10. The piezo stack is directly connected to the needle. This concept allows realizing any partial needle lift between zero and a maximum value, which is given by the maximum elongation of the stack. Because the needle lift is directly proportional to the voltage, it is possible to realize any kind of injection rate shape if the voltage can be controlled accordingly. This system has been designed as a research tool by the Institute of Technical Combustion at the University of Hanover in order to investigate the effect of different injection rate shapes on the combustion process of DI diesel engines (e.g. [54, 78, 75, 81]). The direct needle control has also turned out to be very effective in the case of pulsed injection in HCCI applications [66] (Sect. 6.4). In the case of ramp- or boot-type injections, the rate is controlled by needle seat throttling.

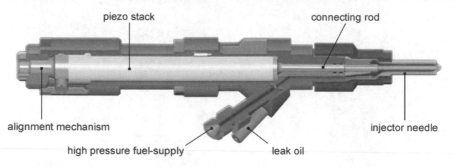

Fig. 6.10. Piezo injector with direct needle control [54]

Seat throttling results in increased turbulence and cavitation inside the nozzle and in a well-dispersed but low-momentum spray with decreased penetration [81]. These sprays are usually not capable of including the complete cylinder charge in the combustion process, and it has been shown that this kind of rate shaping is not optimal. For this reason, the piezo injection system has been extended in order to realize fully flexible pressure-modulated rate shaping, Sect. 6.2.6.

6.2.4 Variable Nozzle Concept

The concept of a variable nozzle is based on the approach of decoupling the trade-off between emission and power. In the case of a conventional nozzle, the size of the nozzle holes depends on the number of holes and on the maximum mass flow rate that has to be injected at full load. In the case of part load however, smaller nozzle holes would be favorable in order to enhance mixture formation and to reduce NO_x and soot emissions. In some operating points of the engine map, the desired small masses can only be injected by partial opening of the nozzle. This results in strong throttling effects at the needle seat and thus in a reduction of spray penetration and air entrainment. The near-nozzle combustion of these sprays often increases soot emissions.

Furthermore, it has been shown that in the case of very small nozzle holes a pre-injection can be omitted because of the small amount of fuel being injected during the ignition delay, while the noise level can be reduced to that of modern injection strategies with pre-injection [33]. Because very small holes are not suitable for injecting larger amounts of fuel into the combustion chamber, the effective cross-sectional area of the injector must be variable, such that the injection rate can be increased during the injection event.

Fig. 6.11 shows a possible design of such a variable nozzle from Bosch [15, 9]. The basic concept consists of two coaxial needles, which can be opened and closed independently of each other. The outer needle is responsible for the injection of small amounts of fuel through small nozzle holes (k- and ks-holes in order to minimize cavitation and discharge variation and to increase penetration and reduce soot), while the inner needle controls the mass flow through a second row of larger injection holes. Depending on the operation point, the appropriate row of

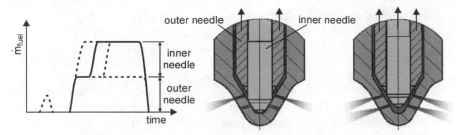

Fig. 6.11. Variable nozzle concept: possibility of rate shaping [9]

nozzle holes can be chosen [15]. A further possible injection strategy is shown in Fig. 6.11: the outer needle is responsible for pre-injection, and then the desired shape of the main injection rate as well as the amount of injected fuel, which strongly depend on the operating point, can be determined by a staggered opening of outer and inner needle.

Due to improved production engineering, it is possible to use different nozzle hole geometries, sizes, and spray directions for the holes of each row. This offers the possibility to adapt the nozzle to combustion chamber geometry and air motion further and to improve the inclusion of the complete cylinder charge in the combustion process.

Furthermore, this type of injector is also well suited for HCCI applications, Sect. 6.4. Depending on the engine operation mode, different spray directions, the narrow arrangement during HCCI-mode at low loads (e.g. inner needle) and the wide arrangement during conventional operation at higher loads (e.g. outer needle), can be realized with the same injection system.

6.2.5 Increase of Injection Pressure

On the one hand, increased injection pressures offer the possibility better to include the complete air charge in the mixture formation and combustion process and thus to increase the specific power (kW/liter) of an engine. However, the more sophisticated the combustion concept and the higher the injection pressure of the basic engine, the less the increase of specific power. On the other hand, increased injection pressures offer the possibility to reduce the nozzle hole sizes. Due to a better mixture formation compared to the basic concept, this effect reduces the soot-NO_x trade-off. In this case, the specific power of the engine is not increased.
In the case of future downsizing concepts with considerably increased supercharging pressures as well as EGR-rates and specific power, injection pressures of more than 200 MPa will be necessary in order to achieve a stable and effective combustion.

However, there is another important aspect that must be kept in mind when discussing the subject of optimum injection pressure: the more the injection pressure is increased, the more energy is needed to drive the injection system. This effect increases fuel consumption. Hence, it is necessary to increase the efficiency of the injection systems by reducing leak flow and by dynamically adjusting the maximum pressure to the actual needs of the engine, depending on the operating point.

A possible way of generating extremely high maximum injection pressures only in the case of demand is the use of a pressure amplifier unit. Fig. 6.12 shows such a hydraulically amplified diesel common-rail injection system from Bosch [14], where the pressure amplifier is directly integrated into the injector body of a common-rail injector. This concept provides maximum injection pressures of more than 220 MPa, allowing to inject large fuel quantities for high specific power through small nozzle holes. Due to the smaller holes, mixture formation can be improved in all operating points of the engine map. Because the pressure amplifier is driven hydraulically by fuel from the rail, this concept requires increased fuel mass flows, but at lower pressure levels. In the case of a transmission ratio of about 1:2, only rail pressures of 130 MPa or even less are necessary. For this reason, only the nozzle unit must be designed for high pressures.

The amplifier is activated by an additional solenoid valve. During the pre-injection and the first phase of the main injection, the injector acts as a standard common-rail system. The fuel from the rail (moderate pressure) takes the way through the no-return valve to the nozzle. Due to the moderate injection pressure during this time, the injected fuel mass as well as the mixture formation process can be reduced in order to decrease noise and NO_x-formation. After the start of the main injection, the additional solenoid valve (valve 1) is opened. Then the amplifier piston moves downwards and compresses the volume below. The no-return valve closes and the injection pressure increases. The maximum pressure is determined by the geometry of the amplifier piston. At the end of the main injection the needle is closed and opened again (valve 2) in order to perform a high-pressure post-injection. After the post-injection the pressure amplifier is deactivated, such that the following injection can start again with the moderate rail pressure.

Varying energizing timings of both solenoid valves, flexible pressure curves from boot to rectangular can be generated. Thus, this kind of future common-rail injection system enables us to realize the favorable pressure-modulated injection, which can only be performed by camshaft driven systems (Sect. 2.2.1) so far.

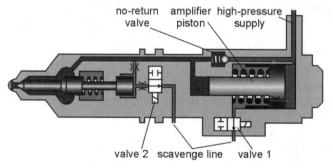

Fig. 6.12. Amplified pressure common-rail system from Bosch [14]

However, the pressure amplification is fixed due to the geometry of the piston, limiting the degree of flexibility. A fully flexible research system is presented in the following section.

6.2.6 Pressure Modulation

Rate shaping is an effective measure to reduce the soot-NO_x trade-off if no EGR is used. However, the potential of fully flexible injection rate shaping cannot be completely realized if the mass flow is only controlled by needle lift and pressure is maximal (Sect. 6.2.3). Better results can be achieved if the injection rate is a function of injection pressure and if the needle opens as fast as possible to avoid throttling effects. Today's cam shaft driven injection systems (Sect. 2.2.1) perform such a pressure-modulated injection, but the increase of pressure during injection is a function of injector geometry and engine speed and cannot be varied independently in order to optimize each point of the engine map. Further on, the realization of multiple injections is complicated. In contrast to this, common rail injection systems allow for performing fully flexible multiple injections, but a modulation of injection pressure during the injection event is not possible. The amplified pressure common rail system, which has been described in Sect. 6.2.5, is the first commercial system that combines both advantages. However, the pressure amplification is fixed due to the geometry of the piston, and this significantly limits the degree of flexibility. For this reason, research systems have been used in order to evaluate the potential of pressure modulation [89, 59, 58, 41]. The maximum degree of flexibility has been obtained by an extension of the directly driven piezo injection system (Sect. 6.2.3), the so-called Twin-CR system Fig. 6.13.

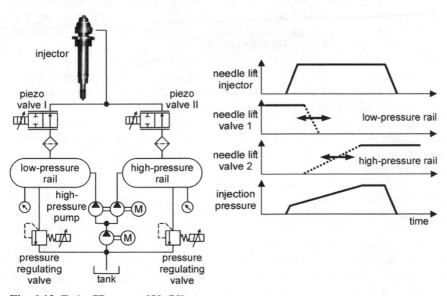

Fig. 6.13. Twin-CR system [50, 75]

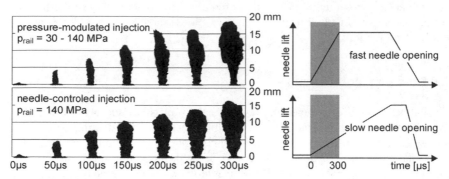

Fig. 6.14. Optical investigation of a ramp-type injection: comparison of pressure-modulated and needle lift controlled rate shaping during the first 300 µs of injection [80]

The two rails are used in order to provide the injector with two different pressure levels, which can be dynamically adjusted by the two pressure control valves. In order to perform a ramp- or boot-type main injection, the additional piezo valve controlling the mass flow from the low-pressure rail to the injector is opened. As soon as the injector is opened, the injection starts at a low pressure level. In order to increase injection pressure, the low-pressure supply is closed while the high-pressure supply is opened. The exact shape of the resulting injection pressure curve can be precisely adjusted by the opening and closing speeds and timings of both additional piezo valves. At the end of injection, the injector needle is closed. Pre- and post-injections can be realized by multiple opening and closing of this needle. Finally, the piezo valve of the high-pressure rail closes, the one of the low-pressure rail opens again, and the system is ready to perform the next injection.

This research injection system has been used to investigate the potential of a pressure-modulated rate shaping in comparison to the one controlled by needle lift. Fig. 6.14 shows optical investigations of a ramp-type injection rate. It is clearly visible that due to throttling and increased cavitation in the case of partial needle lift a stronger disintegration of the spray near the nozzle is achieved, while the resulting spray penetration is reduced. The pressure-modulated injection results in smaller spray angles and increased penetration.

Fig. 6.15 from Seebode et al. [74] finally shows a comparison of needle-controlled and pressure-modulated injection rate shaping and their effects on combustion. Four different injection rate curves, the standard rectangular one, the slow and the fast ramp, and the boot injection were investigated, Fig. 6.15a. As can be seen in Fig. 6.15b, the boot-type injection results in a lower peak pressure due to less premixed fuel-air mixture and a slower heat release at the beginning of combustion, and also due to a later start of combustion. This results in a significant reduction of NO_x-emissions. Compared to the pressure-controlled injection, the enhanced dispersion but lower penetration of the needle-controlled injection (decreased entrainment of cylinder charge, higher local fuel-air ratios) produces increased soot emissions and fuel consumption, Fig. 6.15c. Fig. 6.15d finally compares different shapes of pressure-controlled injections and shows that the

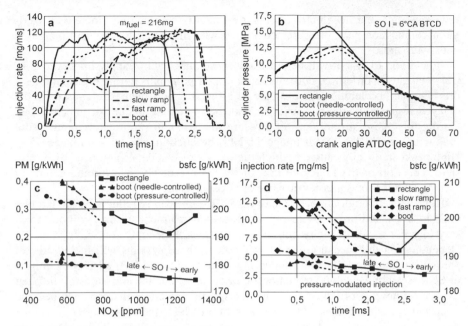

Fig. 6.15. Reduction of soot-NO_x trade-off: comparison of pressure-modulated and needle lift controlled rate shaping [74]. 2.0 liter single cylinder engine, ε: 16:1, 1300 rpm, m_{fuel} = 216 mg (high part load), no EGR

ramp-type injection is best suited for the engine. Further detailed investigations that also include post-injections are published in [75].

Altogether, the pressure-modulated injection is an effective measure to reduce the soot-NO_x trade-off. However, due to the complexity of the injection system (costs) and due to the fact that injection rate shaping is only effective if no EGR is applied, it will perhaps never be used for series production engines.

6.2.7 Future Demands

In future, very flexible high-pressure fuel injection systems with multiple injection and rate shaping capabilities as well as increased injection pressures are necessary in order to realize the optimum rate shaping and injection timing for each single point of the engine map, and to get the best compromise between emission trade-off and fuel consumption. The more the injection pressure is increased, the more the efficiency of the injection system itself becomes important in order to reach a low overall fuel consumption of the engine. New developments on actuators (especially piezo technique), nozzle design, and control strategies are key factors for future diesel injection technology.

Besides the injection system, the EGR-rate, the compression ratio, the shape of the combustion chamber, the air motion, the fuel-air ratio etc. are important measures to improve the combustion process and to achieve a significant reduction of raw emissions. Further improvements are effective intercooling, increased maximum combustion peak pressures combined with higher EGR-rates and increased boost pressures.

In addition to this, it will be more and more important to apply improved control strategies in order to reduce the formation of pollutants during transient operation. During acceleration for example, the fuel-air ratio increases, caused by the delayed response of the turbocharger and the increased amount of injected fuel. During this phase, a transient rail pressure increase can significantly reduce soot production while the disadvantage in NO_x-production can be minimized [52]. High peaks in opacity caused by high fuel-air ratios can be also reduced by a transient reduction of the EGR-rate [52] or by the use of variable turbine geometry (VTG), which allows a faster increase of boost pressure.

However, although all these measures will significantly reduce the engine raw emissions, the additional use of exhaust gas after-treatment systems will be necessary in order to fulfill future emission limitations, especially for heavy vehicles. Depending on the special range of application, these systems will include diesel particulate filters or oxidation catalysts to reduce soot, and NO_x-adsorber catalysts or SCR (selective catalytic reduction) in order to reduce the emissions of nitrogen oxides. In the case of large-bore ship engines, the injection of water may become important in order to reduce the formation of both NO_x and soot.

6.3 DI Gasoline Engines

6.3.1 Introduction

Since Mitsubishi has presented the first modern series production gasoline direct injection (GDI) engine in 1995, the development of these engines has been considerably advanced. The main reason for this trend is that the fuel consumption of today's gasoline engines must be significantly reduced in order to fulfill future demands regarding CO_2 emission. In contrast to the diesel engine, which has an excellent efficiency but suffers from its high soot and NO_x-emissions, the conventional port-fuel-injected (PFI) gasoline engine will be able to fulfill future legislation regarding the emission of soot, NO_x, unburned hydrocarbons, and CO if combined with a conventional three-way catalyst system for exhaust gas after-treatment. The most important challenge however will be the reduction of specific fuel consumption in order to lessen the increasing discrepancy between diesel and gasoline engine and to reduce the overall CO_2 fleet emission.

Among other techniques like cylinder cutoff or the application of variable valve trains for example, the direct injection of gasoline is the measure with the highest individual potential to reduce fuel consumption and thus also CO_2-emissions. Compared to a similar PFI engine, about 15–25% reduction of fuel consumption at part load are theoretically possible [35, 93, 105, 77]. Depending on the operating point of the engine, the direct injection of gasoline offers different advantages compared to the PFI technique.

First of all, the evaporation of fuel inside the cylinder reduces the average temperature of the in-cylinder charge. If the fuel is injected during the intake stroke, this effect may increase volumetric efficiency by up to 10 % [19]. The reduction of temperature reduces the possibility of knock, and the compression ratio can be increased by about 1 to 1.5 units [35], which directly results in an increase of thermal efficiency. The direct injection improves the transient behavior of the engine [19, 105] and may also reduce the emission of unburned hydrocarbons during cold start, because the formation of a liquid film inside the induction system is circumvented and a more precise control of the air-fuel mixture is possible. If engine speed and load are controlled by a conventional throttle, conventional stoichiometric fuel-air mixtures can be used, allowing the application of the well-known durable and highly effective three-way catalyst for exhaust gas after-treatment.

While at full load the above mentioned advantages of direct injection result in only a small reduction of fuel consumption compared to the conventional PFI engine (wide open throttle), the full fuel economy potential can be realized at part load. The main advantage of the direct injection is the fact that in the case of part load throttling can be eliminated and thus pumping losses are minimized, Fig. 6.17. The reduction of load is achieved by a reduction of the injected fuel quantity, while the airflow is not throttled. This approach of qualitative load control is well known from the DI diesel engine. Hence, there is no homogeneous fuel-air mixture inside the whole cylinder any more, but a stratified charge. This stratified charge consists of a region of fuel-rich mixture and pure air or a mixture of air and recycled burnt gases in the remaining volume, Fig. 6.16. This stratified charge is achieved by a late injection during the compression stroke. Combustion and energy release only take part inside the mixture region. Due to stratification, there is a further advantage of the GDI process: because the reactive zone is separated from the wall by the non-reacting part of the cylinder charge, the heat losses to the engine walls are reduced.

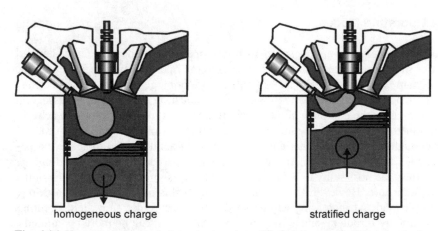

homogeneous charge   stratified charge

Fig. 6.16. Homogeneous (early injection) and stratified-charge mode (late injection)

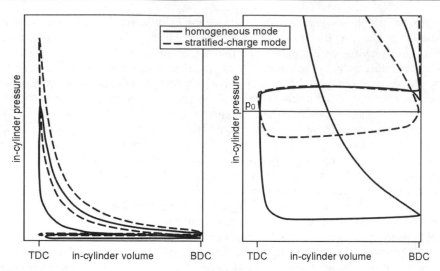

Fig. 6.17. Reduction of throttle losses, stratified-charge combustion

Compared to the diesel engine, the most important difference however is that in the case of diesel fuel auto-ignition occurs, while in the case of gasoline a spark is necessary. For this reason, non-throttled operation at part load requires the fuel-air mixture to be concentrated in an ignitable cloud around the spark plug at the time of ignition. Most of today's stratified-charge GDI engines employ a large-scale air motion (swirl or tumble) as well as specially contoured piston surfaces in order keep the fuel cloud compact and to transport it to the spark plug. The challenge is to control the stratified-charge combustion over the required operating range. Possible techniques as well as their advantages and disadvantages will be discussed in Sect. 6.3.3.

6.3.2 Operating Modes

Dependent on load and engine speed, different operating modes have to be applied in order to realize a stable and satisfactory engine operation within the complete engine map, Fig. 6.18. In the case of full load a homogeneous stoichiometric or even fuel-rich mixture inside the complete combustion chamber is necessary in order to include the complete air charge in the combustion process and to achieve maximum torque. Early injection during the intake stroke is applied in order to have enough time to inject the required large fuel quantities and to achieve a homogeneous fuel-air mixture. At medium and low part load the stratified-charge mode offers the possibility to reduce pumping losses significantly due to the omission of a throttle. However, the operating range of this mode is limited in engine speed and load. At increasing engine speed, the in-cylinder flow field becomes more and more turbulent, and above approx. 3000 rpm it can no more be utilized

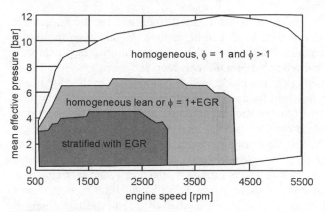

Fig. 6.18. GDI operating modes within the engine map [93]

to keep the mixture cloud compact. Above break mean effective pressures (BMEP) of approx. 5 bar, the injected fuel quantity becomes too large in order to realize a sufficient mixture formation prior to ignition, and soot is produced [35, 69]. Although in the case of stratified-charge the average temperature of the cylinder is reduced due to the overall lean operation, the local temperature of the reaction zone remains high, because here a stoichiometric or slightly rich mixture exists. Furthermore, the GDI engine operates at a higher level of pressure and temperature at unthrottled stratified-charge mode, compared to the PFI engine (throttling reduces mass of in-cylinder charge, Fig. 6.17). This effect also tends to increase NO_x-emissions. For this reason EGR has to be applied in order to reduce local temperatures and NO_x raw emissions.

The increased content of oxygen in the exhaust gases significantly reduces the conversion rates of the well-known three-way catalyst (due to high O_2-content CO and HC are no more available for the reduction of NO_x), such that this durable and cost-effective after-treatment system can no longer be used to realize the necessary reduction of engine-out raw emissions. For this reason, more expensive and complex $DeNo_x$ systems like the NO_x-adsorber catalyst for example have to be used. However, the regeneration of these systems requires additional energy. Depending on the frequency and time span of regeneration, the effective advantage of the GDI engine over the PFI concept can be significantly reduced. Furthermore, only sulfur-free fuels (< 10 ppm) can be used, because otherwise the combustion products (SO_x) will considerably reduce the efficiency of NO_x adsorption.

At increased loads above approx. 5 bar BMEP or engine speeds above approx. 3000 rpm the engine must be operated in homogeneous mode. If no further measures are applied, the fuel-air ratio is stoichiometric and a conventional throttle must be used to control load. A second possibility is to generate a homogeneous but lean fuel-air mixture and to reduce throttling. However, misfiring may occur if the mixture becomes too lean, and the emissions of unburned hydrocarbons as well as fuel consumption may increase. In this case, a strong in-cylinder air motion is favorable, but its generation increases the pumping losses again. Furthermore, the increased content of oxygen in the exhaust gases requires the use of

DeNOx-catalysts again. In order to avoid these difficulties, the homogeneous stoichiometric mode combined with EGR can be applied. In this case, the three-way catalyst can be used again, and the reduction of pumping losses is achieved by replacing a part of the cylinder charge with inert recycled burnt gas.

There is another stratified operating mode, which is important in the case of acceleration. The homogeneous split mode (two-stage injection) is used for several cycles during the transition from stratified to homogeneous mode. The larger part of the fuel quantity is injected early during the intake stroke and forms a homogeneous but lean mixture inside the complete cylinder. Then the remaining fuel quantity is injected during the compression stroke and forms a fuel-rich ignitable zone around the spark plug. This region is ignited by the spark, and its flame then ignites the remaining lean mixture parts. Near the full load limit, ignition must usually be retarded in order to prevent knock. However, due to the late combustion, this measure usually decreases of thermal efficiency. In this case, the application of the homogeneous split injection is advantageous, because the stratified charge is more knock-resistant and allows for the retainment of the early ignition timing. Furthermore, the split injection can also be applied to increase exhaust gas temperatures and to reduce the time of catalyst light-off [18].

Further improvements regarding especially the emission of unburned hydrocarbons can be achieved if the conventional starting strategy with early low-pressure injection is replaced by a late high-pressure stratified injection [46]. In this case the generation of a sufficiently high injection pressure prior to injection must be guaranteed. Then, the significant fuel enrichment necessary in order to compensate for the deteriorated mixture formation due to the low injection pressures can be circumvented. This results in a significant reduction of injected fuel quantity as well as emitted unburned HC emissions during engine start. If this starting concept is combined with the homogeneous split injection for fast catalyst light-off, the cumulative HC emissions can be drastically reduced, such that even SULEV emission legislation can be fulfilled [46].

6.3.3 Stratified-Charge Combustion Concepts

In the case of full load, fuel injection starts during the induction stroke in order to have enough time to inject the desired mass and to achieve a homogeneous stoichiometric mixture inside the complete cylinder. Because in-cylinder pressures at this time are small, low injection pressures are sufficient. In the case of stratified operation however, the fuel is injected during the compression stroke. This late injection timing is necessary in order keep the spray cloud compact and to minimize fluctuations of its position and spatial structure at the time of ignition. The more the fuel cloud can be kept compact, the more effective the stratified combustion and the higher the possible reduction of fuel consumption. Large and extremely lean regions at the border of the cloud will not burn and will increase fuel consumption and emission of unburned hydrocarbons. Because of the increased in-cylinder pressures as well as the very small time interval between

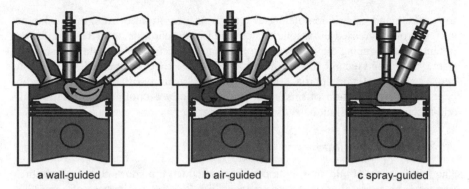

a wall-guided   b air-guided   c spray-guided

Fig. 6.19. Schematic description of the three possible stratified-charge combustion concepts

injection and ignition, the late injection timing requires significantly increased injection pressures in order to achieve a fast and effective mixture formation.

Two often opposing objectives have to be realized if an application to series production engines is regarded: the fuel economy potential of the stratified operation mode shall be maximized, while at the same time a sufficient robustness of the combustion concept has to be achieved in order to avoid any kind of misfiring in the whole operating range. Furthermore, it must also be possible to operate the engine in the homogeneous mode in order to achieve full load.

Regarding the basic arrangement of injector and spark plug, two concepts, the narrow and the wide arrangement, can be distinguished. In the case of the wide arrangement, the spark plug is usually mounted in the center of the cylinder head, while the injection nozzle is at the side. The central arrangement of the spark plug is advantageous for combustion: it allows a fast and effective burning of the mixture cloud due to the formation of a spherical flame front. The relatively large distance between the injector and the spark plug is also advantageous for mixture formation, because the time interval between injection and ignition is usually longer, compared to a narrow arrangement. Hence, a more homogeneous mixture inside the spray cloud can be achieved, and the risk of soot formation due to excessively rich fuel regions is minimized. However, it is absolutely necessary to keep the cloud compact in order to avoid the formation of very lean mixture regions in the outer spray areas, because this effect reduces the ability of stratification and increases HC emissions and fuel consumption.

The three basic approaches of controlling the stratified-charge combustion, the so-called wall-guided, air-guided, and spray-guided techniques, Fig. 6.19, will be discussed in the following.

Wall-Guided Technique

Fig. 6.19a shows the basic arrangement of the injector and the spark plug in the case of the so-called wall-guided technique (wide arrangement). This approach uses a specially shaped piston surface in order to transport the fuel to the centrally arranged spark plug. Because a considerable amount of fuel is injected on the piston surface and cannot completely evaporate until ignition occurs, this technique suffers from increased emissions of unburned hydrocarbons and CO, and the full potential of reducing fuel consumption cannot be reached. For this reason, the pure wall-guided technique is of little importance. However, the wall-guided concept is a very reliable approach regarding the robustness of the combustion concept and the prevention of misfiring. Today, usually a combination of the wall- and the air-guided technique is used.

Air-Guided Technique

The fuel is injected into an in-cylinder airflow, which transports the compact spray plume to the spark plug (wide arrangement, Fig. 6.19b). In the case of a pure air-guided technique, there is no wall wetting. The generation of a stable air motion that keeps the spray plume compact and transports it to the spark plug while enhancing a homogeneous air-fuel mixing inside the cloud, as well as the exact timing of injection, are crucial for the efficiency and reliability of this concept. The in-cylinder airflow is created by a special shape of the inlet ports, and its intensity can usually be controlled by special air baffles in the manifold. Two main in-cylinder air motions are possible, the swirl and the tumble. In the case of a flat cylinder head, the swirl flow is usually utilized, while a pent roof cylinder head also allows the application of a stable tumble flow. In the case of tumble, the spray plume is usually deflected from a shaped cavity in the piston, and the mixture is transported to the spark plug. The swirl component of the in-cylinder motion generally experiences less viscous dissipation than the tumble component, is therefore preserved longer during the compression stroke, and is of greater utility for maintaining mixture stratification. Special piston geometries are used to enforce the effect of air motion and to make sure that the mixture cloud reaches the spark plug at the time of ignition, Fig. 6.20.

If an optimum air motion can be produced in any point of the engine map belonging to the area of stratified operation, fuel consumption can be significantly reduced. However, the generation of a stable airflow that enhances mixture formation inside the spray cloud, keeps it compact at the same time, and transports it to the spark plug, such that ignition can occur at a thermodynamically optimum

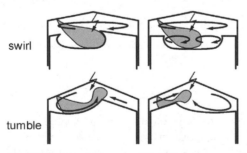

swirl

tumble

Fig. 6.20. Possible combustion chamber geometries as well as in-cylinder air motions, air-guided technique

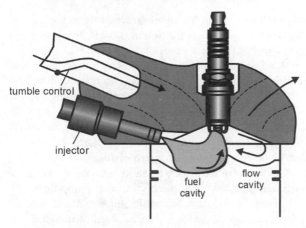

Fig. 6.21. Wall-air-guided technique, FSI combustion concept, Volkswagen [83]

timing, is nearly impossible to realize for all speed and load points within the stratified operation range. Furthermore, the generation of swirl or tumble increases losses due to throttling and thus reduces fuel economy.

The geometries of pistons used for the air-guided technique are all quite complex. Compared to a conventional flat piston, the increased surface results in increased heat transfer to the engine. Furthermore, there are more sharp edges, and due to the complex geometry the combustion chamber volume is less spherical. For these reasons, the knock resistance is deteriorated [19, 100] and the compression ratio often has to be lower than in the case of a flat piston in order to prevent knock at full load.

Today, a combination of the air-guided and wall-guided techniques is the only concept that allows to realize a stable stratified operation in series production GDI engines, Fig. 6.21. The more the formation of a liquid wall film on the piston surface can be circumvented, and the better an optimal in-cylinder air motion can be generated and controlled, the higher the potential efficiency of the concept.

Spray-Guided Technique

The spray-guided technique is the concept that theoretically allows for the attainment of the highest fuel economy. However, this approach is the most complicated to realize, and for this reason it has only been investigated and tested in research engines so far. The spray-guided concept is characterized by a narrow arrangement of the injector and the spark plug, Fig. 6.19c. The spray is directly transported to the spark plug by its kinetic energy. Special combustion chamber and piston geometries are not necessary, and the in-cylinder airflow is also of secondary importance. Usually, strong in-cylinder charge motions are disadvantageous because they may disturb the formation of the desired spray shape. Due to the narrow arrangement, the time between injection and ignition, and thus the time for mixture formation, is extremely small. For this reason, high injection pressures of more than 20 MPa will be necessary to provide enough energy for mixture formation [100] and to avoid the production of soot. The generation of these high injection pressures causes problems regarding system friction and wear, because gasoline has a lower lubricity and viscosity than diesel fuel.

Due to the very short time for mixture formation, the formation of large lean zones at the border of the spray is not possible, and the mixture region can be kept very compact. Because the time of arrival of the spray at the spark plug is only dependent on injection timing and not on complicated air motions, there are no restrictions in ignition timing, and the thermodynamically optimal timing can be realized much easier than in the case of the wall-air-guided concepts. Hence, the spray-guided technique offers the largest possible decrease of fuel consumption at part load. Because the spray does not impact on a wall, and because a strong in-cylinder air motion is not required, heat losses to the engine and pumping losses are the smallest of all three concepts. Further on, more spherical and knock-resistant piston and combustion chamber geometries can be used. Altogether, an additional benefit in fuel consumption of 5% compared to the wall-air-guided technique is expected [100].

Besides of all these advantages, there are still serious challenges, which have to be overcome in order to realize the spray-guided concept in series production engines. The most serious problem is the achievement of the required spatial accuracy and reproducibility of the spray shape for all operating points within the stratified mode region in the engine map. Due to the extreme stratification, the gradient of fuel vapor concentration at the outer spray region is also extremely strong. In order to achieve the existence of an ignitable fuel-air mixture at the position of the spark, the spatial arrangement of injector and spark plug have to be carefully optimized. A small displacement usually results in misfiring or deposition of liquid fuel on the spark plug. If the spark plug is wetted by liquid fuel, soot may be produced, and carbonization as well as extreme thermal stress (thermoshock) will significantly decrease the spark plug's lifetime.

A further problem is the fact that only a very short time interval for inflammation exists, Fig. 6.22. During injection, the fuel-air mixture at the spark plug is too

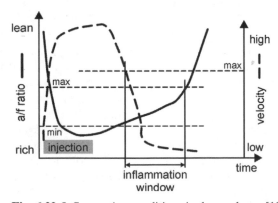

Fig. 6.22. Inflammation conditions in the spark gap [100]

rich, and the flow velocities are too high for inflammation. After the end of injection, the flow velocity decreases, but the mixture quickly becomes too lean for ignition. Hence, ignition timing is strongly dependent on injection timing, and the accurate timing of both events is crucial in order to avoid misfiring.

Depending on the injector concept, the spray quality (shape, SMD, penetration, cone angle) varies more or less with changing in-cylinder pressure (Sect. 2.2.2). This causes serious problems concerning the reliability of ignition at different load points. Because the spray is directly transported to the spark plug by its kinetic energy, the time of ignition is directly dependent on injection timing. In order to avoid premature ignition, injection timing must be increasingly retarded if the load is reduced. The more the injection timing is retarded, the higher the in-cylinder pressure during injection, and the stronger the possible change in spray shape. For this reason, injectors that produce a completely pressure-independent spray shape have to be developed. High-pressure injection combined with multi-hole injectors is a promising approach (see also Sect. 2.2.2). However, these injectors still have to be significantly improved in order to permit a control of spray penetration (dependent on injection timing and in-cylinder pressure) and to avoid wall wetting. Furthermore, mixture formation has to be improved in order to homogenize the fuel-air mixture inside the spray prior to ignition and to suppress the formation of soot at higher loads.

6.3.4 Future Demands

Today the standard gasoline engine uses port fuel injection. This technique has reached a high development status, and future improvements, especially the combination with variable valve trains (VVT), will help further to reduce fuel consumption. However, the more the theoretical advantages of the GDI technique can be realized in series production engines, the more the PFI engine will be substituted by direct injection concepts. Furthermore, many additional measures like turbo-charging or VVT can also be successfully combined with this concept [27]. The main advantages of direct injection are in-cylinder charge cooling due to fuel evaporation (increase of efficiency due to higher possible compression ratio, increase of volumetric efficiency), and throttle-less operation and reduction of heat transfer to the wall during stratified-charge combustion. In addition to this, GDI engines show faster transient response due to a more precise fuel delivery, and lower HC- and CO-emissions are possible during cold start.

Regarding the stratified-charge combustion, the spray-guided technique offers the highest theoretical efficiency, but is the most complicated one to realize. There are two main problems in this concept, which have to be solved. First, the generation of a well-mixed but extremely compact spray plume must be possible in extremely short time intervals in order to achieve a high degree of stratification while preventing the formation of soot. Second, an absolutely reliable way of igniting this mixture must be realized in order to avoid misfiring, emission of unburned hydrocarbons as well as losses in efficiency. If a spark plug is used for ignition, the spray cone angle must be absolutely independent of backpressure.

Fuel injection systems for GDI engines must also have the capability of providing early injection for homogeneous-charge combustion at full load, where a well-dispersed fuel spray is desirable to ensure a homogeneous charge even for the largest fuel quantities. Due to the lower in-cylinder pressures during early injection, the injection system must be capable of adapting the spray penetration in order to avoid wall wetting. Up to now, none of the three possible nozzle concepts, the multi-hole injector, the outwardly opening nozzle, and the inwardly opening swirl injector, have been able to fulfill all of these requirements.

While hollow-cone sprays show a faster mixture formation, the multi-hole nozzle produces a stable spray geometry. Today, all three categories are under development and are being optimized for use in spray-guided applications. Investigations with high-pressure multi-hole injection systems with specially adapted numbers, sizes, and positions of the nozzle holes have been reported by Bosch [100], while Siemens has developed a piezo-actuated outwardly opening injector [4]. Stegemann et al. [79] constructed a piezo-actuated pressure-swirl injector, which is able to adjust spray angle and penetration, and which can also be operated in a way that avoids the formation of the well-known pre-spray.

As far as ignition is concerned, there are considerable problems regarding misfiring as well as thermal stress and carbonization of the spark plug. A possible future technique might be the application of the laser-induced ignition [25]. This approach offers multiple advantages. The life-time of the ignition system is no longer reduced by thermo-shock and carbonization, and the optical system, which is used to bring the laser beam into the combustion chamber, can be easily used to vary the spatial position of the focus point and thus to compensate the variation of spray shape with increasing backpressure, Fig. 6.23. In this case, it is possible to uncouple the process of optimized mixture formation and reliable ignition. Such an ignition system has been successfully investigated by Geringer et al. [25].

Future injection and ignition systems will help to realize the spray-guided stratified-charge combustion in series production engines, to extend the application of this concept to higher load and speed regions of the engine map, and to improve transient engine and emissions behavior in order to exploit the full benefits of the GDI technique. Due to the different operating modes, durable exhaust gas after-treatment techniques must finally be developed in order to ensure high conversion rates for all of the different exhaust gas temperatures and oxygen contents.

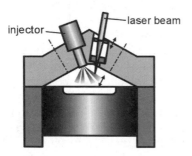

Fig. 6.23. Laser-induced ignition [25]

6.4 Homogeneous Charge Compression Ignition (HCCI)

6.4.1 Introduction

Besides the further development and enhancement of both the conventional gasoline and the conventional diesel engine processes, which have been discussed in the two previous sections, one combined combustion concept, the so-called homogeneous charge compression ignition (HCCI) combustion, promises further improvements. This approach theoretically combines the advantages of both conventional processes: low raw emissions and high fuel economy. It can be realized by either modifying the gasoline or the diesel engine. In both cases, the combustion is initiated by auto-ignition of an overall lean and homogeneous fuel-air mixture.

Regarding the conventional spark-ignited gasoline engine, efficiency and power are largely limited by knock, which necessitates the use of less than optimal compression ratios and limit thermal efficiency. Furthermore, modern stratified direct injection concepts suffer from high air excess ratios at part load that complicate an efficient exhaust gas after-treatment. In the case of HCCI, two characteristics of the diesel combustion process are introduced: the compression ratio is increased (improved thermal efficiency) to achieve auto-ignition, and load is controlled by the quality of the fuel-air ratio. Hence, the engine operates without throttle, and the homogeneous mixture becomes extremely lean at low load.

Regarding the conventional diesel combustion process, the existence of fuel-rich and lean zones results in the formation of soot and NO_x during most of the combustion time, and a sufficient and simultaneous reduction of both pollutants is not possible because of the well-known NO_x-PM-trade-off. In the case of HCCI combustion, this process is changed in such a way that the lean fuel-air mixture is homogenized prior to auto-ignition, such that the strongly heterogeneous combustion can be circumvented. Because no fuel-rich zones exist, the formation of soot during combustion is suppressed, resulting in a non-luminous flame. On the other hand, the high air excess ratio results in low overall and local temperatures, suppressing also the formation of thermal nitric oxides, Fig. 6.24. The realization of this new combustion concept is not limited to special fuels, and besides diesel fuel and gasoline also alternative fuels like natural gas, methanol and hydrogen etc. can be used.

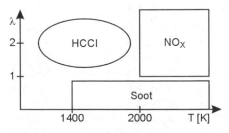

Fig. 6.24. HCCI-combustion: simultaneous reduction of soot and NO_x

Because there is no difference in local and overall lambda and temperature values in the case of the perfect HCCI-process, the compression of the completely homogeneous mixture leads to a simultaneous auto-ignition of the whole charge. Usually, a perfect homogenization prior to ignition cannot be realized in engine applications (Sect. 6.4.4), such that there will always be a certain degree of inhomogeneity. Nevertheless, in contrast to the conventional diesel combustion, the reaction rate is not controlled by turbulent mixing (diffusion burning), and in comparison to the gasoline engine, there is no distinct flame front that propagates through the volume. Since combustion reactions are not initiated by a spark, extremely lean mixtures can be burnt. This results in a very fast overall heat release.

Compared to the SI engine, where only a small part of the total mixture is part of the flame front and takes part in the combustion process at the same time (the entire heat of this part must be released), the entire charge reacts at the same time in the case of HCCI combustion. However, the local temperatures are significantly lower than that of the flame front in the SI engine or that of the stoichiometric zones of the diesel spray combustion, and compared to the SI combustion, the local HCCI combustion releases the heat from the entire charge more slowly and uniformly. Because the whole charge is involved simultaneously in the combustion process, HCCI heat release has a smaller overall duration than the conventional ones.

Many names have been given to combustion concepts with HCCI characteristics, e.g. Active Thermo-Atmosphere Combustion (ATAC, [62]) and Compression-Ignited Homogeneous Charge (CIHC, [57]), Premixed Lean Diesel Combustion (PREDIC, [88]), Premixed Charge Compression Ignition (PCCI, [7]), Activated Radical Combustion (AR, [38]), Controlled Auto-Ignition (CAI, [61]), Homogeneous charge Intelligent Multiple Injection Combustion System (HIMICS, [13]), Uniform Bulky Combustion System (UNIBUS, [103]) or Modulated Kinetics (MK, [43]). Nevertheless, the worldwide accepted expression, which was formed by Thring [91], is HCCI. In Europe, CAI is also used for gasoline engines, for which a higher inhomogeneity is needed to control ignition, and where the expression "homogeneous charge" is less precise. In the following, HCCI will be used for both engines, and CAI will only be utilized if special characteristics of the gasoline HCCI concept are discussed.

The first HCCI engines were two-stroke engines [60, 62]. The main target of these investigations was to eliminate misfire and to stabilize the combustion process at part load. HCCI operation, when optimized, has been shown to provide efficient and stable operation. In Fig. 6.25, a comparison of the cylinder pressure histories for SI and CAI operation is given. The diagrams clearly show the elimination of cycle-to-cycle variations of cylinder pressure and the positive effect on combustion stability, which has also been reported by several other authors (e.g. [37, 60, 62, 67]). Another successful two-stroke CAI-concept is the Activated Radical Combustion (AR). Honda used this combustion concept for motorcycles. In this case the HCCI-process was used to improve the stability of combustion and to reduce HC-emissions and fuel consumption at part load. High EGR-rates of up to 80% were used. At higher loads and at full load, the motor was driven as a conventional SI engine.

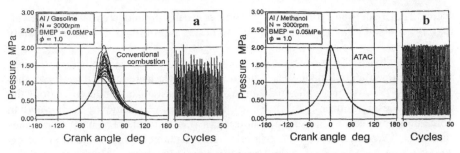

Fig. 6.25. Comparison of cylinder pressure histories. **a** SI operation (gasoline), **b** HCCI operation (labeled ATAC) with methanol [37]

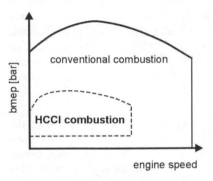

Fig. 6.26. Dual-mode operation

Najt and Foster [57] investigated the HCCI-concept for four-stroke-engines for the first time. The authors used a machine with variable compression ratio and heating of the intake air, and the fuels were mainly blends of heptane and octane. It was shown that high compression ratios allow a decrease in intake air temperature but result in excessive heat release and knocking. Thring [91] used a four-stroke engine with variable compression ratio in order to investigate the behavior of gasoline and diesel HCCI combustion. As a consequence of the limited operating range he suggested a hybrid combination of HCCI (part load) and conventional combustion (high and full load). Further basic investigations have been performed by Stockinger et al. [85] and Ayoma et al. [7].

Altogether, these early investigations have already shown that an important reduction of NO_x and soot raw emissions could be achieved, while in most cases the CO and HC emissions increased because of the low combustion temperatures. Furthermore, it has turned out that HCCI combustion is only possible at moderate loads and engine speeds, Fig. 6.26, due to uncontrollable and premature knock-like combustion at higher loads and partial burning at very low loads or high engine speed (Sects. 6.4.4 and 6.4.5). In the case of higher loads and especially at full load, conventional diesel spray combustion or gasoline SI-combustion have to be used. This so-called dual-mode concept is only attractive for applications with a high degree of part load. However, it is expected that the further development of special HCCI-fuels and combustion phasing control will help to expand the area of HCCI-combustion in the engine map and to better exploit the full benefit of HCCI combustion.

Several dual-mode concepts have already been realized: Gatellier et al. [24] have developed a dual-mode diesel engine that is able to reach near zero particulate and NO_x emissions while maintaining the performance standards of conventional DI diesel engines at full load. 50 kW/l have been achieved with conventional limiting factors and engine parameters. Duret et al. [17] have realized a similar 60 kW/l dual-mode HCCI diesel engine with early direct injection during the HCCI mode. The authors use a specially adapted narrow angle direct injection system (NADI), a compression ratio of 14:1 and zero swirl. The engine emits ultra low NO_x and soot at part load while retaining the high fuel economy of the full load performance of conventional diesel engines. Kimura et al. [43] used late direct injection during MK-HCCI combustion at part load and conventional diesel combustion at full load. This dual-mode concept is already realized in production engines since 1998 [43].

6.4.2 HCCI Chemistry

The HCCI oxidation chemistry determines the auto-ignition timing, the heat release rate, the reaction intermediates and the final products of combustion. HCCI-combustion of most fuels, especially typical diesel fuels, displays a two-stage heat release as shown in Fig 6.27. Only high octane number gasoline fuels perform a single-stage combustion [31].

The first stage of heat release is associated with low temperature kinetic reactions (low temperature oxidation, LTO) and the second and much stronger one (main reaction) is the high temperature oxidation (HTO). The time delay between LTO and HTO is attributed to the negative temperature coefficient regime (NTC) [67, 106, 107]. About 7–10% of the energy is released during LTO [106], the rest is released during HTO.

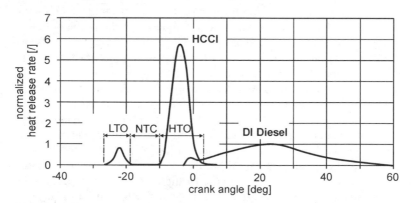

Fig. 6.27. Two-stage heat release of HCCI diesel combustion, schematic diagram

The major difference between the conventional diesel combustion and the HCCI process lies in the composition of fuel vapor at ignition. Because of its very heterogeneous fuel-air mixture, conventional combustion starts at those local mixture regions having the best ignition conditions, and then the heat generated from the combustion of these local sections enhances ignition of the neighboring regions. After ignition, the conventional diesel combustion begins with a very fast high temperature oxidation of the premixed air-fuel quantity, which has been generated during ignition delay (premixed peak). The rate of heat released by the subsequent diffusion burning depends on the velocity of the turbulent mixing of unburned or partially burnt fuel and air. In Fig 6.27, the premixed peak near TDC and the subsequent comparatively slow diffusion burning are shown.

In the case of HCCI combustion on the other hand, the whole cylinder charge reacts simultaneously. The whole charge is firstly oxidized according to the LTO chemistry and then, after a second delay (NTC region), passes to the high temperature oxidation.

In order to achieve a homogeneous air-fuel mixture, the time available for mixing must be maximized, and fuel injection must occur at early crank angles. However, this often leads to an early start of LTO and HTO reactions. Because the whole charge reacts simultaneously, the heat is released in a very short time, resulting in a considerable pressure rise, increase of noise ($dp/d\varphi$) and peak pressure compared to the conventional diesel combustion.

Here, three main challenges of HCCI combustion, the formation of a homogeneous air-fuel mixture, and the prevention of premature ignition timing and the prevention of excessive heat release rates, which limit the use of HCCI combustion to part load today, become already apparent.

The theoretical and practical roots of HCCI-combustion concepts are ultimately credited to the Russian scientist Nikolai Semenov, who began pioneering work in the field of ignition in the 1930s. Two relevant aspects of his work are the thermal and the chemical ignition theories [44, 63, 64]. The thermal ignition theory postulates that the combustion process can be initiated only, if sufficient molecular collisions are occurring, i.e. sufficient temperature and pressure conditions must be existent. The chemical or chain theory of ignition hypothesizes, that combustion involves a process of branching chemical reaction chains, initiated at active chemical centers. Spontaneous ignition occurs if the number of chains being initiated exceeds the number of chains being terminated. In the 1970s, skeletal kinetic models, based on degenerate-branched-chain and class chemistry concepts, were developed for the prediction of auto-ignition delay time in engines [32] and formed the basis for the development of further extended chemical kinetic models. In the following, the most important chemical reactions during LTO and HTO of alkanes, which are used in the chemical kinetic models, are explained.

Low Temperature Oxidation of Alkanes

The time of occurrence and the following heat release of the first stage ignition depend on the fuel molecular size and structure. The LTO reactions generally occur at temperatures below approx. 850 K [31, 92]. The most important steps during LTO are the following, further details are given in [13, 11, 10, 96, 98].

The reaction starts with the generation of hydrocarbon radicals $R^\bullet$. Fuel (RH), (R: rest molecule, H: hydrogen atom) reacts with oxygen (O_2) in an endothermic reaction:

$$RH + O_2 \rightarrow R^\bullet + HO_2^\bullet. \tag{6.1}$$

This reaction is quite slow, and as soon as enough radicals exist, the fuel molecules are attacked by the following type of reaction, which is important during the whole combustion process:

$$RH + X^\bullet \rightarrow R^\bullet + XH. \tag{6.2}$$

$X^\bullet$ is an arbitrary radical, preferably the $OH^\bullet$ radical [26]. The following two-step mechanism finally starts the low temperature oxidation:

Step 1: Together with oxygen the already existent hydrocarbon radicals form peroxy radicals ($RO_2^\bullet$)

$$R^\bullet + O_2 \Leftrightarrow RO_2^\bullet \text{ (first } O_2\text{-addition).} \tag{6.3}$$

Then, an internal H-atom abstraction

$$RO_2^\bullet \rightarrow HO_2R^{\bullet\bullet} \text{ (internal H-atom abstraction)} \tag{6.4}$$

occurs. An external H-atom abstraction is also possible. This reaction would result in a subsequent chain branching and thus could lead to ignition, while the internal H-atom abstraction does not. However, the external H-atom abstraction is much slower than the internal one [11, 10], and thus reaction 6.4 is dominant and ignition does not occur.

Step 2: A second O_2-addition,

$$HO_2R^{\bullet\bullet} + O_2 \Leftrightarrow HO_2R^{\bullet\bullet}O_2 \text{ (second } O_2\text{-addition),} \tag{6.5}$$

and the subsequent external or internal H-atom abstraction finally result in ignition:

$$HO_2R^{\bullet\bullet}O_2 + RH \rightarrow HO_2R'O_2H + R^\bullet \text{ (external H-atom abstraction)} \tag{6.6}$$

$$HO_2R'O_2H \rightarrow HO_2R^{\bullet\bullet}O + OH^\bullet \text{ (chain branching)} \tag{6.7}$$

$$HO_2R^{\bullet\bullet}O \rightarrow OR'O + OH^\bullet \text{ (chain propagation)} \tag{6.8}$$

or

$$HO_2R^{\bullet\bullet}O_2 \rightarrow HO_2R^{\bullet\bullet\bullet}O_2H \text{ (internal H-atom abstraction)} \tag{6.9}$$

$$HO_2R^{\bullet\bullet\bullet}O_2H \rightarrow HO_2R^{\bullet}O + OH^\bullet \text{ (chain propagation)} \tag{6.10}$$

$$HO_2 R''O \rightarrow OR'''O + OH^\bullet \text{ (chain branching).} \qquad (6.11)$$

R, R' and R'' are fuel molecules and "$^\bullet$" denotes fuel radicals according to Warnatz et al. [97]. The $OH^\bullet$ radicals produced by these reactions oxidize the hydrocarbons (reaction 6.2),

$$RH + OH^\bullet \rightarrow R^\bullet + H_2O, \qquad (6.12)$$

and the increasing reaction rate of these exothermic reactions (reactions 6.3 to 6.12) result in the first heat release (LTO). However, backward reactions become dominant as temperature increases due to heat release and further compression. In order to show that the reactions in both directions are important (especially in Eqs. 6.3 and 6.5), the "$\Leftrightarrow$" symbol is used. With increasing temperature, the formation of oxidizer ($OH^\bullet$) degenerates, leading to the degeneration of the first-stage heat release. This mechanism is called degenerate chain branching. It is responsible for the increase of ignition delay with rising temperature (Negative Temperature Coefficient (NTC), Fig. 6.28) and explains the existence of the NTC region between LTO and HTO during the two-stage HCCI combustion.

NTC Region

At temperatures between approx. 800 K and 1000 K, the fuel radicals from reaction 6.2 feed the following two reactions,

$$R^\bullet + O_2 \rightarrow \text{alkene} + HO_2^\bullet, \qquad (6.13)$$

$$HO_2^\bullet + HO_2^\bullet \rightarrow H_2O_2 + O_2, \qquad (6.14)$$

and result in an accumulation of H_2O_2, which remains relatively inert as long as the temperature is below about approx. 1000 K [31].

Temperature increases due to compression, and above 900-1000 K the chain branching reaction

$$H_2O_2 + M \rightarrow OH^\bullet + OH^\bullet + M \qquad (6.15)$$

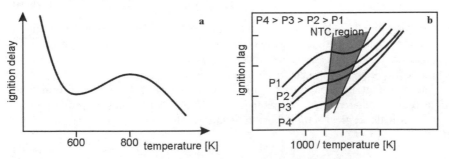

Fig. 6.28. Negative temperature coefficient for hydrocarbon ignition. **a** [84], **b** [103]

quickly produces large numbers of OH$^{\bullet}$ radicals and thus initiates the second-stage heat release process via reaction 6.12.

This temperature is independent of the fuel. Any engine parameter that makes the in-cylinder charge earlier reach the H_2O_2 decomposition temperature (e.g. higher compression ratio, earlier or increased heat release of first stage ignition, increased intake temperature) will advance the start of main combustion.

High Temperature Oxidation of Alkanes

At temperatures above 900 K, molecules with long chains break up into short C_1- and C_2-hydrocarbons. First, alkyl radicals are formed by abstraction of H-atoms (reaction 6.2):

$$RH + X^{\bullet} \rightarrow R^{\bullet} + HX . \tag{6.16}$$

$X^{\bullet}$ represents radicals like H$^{\bullet}$, O$^{\bullet}$, OH$^{\bullet}$ and HO$_2^{\bullet}$ [30]. The most important chain branching reaction responsible for the production of radicals during HTO (> 1100 K) is

$$H^{\bullet} + O_2 \rightarrow O^{\bullet} + OH^{\bullet} . \tag{6.17}$$

After H-atom abstraction the high temperatures lead to thermal break-up,

$$R^{\bullet} \rightarrow R''^{\bullet} + alkene . \tag{6.18}$$

The new alkyl radicals $R''^{\bullet}$ undergo further break-up, and the alkenes C_nH_{2n} are spilt off by the radicals O$^{\bullet}$, OH$^{\bullet}$, HO$_2^{\bullet}$ [30]. The further decomposition results in mainly $CH_3^{\bullet}$ and $C_2H_5^{\bullet}$ radicals, and the subsequent reactions lead to the formation of formaldehyde (CH_2O) and to the burning of C_2 hydrocarbons (C_2H_6, C_2H_5, C_2H_4, C_2H_3, C_2H_2). Detailed reaction paths can be found in [95, 96]. Altogether, the decomposition of fuel in C_1- and C_2-hydrocarbons is not dependent on the molecular weight of the fuel itself, and after initiation of the high temperature reactions their velocity is influenced not much by the fuel.

The HTO leads to the formation of CO, which finally oxidizes to CO_2. This final reaction releases a significant amount of energy and thus is important for high fuel economy. The most important reaction [26, 30] is

$$CO + OH^{\bullet} \rightarrow CO_2 + H^{\bullet} , \tag{6.19}$$

the reaction rate of which is only important above 1100 K. Below this temperature, there is only a partial oxidation of CO. This explains the increased CO emissions of the HCCI-process, the combustion temperatures of which are much lower compared to the conventional diesel process.

The two-stage ignition occurs, if diesel or diesel-like fuels are used. In the case of gasoline, methane or natural gas there is no LTO [2, 31], and the ignition timing depends on the critical temperature that is directly reached by compression [31, 104].

Extended Chemical Models

The chemical kinetics of auto-ignition and combustion is extremely complex. In the 1970s the first reduced kinetic models were developed for prediction of auto-ignition delay time in an engine [32]. Since the 1980s, more and more detailed chemical kinetic models have been published. However, there are hundreds of species and reactions that have to be modeled, even if simple hydrocarbons like butane [29] are used. In the case of fuels with higher molecular weight such as n-heptane [13], the modeling becomes even more complex. Despite the complexity, several approaches using detailed chemical mechanisms have been reported. Some authors use single-zone models [3, 31, 42], others multi-zone ones along with CFD [2, 45]. For example, Groenendijk et al. [31] used a detailed single-zone model (857 species and 3606 reactions for iso-octane and 544 species and 2446 reactions for n-heptane). These models are very useful for fundamental analyses of the effect of fuel or EGR composition on ignition timing and combustion.

Nevertheless, besides a detailed description of the relevant chemistry, an accurate consideration of three-dimensional turbulence and inhomogeneities inside the combustion chamber is also of fundamental importance for the HCCI combustion process. However, the combination of CFD and detailed chemistry results in very time-expensive and often impractical calculations. From this point of view, a combination of reduced chemical models and CFD seems to be the best solution. The reduced chemical models must be able to map the important reactions required to calculate the main features like ignition timing, heat release, temperature, and pressure histories and fuel consumption. Li et al. [51] have developed a reduced kinetic model for primary reference fuels (PRF). Zheng et al. [106] used this model in order to simulate HCCI first-stage ignition. The model includes 29 reactions and 20 active species. The simulations of Zheng et al. [106] agree well with experimental data concerning pressure, ignition timing, and first-stage ignition heat release. Zheng et al. [107] successfully extended this model through the entire HCCI combustion process. The new model includes 69 reactions and 45 active species and combines the chemistry of the low, intermediate, and the high temperature regions.

6.4.3 Emission Behavior

As already described, one of the most important benefits of HCCI combustion is the enormous reduction of NO_x raw emissions of 90–98% in comparison to conventional combustion [17, 24, 43, 61]. As far as homogeneity is concerned, a large degree of mixture inhomogeneity can be tolerated without resulting in increased NO_x formation [28]. The biggest part of the very low total NO_x emissions is due to the NO formation mechanism from N_2O [16, 53, 97]. This mechanism is important for combustion processes with high air excess ratios, low temperatures, and high pressures [73].

Today, HCCI combustion is limited to part load operation, where the mixture equivalence ratios and thus also the combustion temperatures of the homogeneously mixed lean charges are low. With an increase in load, the peak combustion temperatures rise and the advantage over the DI diesel engine decreases. Figure 6.29 shows the result of a calculation performed with a quasi-dimensional engine model, which was used to estimate the NO_x emissions from HCCI diesel combustion in comparison to conventional DI diesel combustion with and without EGR. The model requires the crank-angle resolved heat release (Fig. 6.29a) as input and calculates the NO_x formation rate according to the thermal (Zeldovich) and the prompt NO formation mechanism. The compression ratio of all engines is 16:1. The results show a large reduction of NO_x at part load, but a diminishing advantage of HCCI versus DI-diesel at high loads due to increasing peak temperatures, caused by an increasingly fuel-rich homogeneous mixture.

Combustion phasing is another important parameter influencing NO_x formation during HCCI combustion. Premature combustion results in an increased formation of NO_x due to a significant increase in peak pressure and peak temperature.

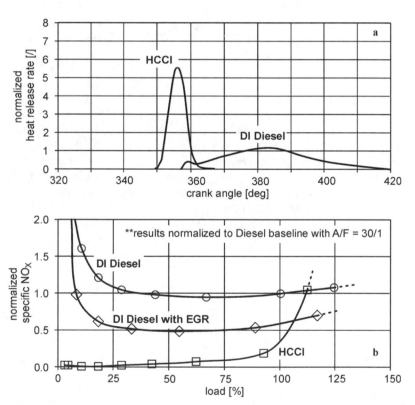

Fig. 6.29. Predicted NO_x emissions versus engine load for typical HCCI and DI-diesel combustion [87]. **a** assumed heat release, **b** NO_x emissions versus load, A/F: air-fuel ratio.

Another important benefit in the case of HCCI diesel combustion is the possibility of reducing NO_x and particulate matter simultaneously [17, 24, 43, 65]. The formation of soot requires fuel-rich zones ($\phi > 1.7$) and temperatures above 1400 K [8, 73], which are not present due to the homogeneous lean mixture. However, fuel deposition on the walls may result in poor mixture preparation and in local fuel-rich regions that are subject to incomplete combustion and produce soot. This can especially happen during early direct injection, because pressure and temperature inside the cylinder are low and fuel penetration is increased.

Compared to conventional combustion concepts, HCCI combustion usually results in significantly higher HC emissions [61, 67, 73, 87]. The HCCI diesel experiments of Schlotz [73] for example showed approximately five times more HC emissions in comparison to the conventional diesel combustion. This is caused by the low combustion temperatures due to the lean mixtures and/or the high EGR rates. EGR is usually needed to lengthen the combustion duration in order to avoid extreme heat release rates. In the case of excessive EGR this can result in misfiring and thus in a significant increase of HC emissions. On the other hand, EGR can also help to reduce the amount of unburned hydrocarbons, because they get a second chance to take part in the combustion process.

In the case of early direct injection, fuel deposition on the cylinder walls and especially in the top-land may also result in an excessive increase of HC emissions. In this case, the fuel injection system must be specially adapted to combustion chamber geometry and gas density, Sect. 6.4.4. The lower the volatility of a fuel, the more serious the problem of wall wetting.

The reduced combustion temperatures are responsible for partial burning and for decreased post-combustion oxidation rates within the cylinder, especially near the walls. For this reason, HCCI combustion also typically results in higher CO emissions than conventional diesel or spark-ignited combustion. The HCCI diesel experiments of Schlotz [73] show approximately ten to twenty times more CO emissions compared to the conventional diesel combustion.

The amount of HC and CO emissions directly influences fuel consumption, because both components contain chemical energy that has not been released during combustion. If a proper phasing of combustion relative to the engine cycle is realized, the HCCI process can approximate the ideal Otto cycle (combustion at constant volume) because of the high heat release rates and the very short combustion duration. If there is no partial burning or misfiring, this generally results in high efficiency [20]. Furthermore, the low combustion temperatures result in a reduced loss of heat to the engine. In the case of the gasoline engine, an additional advantage in efficiency is provided by the omission of throttle losses. The theoretical potential of reducing fuel consumption at partial load is comparable to that of the spray-guided direct injection technique [94, 23]. HCCI fuel efficiencies comparable to those of conventional diesel combustion at part load have been reported by several authors [7, 34]. However, if fuels with low volatility and high ignitability like diesel are used, this beneficial effect of HCCI combustion cannot always be realized because of insufficient mixture preparation, fuel wall impingement, partial burning, and poor combustion phasing (e.g. premature ignition).

6.4.4 Basic Challenges

As already mentioned in the previous sections, there are many obstacles that must be overcome in order to realize the advantages of HCCI combustion in modern engines. The three main challenges are

- the homogenization of fuel, air, and recycled burnt gases prior to ignition,
- the control of ignition and combustion timing, and
- the control of heat release rates.

The difficulties associated with these three main challenges are described in the following.

Homogeneous Mixture Formation

An effective mixture formation and the avoidance of fuel-wall interactions is crucial for achieving high fuel efficiency, reducing HC and PM emissions, and preventing oil dilution. Deposition of fuel on the walls has been proven detrimental to HC emissions even for moderately volatile fuels like gasoline [86]. Regarding the mixture formation and homogenization of fuel and air, two main categories, the external and the internal mixture formation, can be distinguished, see Fig. 6.30.

The simplest way of achieving a homogeneous in-cylinder mixture is the introduction of fuel upstream of the intake valves (external mixture formation). The mixture enters the cylinder during the intake stroke. This method is also known as port injection or fumigation. The turbulence created by the intake flow supports further homogenization. Because the air-fuel mixture experiences the whole time-temperature development, the port injection belongs to the early homogenization concepts. A drawback of this strategy is that injection timing cannot be used to influence the start of ignition. In the case of heavy fuels with reduced volatility, the port injection results in poor evaporation as well as increased wall impingement, HC and CO emissions, fuel consumption and oil dilution. This injection is mainly attractive for gaseous and liquid fuels with high volatility, but not for diesel fuel.

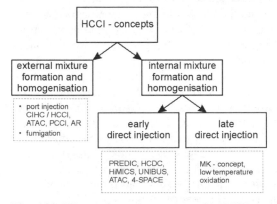

Fig. 6.30. Mixture formation concepts for HCCI-engines [56]

In the case of internal mixture formation, the fuel is directly injected into the cylinder. Two concepts, that of early and of late injection, are possible. Early injection is the mostly used method for HCCI diesel applications and uses a long ignition delay along with low temperatures to homogenize the fuel mixture. A part or even the total amount of the fuel is injected noticeably before top dead center, see Fig. 6.31. In the case of diesel injection, the poor volatility of the fuel and the low gas density of the air inside the cylinder can result in considerable wall wetting. For these cases, new and highly flexible injections systems have to be designed, which have to be specially adapted to the variation of combustion chamber geometry as well as in-cylinder pressure and temperature during injection. Despite the problems of the in-cylinder injection, direct injection is expected to become the preferred method for HCCI engines in the future [87].

Some researchers have already spent significant effort on developing low-penetration injectors and minimizing wall impingement [24, 17, 34, 39, 73]. A suitable injection system must have a high degree of flexibility in order to allow an adaptation of the injection strategy to the varying boundary conditions during injection. High-pressure injection in combination with a large number of small nozzle holes is generally used in order to increase spray disintegration and to include the complete cylinder charge in the mixture formation process while avoiding wall film formation. A further adaptation of the spray penetration can be realized by splitting the injection event into several pulses having different durations. Figure 6.31 shows an example of such a high-pressure pulsed injection strategy. The shorter the pulse duration, the less the momentum of the liquid, resulting in reduced penetration. The area below the curves represents the fuel mass belonging to each pulse. The low gas density at the beginning of injection requires short pulses with reduced injection velocities, and the time interval between the pulses is relatively large. As the piston moves up, density and temperature in the cylinder increase and penetration is reduced. The pulse durations can be prolonged, while the time intervals between subsequent pulses are decreased. At the end of the pulsed injection the distance between nozzle and piston reduces significantly, and the mass injected per pulse must be reduced again in order to prevent fuel deposition on the piston. A piezo common rail injection system capable of performing these fully variable pulsed injections has been developed by Meyer et al. [54].

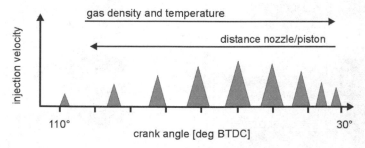

Fig. 6.31. Pulsed injection strategy for early in-cylinder injection

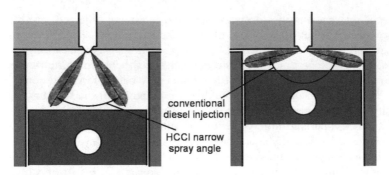

Fig. 6.32. Adaptation of spray angle for early in-cylinder injection

In the case of early in-cylinder injection, the spray direction must be adapted as well. In comparison to the conventional diesel injection near TDC, the volume between the nozzle and the piston is significantly larger. In order to achieve an adequate mixing of fuel and gas and to prevent fuel deposition on the cold liner, the angle between the spray plumes must be decreased, see Fig. 6.32. In order to realize both combustion modes in an engine (dual-mode), the use of a variable nozzle concept, Sect. 6.2.4, would be highly favorable.

Figure 6.33 shows an example of a numerical investigation of the mixture formation in a 2.0 liter single-cylinder HCCI diesel engine with compression ratio 14:1 and zero swirl. A 13-hole common rail injector (hole diameter: 0.12 mm, one central hole, two rows of six holes with different spray directions) is used. Between 110° BTDC and 30° BTDC, nine pulses with a total mass of 70 mg, as shown in Fig. 6.31, are injected. The black dots represent the liquid droplets, the shaded background represents the ratio of air and evaporated fuel.

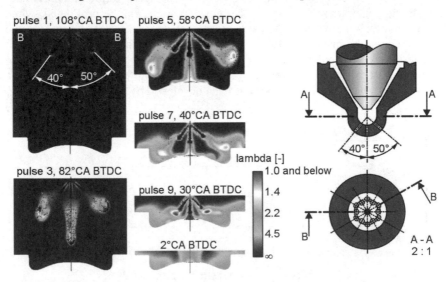

Fig. 6.33. Numerical investigation of mixture formation for early in-cylinder injection

Liquid deposition on the liner can be avoided because of the short pulses and the narrow spray angles. However, these smaller angles result in fuel deposition on the piston at the end of the pulsed injection, and the use of a variable nozzle geometry would again be highly favorable. Because the temperature of the piston is significantly higher than the one of the liner, fuel deposition on the piston is less critical. Near TDC, a very lean but not completely homogeneous mixture can be achieved. Although the NO_x emissions are not influenced very much by some degree of inhomogeneity, partial burning may occur in the very lean regions, which can result in increased HC and CO emissions.

A similar narrow spray angle and multi-hole HCCI injection system for early direct injection has been developed and tested by Gatellier et al. [24]. Multi-stage injection with up to eight pulses per cycle is used in order to reduce spray penetration. Engine and injection system are developed for dual-mode operation.

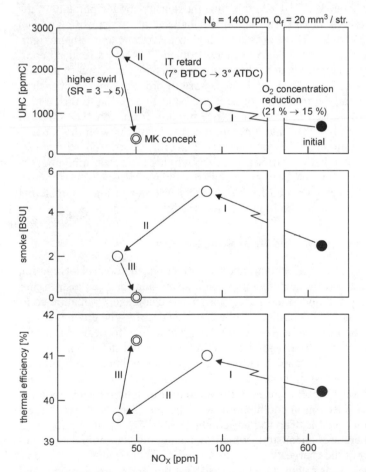

Fig. 6.34. Effects of EGR, retarded injection timing (IT) and increased swirl on exhaust emissions and thermal efficiency, Nissan MK-concept [43]

This means that the same injection system can be used for HCCI combustion at part load and for conventional diesel combustion at full load. Another dual-mode HCCI diesel engine has been developed by Duret et al. [17]. The authors also utilized a specially adapted narrow-angle direct injection system (NADI) with 10 holes and angles of 25° to 35° between needle axis and hole axis. An engine with a compression ratio of 14:1 and zero swirl was used. The injection was split into four pulses, starting at about 120°–110° BTDC and ending at 60°–35° BTDC in the case of HCCI combustion. Ultra low NO_x and soot emissions at part load are reported, while the engine also retained the high fuel economy during conventional combustion at full load.

The second direct injection strategy is late in-cylinder injection, which was chosen for example by Kimura et al. [43] (Nissan MK-concept, Fig. 6.34) and by Shimazaki et al. [76]. In the case of the MK-concept, the late direct injection of diesel fuel starts at about 3° ATDC. At this time the density and temperature of the cylinder charge are high, mixture formation is enhanced, and a deposition of fuel on the walls can be avoided. However, due to high pressure and temperature, ignition delay is short. Only with the use of heavy cooled EGR and a reduced compression ratio (16:1) can ignition delay be prolonged such that injection can be finished prior to ignition. This is absolutely necessary in order to avoid a diffusion-limited combustion and the formation of soot and NO_x. In addition to this, a high swirl ratio (SR: 5.0) is used to further enhance mixture formation. The onset of heat release is clearly after TDC and thus significantly later than with the conventional combustion, resulting in a reduction of peak pressure and combustion noise, but also in a decrease of thermal efficiency. Although there is a clear inhomogeneity of the mixture, very low NO_x emissions can be achieved. Because it is impossible to achieve a further increase of ignition delay in order to inject larger fuel quantities, the MK-concept is also limited to low loads.

Control of Ignition Timing and Heat Release Rate

Besides the problem of homogeneous mixture formation, the control of ignition timing, which determines the main combustion phasing and thus has a strong influence on efficiency and operating range, is a serious challenge. Premature combustion can result in heavy knock-like combustion that destroys the engine.

A stable combustion can be realized at low and partial load for lean fuel-air ratios and/or large amounts of EGR. An increase in load towards stoichiometric values results in a significant increase in heat release rates and in knock-like combustion. Furthermore, the emission benefits vanish (e.g. [7, 85, 87]). Unlike in spark ignition or conventional diesel engines, a direct control of ignition timing via the spark or the start of injection in combination with the very short and well-known ignition delay is not possible. The start of combustion is significantly influenced by the low temperature chemistry, which depends on the complete time-temperature history of the charge.

Most of the applications with diesel fuel suffer from premature ignition, and cooled EGR or reduced compression ratios are used in order to increase ignition delay.

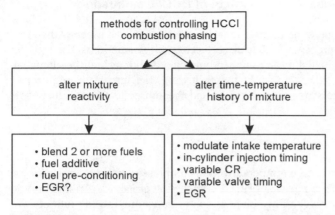

Fig. 6.35. Methods for controlling HCCI combustion phasing [87]

In the case of gasoline applications, the reduced ignitability of gasoline and the generally lower compression ratios of these engines often require measures like intake air heating or non-cooled internal EGR to achieve a reliable ignition. The main reason for the difficulty in controlling the start of combustion exactly is the long time interval during which the low temperature reactions occur. A slight variation of the boundary conditions can easily result in significant variations of the main combustion phasing. HCCI combustion phasing is affected by

- the auto-ignition properties of the fuel,
- the fuel-air ratio,
- the volatility of the fuel,
- the EGR rate, the temperature and the reactivity of the recycled gas,
- the mixture homogeneity,
- the compression ratio of the engine,
- the intake temperature, and
- the heat transfer to the engine.

In order to control combustion phasing, two main groups of approaches can be distinguished [87], see Fig. 6.35. The first group are methods, the purpose of which is to alter the time-temperature history of the mixture. It includes fuel injection timing, variation of intake air temperature, variation of compression ratio (CR) and variable valve timing. The second group attempts to control the reactivity of the charge by varying the properties of the fuel, the fuel-air ratio or the amount of oxygen by EGR. A detailed description of the influence of each method on the HCCI combustion process is given in the following section.

6.4.5 Influence Parameters and Control of HCCI Combustion

The key challenge for operating an engine in HCCI auto-ignition mode is the control of the combustion process without direct access to the reactions. The start of chemical reactions is initiated by the thermodynamic conditions and the chemical composition of the cylinder charge. In the following, the important engine parameters influencing the start of combustion and the subsequent heat release are discussed in detail.

Compression Ratio

Being directly related to charge temperature, the compression ratio is important in determining the rate of heat release in HCCI combustion processes. An increase of compression ratio raises the end-of-compression temperature and the rate of low temperature reactions. This advances the overall ignition process and allows inlet charges of lower temperature to be successfully combusted (CAI process). In the case of HCCI diesel engines the conventional compression ratios usually must be reduced in order to delay the start of combustion and to prevent excessive heat release rates. Nevertheless, the reduction of compression ratios should be moderate, because a reduction of compression ratio decreases thermal efficiency.

Figure 6.36 from Ryan's study of diesel fuelled HCCI engines [71] shows predicted values of start of combustion versus compression ratio. With a constant EGR rate of 40% ignition advances about 200 degrees if the compression ratio is changed from 8:1 to 12:1. Figure 6.37 shows similar results from Velji et al. [94] Excessive compression ratios result in knock-like combustion with high temperatures and increased NO_x emissions.

A variable compression ratio by variable valve timing is an effective but expensive method for solving this conflict. This technology can be used for dual-mode engines to realize the high compression ratios needed for high thermal efficiency in the case of conventional combustion at full load and to enable lower compression ratios in the case of HCCI combustion.

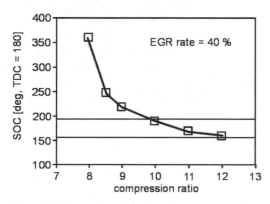

Fig. 6.36. Start of combustion versus compression ratio [71]

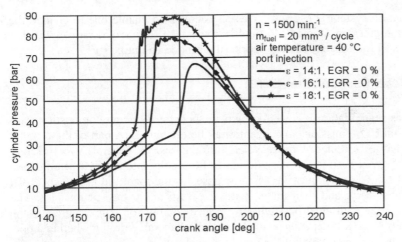

Fig. 6.37. Effect compression ratio on HCCI combustion [94]

Variable Valve Timing

In contrast to HCCI diesel applications, in which EGR-cooling is needed in order to suppress a premature ignition start, lean homogeneous gasoline-air mixtures need additional heating in order to auto-ignite. The required high temperatures and the charge dilution for decelerating the heat release are usually obtained by the application of non-cooled internal EGR, which can be provided efficiently only by variable valve trains. The residual gas can be supplied either by remaining in the combustion chamber due to early exhaust valve closing, or by redrawing from the exhaust system during the induction stroke, Fig. 6.38.

Although simple systems with cam phasers and valve lift shift tappets can fulfill the steady-state HCCI requirements, the dynamic behavior is limited. For fully transient operation a cycle-to-cycle and cylinder individual control of charge composition is advantageous. This is possible with a fully variable electro-hydraulic

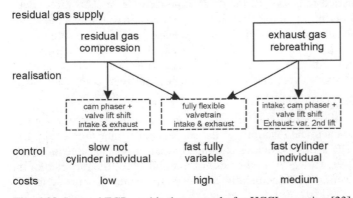

Fig. 6.38. Internal EGR: residual gas supply for HCCI operation [23]

valve actuation system [17, 23, 6]. These systems also enable one to change the effective compression ratio and to modify the intake flow (swirl ratio), which is also beneficial for conventional operation (dual-mode engine).

Exhaust Gas Recirculation (EGR)

Exhaust gas recirculation is a well-known method for reducing combustion temperatures and NO_x-emissions in conventional diesel engines. A part of the original cylinder charge is replaced by recycled burnt gases. In general, two different EGR-concepts are used today, the external EGR and the internal EGR. In the case of external EGR, a portion of the exhaust gases in the manifold is branched off, cooled if necessary, and then mixed with the fresh air in the suction part. The cooling offers an additional possibility to reduce combustion temperatures. In the case of internal EGR it depends on the inlet and outlet valve timing whether a portion of the hot exhaust gases of the previous cycle stays in the cylinder (e.g. [49, 102]) or is sucked back from the manifold (e.g. [6, 23, 49, 102]). In the case of HCCI combustion, EGR has four important effects:

Heat capacity effect: During compression and combustion, the inert burnt gases must be heated up together with the rest of the in-cylinder charge. Because the total heat capacity of the charge is higher with burnt gases due to the higher specific heat capacity values of carbon dioxide (CO_2) and water vapor (H_2O), lower end-of-compression and combustion temperatures are achieved, and heat release rates as well as maximum pressure rise $(dp/dt)_{max}$ are reduced [48, 104]. The heat capacity effect extends the combustion duration if large amounts of EGR are used.

Charge heating effect: if hot burnt gases are mixed with cooler inlet air, the temperature of the inlet charge increases. The heating effect is important for non-cooled EGR applications and is mainly responsible for advanced auto-ignition timing. It also increases the heat release rate and the value of maximum pressure rise $(dp/dt)_{max}$, and shortens the combustion duration [104].

Dilution effect: The introduction of burnt gases into the cylinder replaces a part of the inlet air and causes a reduction of the oxygen concentration. This effect does not affect auto-ignition timing in the case of CAI combustion, but it extends combustion duration and also slows down the heat release rate if large amounts of EGR are used [104, 31]. Experiments of Tsurushima et al. [92] with diesel-like fuels (two-stage ignition) also show a decrease of oxidation rates, but the start of both, LTO and HTO, is delayed, and the time interval between both reactions is expanded.

Chemical effect: Finally, active combustion products present in the burnt gases can participate in the chemical reactions of the subsequent combustion cycle. External and cooled EGR usually provides chemically inert gases and thus does not contribute to this effect. However, in the case of internal EGR, the recycled burnt gases may contain short-lived chemically active components that result in advanced auto-ignition [38, 102, 48].

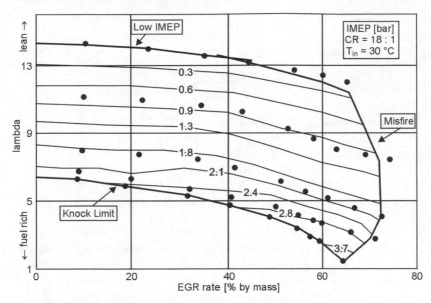

Fig. 6.39. Successful HCCI operational range and IMEP map. Parameter: IMEP, compression ratio 18:1, intake temperature 30°C, n-heptane [65]

The influence of the EGR-effects mentioned above depends on the HCCI-sub-concept that is applied. In the case of HCCI diesel applications, external EGR with additional cooling is customarily used in order to prolong ignition delay, while in the case of CAI combustion (gasoline-HCCI) internal non-cooled EGR is mainly used in order to increase the temperature to the required auto-ignition level. A very extensive study of the HCCI-process on a four-stroke diesel engine ($\varepsilon = 18{:}1$, port injection system, external EGR, fuel: n-heptane) has been carried out by Peng et al. [65]. Similar investigations for a gasoline CAI process have been reported by Oakley et al. [61]. The investigations of Peng et al. give a good overview about the general effect of EGR on knock limit, engine load, combustion stability and emissions.

Figure 6.39 shows the successful HCCI operating region and the values of the indicated mean effective pressure (IMEP) as a function of EGR rate and the overall air-fuel ratio of the cylinder charge. At constant EGR-rate, the HCCI combustion is limited by its low load limit (lean mixture) and by the knocking combustion at higher loads (richer mixture). Starting from low IMEP and increasing load, the amount of fuel injected per cycle must be increased, resulting in lower air-fuel ratios (decreasing lambda values) and richer mixtures inside the cylinder. In the case of knocking combustion, too much heat is released during the very fast combustion which results in excessive peak pressures and can damage the engine. Near this limit, high local temperatures result in an increase of NO$_x$-emissions [23, 24, 31]. An increase of EGR leads to a decrease of combustion temperature, later combustion timing and a reduced heat release rate.

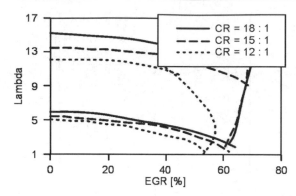

Fig. 6.40. Effect of compression ratio on the operational range of HCCI diesel combustion [17]

This allows for an increase in the amount of fuel injected per cycle and thus in load and for a shift in the knock limit towards smaller lambda values. On the other hand, EGR also reduces the low load limit in terms of lambda: the more recycled burnt gases are present, the lower the combustion temperature. These low temperatures can result in partial burning, which again results in increased HC and CO emissions and in a reduction of efficiency. The third limit of the HCCI-operation region occurs at EGR-rates of about 70%. Reaction rates and ignition timing are so much reduced and retarded that misfiring occurs and HC-emissions increase again.

Similar investigations have been performed by Duret et al. [17]. Figure 6.40 shows the effects of a variation of compression ratio. The operational range of HCCI diesel combustion increases with increasing compression ratio and the misfire limit is shifted to higher EGR ratios. However, the maximum IMEP value is reduced due to earlier ignition, faster combustion and thus earlier onset of knocking combustion at leaner mixtures.

Inlet Air Temperature

Modulating the intake air temperature in order to control the start of ignition is the most popular method in laboratory experiments. Higher intake temperatures advance the start of combustion and vice-versa. Figure 6.41 shows results from Velji et al. [94] for a blend of 80% methane and 20% diesel and different air temperatures. Besides a large loss in volumetric efficiency in the case of high intake air temperatures, the range of crank angle, over which the ignition timing can be influenced, is quite limited. In the case of non-cooled EGR, the variation of the intake temperature is usually one of the multiple effects of a change of the EGR-rate.

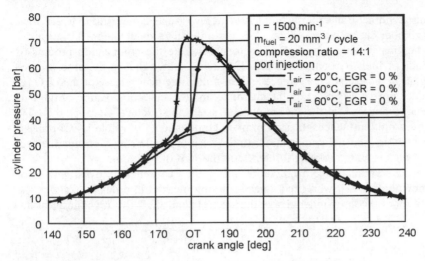

Fig. 6.41. Effect of inlet air temperature on HCCI combustion [94]

Air Excess Ratio

As already described together with the effects of EGR, progressively richer mixtures translate into advancing ignition and faster energy release. At a certain limit, depending on the fuel and further boundary conditions, combustion stability degrades, knock appears and the NO_x emissions increase [23, 44, 31]. On the other hand, leaner mixtures deliver retarded ignition and slower energy release. Extremely lean mixtures result in incomplete combustion at low temperatures leading to increased emissions of CO and HC. Supercharging can increase the IMEP of the engine under HCCI operation if combined with heavy EGR. However, due to the low exhaust gas temperatures, the energy provided by the exhaust gas turbine of the turbocharger is often not sufficient to realize high boost pressures [73].

Engine Speed

In conventional engines, the increasing engine speed results in increased turbulence and burn rates, and thus no measurable effect of speed can be reported. With HCCI systems however, the mixture homogeneity is already completed prior to combustion. This is the reason why reaction rates remain relatively unchanged with varying engine speed and thus the time available for reactions is reduced with increasing engine speed. High engine speed can therefore result in misfire and in reduced power output and efficiency [67, 91].

Fuel Composition

Regarding the HCCI combustion process, fuels can be divided into two main groups, the diesel-like fuels and the gasoline-like ones. Diesel-like fuels (increased ignitability, high cetane number) perform a two-stage combustion, while low-cetane number fuels like gasoline show a single-stage combustion. Using a detailed chemical kinetic model in order to investigate the combustion process of both kinds of fuels, Groenendijk et al. [31] have shown that in the case of two-stage ignition the time of occurrence and the following heat release of the LTO reaction does not only depend on the chemical and thermal composition of the cylinder charge but also on the fuel molecular size and structure. Because the LTO reaction strongly influences the start of main combustion, the whole HCCI process is influenced by the fuel composition. In the case of single-stage ignition the fuel composition has no remarkable influence on the start of ignition.

These observations have been confirmed by the experimental investigations of Tsurushima et al. [92], who used a gas sampling method in order to examine the composition of the in-cylinder charge during LTO and HTO for three diesel-like

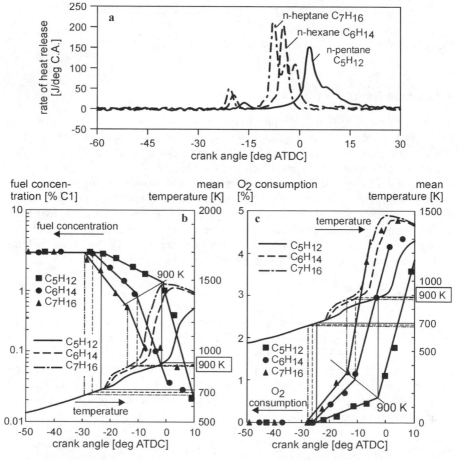

Fig. 6.42. Comparison of oxidation process for different single-component fuels (data from [92]). **a** rate of heat release, **b** fuel consumption, **c** O_2 consumption

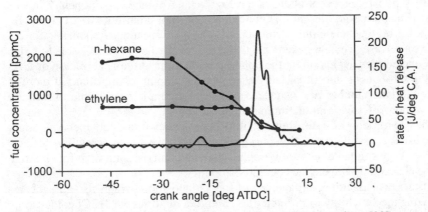

Fig. 6.43. Comparison of oxidation process for different fuels (data from [92])

straight-chain saturated hydrocarbons (n-pentane, n-hexane, n-heptane) with different chain lengths. The engine used in this study was a 2.0 liter four-valve naturally aspirated single-cylinder engine with a compression ratio of 16.5:1. The complete homogenization of the mixture was performed in an external vaporizer, and intake temperature (100°C) as well as equivalence ratio and O_2-concentration were kept constant for all experiments. All fuels show the characteristic two-stage combustion with LTO and HTO, Fig. 6.42a. The ignition and combustion behavior depends on the fuel molecular size: the longer the carbon chain and thus the higher the ignitability (longer molecules break up more easily into radicals), the earlier the start of LTO and the more energy is released. For this reason, the mean temperature in the combustion chamber earlier reaches the value of approximately 900 K, which is needed to initiate the HTO.

A more detailed description of these effects is given in Figs. 6.42b and 6.42c. Figure 6.42b shows the fuel concentrations and mean fuel temperatures versus crank angle, and Fig. 6.42c presents the consumption of oxygen. The fuel concentrations start to decrease at about 700 K. The longer the carbon chain, the earlier the fuel breaks up into radicals and the faster the reactions. This results in an earlier and faster decrease of fuel concentration combined with an earlier and faster consumption of oxygen, and a higher amount of heat is released as proven by the temperature curves. The investigations show that for n-paraffin fuels, the fuel having a longer straight chain is more likely to be oxidized in the low temperature range up to 900 K. The curves of fuel and oxygen concentration change their respective gradients at a temperature of 900 K, which is the start of HTO. From this point on, the gradients are the same for all fuels, and the effect of molecular size is negligible.

The investigations suggest that it is possible to control the oxidation rate in the low temperature range by modifying the fuel. For this reason, Tsurushima et al. [92] have also performed investigations with a mixture of n-hexane and ethylene. The resulting heat release again shows the well-known two-stage behavior, Fig. 6.43, but the behavior of both fuel components during the first stage (LTO) is different. While the concentration of the low cetane number component ethylene keeps almost constant during LTO (ethylene does not contribute to LTO heat release), the high cetane number component n-hexane is mainly responsible for the first heat release. Thus, n-hexane plays the role of the igniter, being able to ignite itself and ethylene. This offers the opportunity to control the heat release of LTO and the subsequent start of HTO by changing the amount of n-hexane. If more of the low cetane number fuel is added, it may for example be possible to increase load without altering ignition timing.

The experiments of Tsurushima et al. [92] have shown that fuel blending might be a very successful combustion phasing control. However, its feasibility for production is questionable because it implies that the composition of the fuel mixture can be changed during engine operation.

Further investigations of the effect of fuel composition, which also aim at characterizing the suitability of different fuels and fuel mixtures for HCCI combustion, have been published by Ryan et al. [72] and Montagne and Duret [55].

Homogeneity

In the case of diesel HCCI combustion, the degree of homogeneity of the charge prior to ignition should be as high as possible in order to prevent the formation of soot, HC and CO emissions. As far as nitric oxides are concerned, a large degree of mixture inhomogeneity can be tolerated without resulting in increased NO_x formation [87].

Regarding CAI gasoline combustion, the example in Fig. 6.44 shows that in the case of realistic values of intake air temperature as well as EGR rate and temperature, a complete homogeneity results in an end-of-compression temperature below the required auto-ignition temperature of about 1000°C. Further measures like intake air heating are necessary. According to ref. [101] the intake air must be heated up to approximately 150°C in order to achieve the required end-of-compression temperatures of 1000–1200°C.

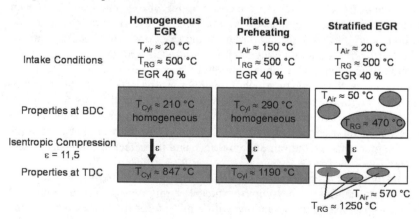

Fig. 6.44. CAI combustion (gasoline): influence of time-temperature history and homogeneity on the ignitability of the mixture [101]

Partly homogenized mixtures on the other hand have a higher probability to ignite earlier if hot EGR zones adjoin to fuel-rich reactive zones. In this case, the ignition occurs at overall lower but sufficiently high local temperatures without heating of the intake air. Ignition occurs first in a relatively small area, but because the whole charge is near the ignition limit, there is no conventional flame combustion. The energy release of the first local ignition initiates a multitude of further ignitions in the whole combustion space resulting in a homogeneous combustion, the energy release rates of which can be controlled by the degree of homogenization [94].

6.4.6 Transient Behavior – Control Strategies

As soon as steady-state conditions are achieved, the HCCI combustion is remarkable stable, but small challenges in the boundary conditions have a significant negative impact on the engine behavior. Unfortunately, in real engine operation the parameters described in the previous section, which are used to control the HCCI combustion process, are interacting strongly. Conventional single input – single output control strategies cannot be applied for a sufficient control under transient conditions any more [23]. Besides conventional and relatively slow mass flow, air excess ratio, and temperature sensors, a real-time combustion signal is needed in order to control the combustion from cycle to cycle and to allow a transient variation of speed and load in the HCCI operation region. One of the most challenging tasks is the mode transition between HCCI and conventional combustion. In this case, the change of the relevant thermodynamic values is unsteady from one cycle to the other, and a model-based combustion control with a precise prediction of charge composition and thermodynamic conditions is required. In the case of a dual-mode CAI engine for example, the transition from SI to HCCI mode must be realized if load is reduced. Due to the typically high exhaust gas temperatures there will be a very advanced combustion with an unfavorable maximum pressure rise in the first HCCI cycle. The sudden change from one mode to the other has to be smoothened by a model-based transition functionality.

6.4.7 Future HCCI Engine Applications

HCCI engines have demonstrated their potential to realize very low emissions of NO_x and particulate matter (PM), as well as high thermal efficiency. However, in order to realize these advantages in modern engines, numerous problems still have to be solved. Although full-time HCCI engines have the biggest theoretical potential to exploit the benefits of HCCI-combustion, it is still in question if this combustion concept will ever function at full load. Today, HCCI applications are limited to part load, but it is expected that the further development of special tailored HCCI-fuels and combustion phasing control might help to expand the area of HCCI-combustion in the engine map. Some authors even predict that future development will result in a so-called Combined Combustion System (CCS) [5, 82], such that there will be no difference between HCCI and CAI combustion any more. The most realistic concept for near-term applications will be the dual-mode concept, which takes advantage of the HCCI or CAI benefits at low loads, and operates on either spark-ignition or conventional diesel combustion at full load. Variable compression ratios due to variable valve actuation will help to realize high compression ratios for high thermal efficiency in the case of conventional diesel combustion at full load and lower compression ratios in the case of HCCI combustion at part load. An appropriate model-based combustion control will help control the HCCI process during transient engine operation and to smooth the transition from one mode to the other.

Compared to the conventional diesel or gasoline engines, a further disadvantage of the HCCI combustion is the increased emission of HC and CO due to low temperatures, incomplete combustion, and flame quenching near the walls. However, these emissions are considered less critical, because they can be reduced very effectively with conventional oxidation catalysts.

Cold start, noise, and lifetime are further problems that have to be solved. Nevertheless, it is believed that the use of advanced technologies like

- fully variable valve actuation,
- electric assisted turbochargers in order to supply sufficient pressure at low exhaust gas temperatures,
- multiple injection strategies,
- variable nozzle concepts, and
- model-based cycle-to-cycle combustion control

will help to make practical applications of HCCI engines realistic in the near future.

References

[1]   Aabo K, Kjiemtrup N (2004) Latest on Emission Control Water Emulsion and Exhaust Gas Re-Circulation. CIMAC Congress 2004, Kyoto, paper 126
[2]   Aceves SM, Flowers DL, Westbrook, CK, Smith JR, Pitz WJ (2000) A Multi-Zone Model for Prediction of HCCI Combustion and Emissions. SAE paper 2000-01-0327
[3]   Aceves SM, Smith JR, Westbrook CK, Pitz WJ (1999) Compression Ratio Effect on Methane HCCI Combustion. ASME Journal of Engineering for Gas Turbines and Power, vol 121, pp 569–74
[4]   Achleitner E, Berger S, Frenzel H, Klepatsch M, Warnecke V (2004) Benzin-Direkteinspritzsystem mit Piezo-Injektor für strahlgeführte Brennverfahren. Motortechnische Zeitschrift (MTZ), 65 (5/2004), pp 338–349
[5]   Affenzeller J, Kriegler W, Lepperhoff G, Owen N, Gruson JF, Blaich M (2003) FURORE, Future Road Vehicle Research, a Roadmap for the Future. European Automobile Engineers Cooperation (EAEC), paper C120

[6]   Allen J, Law D (2001) Advanced Combustion Using a LOTUS Active Valve Train, Internal Exhaust Gas Recirculation Promoted Auto-Ingition. In: Duret P (ed) A New Generation of Engine Combustion Processes for the Future?, Paris, pp 85–100

[7]   Aoyama T, Hattori Y, Mizuta J, Sato Y (1996) An Experimental Study on Premixed-Charge Compression Ignition Gasoline Engines. SAE paper 960081

[8]   Böhm H Hesse D, Jander H, Lüers B, Pietscher J, Wagner HG, Weiss M (1988) The Influence of Pressure and Temperature on Soot Formation in Premixed Flames. 22nd Symposium (International) on Combustion, The Combustion Institute, pp 403–411

[9]   Bonse B, Dittus B, Giersch J, Kerst A, Kügler T, Schuhmacher H, Wintrich T (2003) Diesel Injection Nozzle – Innovations with Opportunities to Reduce Emissions, Fuel Consumption and Noise. 5. Int Stuttgarter Motorensymposium, Stuttgart, pp. 1933

[10]   Chevalier C, Louessard P, Müller UC, Warnatz J (1990) A Detailed Low-Temperature Reaction Mechanism of n-Heptane Auto-Ignition. Proceedings of the 2nd Symposium on Diagnostics and Modeling of Combustion in Reciprocating Engines, Comodia 90, pp 93–97

[11]   Chevalier C, Warnatz J, Melenk H (1990) Automatic Generation of Reaction Mechanisms for the Description of the Oxidation in Higher Hydrocarbons. Berichte der Bunsen-Gesellschaft für Physikalische Chemie vol 94, pp 1362–1367

[12]   Chmela F, Jager P, Herzog P, Wirbeleit F (1999) Emissionsverbesserung an Dieselmotoren mit Direkteinspritzung mittels Einspritzverlaufsformung. Motortechnische Zeitschrift (MTZ) 60, pp. 552–558. English Version: Reducing Exhaust Emissions of Direct Injection Diesel Engines via Injection Rate Shaping. MTZ worldwide, 60 (9/1999), pp. 5–8

[13]   Curran HJ, Gaffuri P, Pitz WJ, Westbrook CK (1998) A Comprehensive Modeling Study of n-Heptane Oxidation. Combust. Flame 114, pp 149–177

[14]   Dohle U, Dürnholz M, Kampmann S, Hammer J, Hinrichsen C (2004) 4th Generation Diesel Common-Rail Injection System for Future Emission Legislation. FISITA World Automotive Congress, Barcelona, paper F2004V271

[15]   Dohle, U (2003) Innovative Injection Technology for Future Diesel Passenger Cars. 12th Aachener Motorenkolloquium, Aachen, pp. 107–123

[16]   Drake MJ, Ratcliffe JW, Blint RJ, Carter CD, Laurendeau NM (1990) Measurements and Modelling of Flamefront NO Formation and Superequilibrium Radical Concentrations in Laminar High-Pressure Premixed Flames. 23th Symposium (International) on Combustion, The Combustion Institute, Pittsburgh, pp 387–395

[17]   Duret P, Gatellier B, Miche M, Montreiro L, Zima P, Marotaux D, Blundell D, Ganser M, Zhao H, Perozzi M, Araneo L (2003) Innovative Diesel HCCI Combustion Process for Passenger Cars: the European SPACE LIGHT Project. EAEC Congress, paper C108

[18]   Ehlers G, Pfalzgraf B, Wurms R (2003) Homogeneous Split – a Highly Efficient Strategy to Reduce the Exhaust Emissions of FSI Engines. 9th Symposium The Working Process of the Internal Combustion Engine, Institute for Internal Combustion Engines and Thermodynamics, Graz University of Technology, Austria

[19]   Eichlseder H, Baumann E, Müller P, Neugebauer S (2000) Chancen und Risiken von Ottomotoren mit Direkteinspritzung. Motortechnische Zeitschrift (MTZ), 61 (3/2000), pp 144–152. English version: Potential and Risks of Gasoline Direct Injection Engines for Passenger Car Drivelines. MTZ worldwide, 61 (3/2000), pp 2–5

[20]   Epping K, Aceves S Bechtold R, Dec J (2002) The Potential of HCCI Combustion for High Efficiency and Low Emissions. SAE paper 2002-01-1923

[21] Fessler H, Langride S, Eckhardt T, Gstrein W (2003) Prospects for the Diesel Engine with stricter emission laws. 9th Symposium The Working Process of the Internal Combustion Engine, Institute for Internal Combustion Engines and Thermodynamics, Graz University of Technology, pp. 1–26

[22] Flynn PF, Durrett, RP, Hunter, GL, Loye, AO, Akinyem OC, Dec, JE, Westbrook, CK (1999) Diesel Combustion: An Integrated View Combining Laser Diagnostics, Chemical Kinetics, and Empirical Validation. SAE paper 1999-01-0509

[23] Fuerhapter A, Piock WF, Fraidl GK (2003) CSI – Controlled Auto Ignition – the Best Solution for the Fuel Consumption Versus Emission Trade Off?. SAE paper 2003-01-0754

[24] Gatellier B, Walter B (2002) Development of the High Power NADI[TM] Concept Using Dual Mode Diesel Combustion to Achieve Zero NOx and Particulate Emissions. THIESEL 2002 Thermo- and Fluid-Dynamic Processes in Diesel Engines, Valencia, 10th-13th September 2002, pp 131–143

[25] Geringer B, Klawatsch D, Graf J, Lenz HP, Schuöcker D, Liedl G, Poick WF, Jetzinger M, Kapus P (2004) Laserzündung, ein neuer Weg für den Ottomotor. Motortechnische Zeitschrift (MTZ), 65 (3/2004), pp 214–219. English version: Laser Ignition, New Potential for the Petrol Engine. MTZ worldwide 65 (3/2004), pp 24–26

[26] Glassman I (1996) Combustion. Third Edition, Academic Press, San Diego

[27] Golloch R (2005) Downsizing bei Verbrennungsmotoren. Springer-Verlag, Berlin, Heidelberg, New York, ISBN 3-540-23883-2

[28] Gray III AW, Ryan III TW, Roberts CE, Dodge LG (1998) Homogeneous Charge Compression Ignition (HCCI) Emissions Formation. 31st ISATA Symposium, Düsseldorf, Germany, paper no 98ATE31

[29] Green RM, Cernansky NP, Pitz WJ, Westbrook CK (1987) The role of Low Temperature Chemistry in the Autoignition of n-Butane. SAE paper 872109

[30] Griffith JF, Barnard JA (1995) Flame and Combustion. Third Edition, Chapman Hall, London

[31] Groenendijk A, Müller E (2002) Mixture Formation and Combustion Control for Low Emission DI Diesel Combustion with HCCI-Characteristics. THIESEL 2002, Valencia, 10th-13th September 2002, pp 145–157

[32] Halstead MP, Kirsch LJ, Prothero A, Quinn CP (1975) A Mathematical Model for Hydrocarbon Autoignition at High Pressures. Proc. Roy. Soc., A346, pp 515–538

[33] Hammer J, Dürnholz M, Dohle U (2004) Entwicklungstrends bei Einspritzsystemen für PKW-Dieselmotoren. Dieselmotorentechnik 2004, pp. 36–52, TA Esslingen

[34] Harada A, Shimazaki N, Satoru S, Miyamoto T, Akagawa H, Tsujimura K (1998) The Effects of Mixture Formation on Premixed Lean Diesel Combustion. SAE paper 980533

[35] Heil B, Weining HK, Karl G, Panten D, Wunderlich K (2001) Verbrauch und Emissionen, Reduzierungskonzepte beim Ottomotor, Teil 2. Motortechnische Zeitschrift (MTZ), 62 (12/2001), pp 1022–1035. English version: Concepts for Reducing Fuel Consumption and Emissions in Spark-Ignited Engines, Part 2. MTZ worldwide 62 (12/2001), pp 17–23

[36] Hulpi J (2004) Humidification Methods for Reduction of NO[x] Emissions. CIMAC Congress 2004, Kyoto, paper 112

[37] Iida N (1994) Combustion Analysis of Methanol Fueled Active Thermo Atmosphere Combustion (ATAC) Engine Using a Spectroscopic Observation. SAE paper 940684

[38] Ishibashi Y, Asai M (1996) Improving the Exhaust Emission of Two-Stroke Engines by Applying the Activated Radical Combustion. SAE paper 960742

[39] Iwabuchi Y, Kawai K, Shoji T, Takeda Y (1999) Trial of New Concept Diesel Combustion System – Premixed Compression-Ignited Combustion. SAE paper 1000-01-0185

[40] Jorach RW, Doppler H, Altmann O (2000) Schweröl-Common-Rail- Einspritzsysteme für Großmotoren. Motortechnische Zeitschrift (MTZ) 61, pp. 854-861. English Version: Heavy Fuel Common Rail Injection Systems for Large Engines. MTZ worldwide, 61 (12/2000) 4, pp. 10–13

[41] Kammerdiener T, Bürgler L (2000) Ein Common-Rail Konzept mit druckmodulierter Einspritzung. Motortechnische Zeitschrift (MTZ) 61, pp 230–238. English Version: A Common Rail Concept with Pressure-Modulated Fuel Injection. MTZ worldwide, 61 (4/2000), pp. 7–11

[42] Kelly-Zion PL, Dec JE (2000) A Computational Study on the Effect of Fuel-Type on Ignition Time in HCCI Engines. Proc. Combust. Inst. 28, paper no. 4E11

[43] Kimura S, Aoki O, Kitahara Y, Aijoshizawa E (2001) Ultra-Clean Combustion Technology Combining a Low-Temperature and Premixed Combustion Concept for Meeting Future Emission Standards. SAE paper 2001-01-0200

[44] Kowalewicz A (1984) Combustion Systems of High Speed Piston I.C. Engines. Wydawnictwa Komunikacji, Warsaw

[45] Kraft M, Maigaard P, Mauss F, Christensen M, Johansson B (2000) Investigation of Combustion Emissions in a HCCI Engine – Measurements and a New Computational Model. Proc. Combust. Inst. 28, paper no. 4E12

[46] Kufferath A, Samenfink W, Gerhardt J (2003) Die neue Emissionsstrategie der Benzin-Direkteinspritzung. Motortechnische Zeitschrift (MTZ), 64 (11/2003), pp 916–923. English version: The new Emissions Strategy of Gasoline Direct Injection. MTZ worldwide 64 (11/2003), pp 6–9

[47] Lavoie GA, Heywood JB, Keck JC (1970) Experimental and Theoretical Study of Nitric Oxide Formation in Internal Combustion Engines. Combustion Science and Technology, vol 1, pp 313–326

[48] Law D, Allen J (2002) On the Mechanism of Controlled Auto Ignition. SAE paper 2002-01-0421

[49] Law D, Kemp D, Allen J, Kirkpatric G, Copland T (2000) Controlled Combustion in an IC-Engine with a Fully Variable Valve Train. SAE paper 2000-01-0251

[50] Lettmann H, Seebode J, Merker GP (2004) Numerical and Experimental Evaluation of DI-Diesel Soot and Nitrogen Oxide Emissions in Respect of Different Injection Strategies. FISITA World Automotive Congress, Barcelona, paper F2004V240

[51] Li H, Miller DL, Cernansky NP (1996) Development of Reduced Kinetic Model for Prediction of Preignition Reactivity and Autoignition of Primary Reference Fuels. SAE paper 960498

[52] Mahr, B (2002) Future and Potential Diesel Injection Systems. THIESEL 2002 Conference on Thermo- and Fluid-Dynamic Processes in Diesel Engines

[53] Malte PC, Pratt DT (1974) Measurement of Atomic Oxygen and Nitrogen Oxides in Jet-Stirred Combustion. 15th Symposium (International) on Combustion, The Combustion Institute, Pittsburgh, pp 1067–1070

[54] Meyer S, Krause A, Krome D, Merker GP (2002) Ein flexibles Piezo-Common-Rail System mit direktgesteuerter Düsennadel (in German). Motortechnische Zeitschrift (MTZ) 2/2002 Jahrgang 63, pp 86–93, also available in English (MTZ Worldwide)

[55] Montagne X, Duret P (2001) What will be the Future Combustion and Fuel-Related Technology Challenges?. In: Duret P (ed) A New Generation of Engine Combustion Processes for the Future?, Paris, pp 175–180

[56] Müller E, Groenendijk A, Raatz T (2001) Homogene Dieselverbrennung – Die Lösung des NOx-Partikel-Problems (in German). Berichte und Informationen 4. Dresdner Motorenkolloquium 2001, pp 32–44

[57] Najt PM, Foster DE (1983) Compression-Ignited Homogeneous Charge Combustion. SAE paper 830264

[58] Namekawa S, Nakano R, Ishida H (2004) Development of New Common Rail Fuel Injection System for the Latest Development MHI MARK-30B Engine. 24th CIMAC-Congress, Kyoto, paper 113

[59] Nishimura T, Satoh K, Takahashi S, Yokota K (1998) Effects of Fuel Injection Rate on Combustion and Emission in a Diesel Engine. SAE paper 981929

[60] Noguchi M, Tanaka Y, Tanaka T, Takeuchi Y (1979) A Study on Gasoline Engine Combustion by Observation of Intermediate Reactive Products during Combustion. SAE paper 790840

[61] Oakley A, Zhao H, Ladommatos N, Ma T (2001) Experimental Studies on Controlled Auto-Ignition (CAI) Combustion of Gasoline in a 4-Stroke Engine. SAE paper 2001-01-1030

[62] Onishi S, Jo SH, Shoda K, Jo PD, Kato S (1979) Active Thermo-Atmosphere Combustion (ATAC) – A New Combustion Process for Internal Combustion Engines. SAE paper 790501

[63] Oppenheim AK (1984) The Knock Syndrome – its Cures and its Victims. SAE paper 841339

[64] Oppenheim AK (2003) Quest for Controlled Combustion Engines. SAE paper 880572

[65] Peng Z, Zhao H, Ladommatos N (2003) Effects of Air/Fuel Rates on HCCI Combustion of n-heptane, a Diesel Type Fuel. SAE paper 2003-01-0747

[66] Pöttker S, Eckert P, Delebinski T, Baumgarten C, Oehlert K, Merker GP, Wagner U, Spicher U (2004) Investigations of HCCI Combustion Using Multi-Stage Direct-Injection with Synthetic Fuels. SAE paper 2004-01-2946

[67] Pucher GR, Gardiner DP, Bardon MF, Battista V (1996) Alternative Combustion Systems for Piston Engines Involving Homogeneous Charge Compression Ignition Concepts – A review of Studies Using Methanol, Gasoline and Diesel Fuel. SAE paper 962063

[68] Ricaud JC, Lavoisier F (2002) Optimizing the Multiple Injection Settings on an HSDI Diesel Engine. THIESEL 2002 Conference on Thermo- and Fliud Dynamic Processes in Diesel Engines

[69] Robert Bosch GmbH (2003) Ottomotor-Management. Robert Bosch GmbH, Gelbe Reihe, 2nd Edition, May 2003

[70] Robert Bosch GmbH (2004) Diesel-Speichereinspritzsystem Common Rail. Robert Bosch GmbH, Gelbe Reihe, 3rd Edition, October 2004

[71] Ryan III TW, Callahan TJ (1996) Homogeneous Charge Compression Ignition of Diesel Fuel. SAE paper 961160

[72] Ryan III TW, Matheaus A (2002) Fuel Requirements for HCCI Engine Operation. THIESEL 2002 Thermo- and Fluid-Dynamic Processes in Diesel Engines, Valencia, 10th-13th September 2002. pp 361–375

[73] Schloz E (2003) Untersuchungen zur homogenen Dieselverbrennung bei innerer Gemischbildung. Dissertation, Universität Karlsruhe

[74] Seebode J, Merker GP, Lettmann H (2004) Injection Strategies under the Influence of Pressure Modulation and Free Rate Shaping in Modern DI-Diesel Engines. CIMAC Congress 2004, Kyoto, paper 47

[75] Seebode, J (2004) Dieselmotorische Einsptitzratenformung unter dem Einfluss von Druckmodulation und Nadelsitzdrosselung. Ph.D. Thesis, Institute of Technical Combustion, University of Hanover, Germany

[76] Shimazaki N, Tsurushima T, Nishimura T (2003) Dual Mode Combustion Concept with Premixed Diesel Combustion by Direct Injection Near Top Dead Center. SAE paper 2003-01-0742

[77] Spicher U, Reissing J, Kech JM, Gindele J (1999) Gasoline Direct Injection (GDI) Engines – Development Potentialities. SAE paper 1999-01-2938

[78] Stegemann J (2004) Dieselmotorische Einspritzverlaufsformung mit piezoaktuierten Experimentaleinspritzsystemen. Ph.D. Thesis, Institute of Technical Combustion, University of Hanover, Germany

[79] Stegemann J, Tremel O, Rölle T, Pape J, Merker GP (2005) Variable Einspritzsysteme als Entwicklungstools zur Optimierung von Ottomotoren mit Direkteinspritzung. Motortechnische Zeitschrift (MTZ), 66 (6/2005), pp 916–923. English version: Variable Injection Systems Used as Development Tools to Optimise Direct-Injection Spark-Ignition Engines. MTZ worldwide 66 (6/2005)

[80] Stegemann J, Meyer S, Rölle T, Merker GP (2004) Einspritzsystem für eine vollvariable Verlaufsformung. Motortechnische Zeitschrift (MTZ) 65, pp. 114–118. English Version: Injection System for Fully Variable Control of the Shape. MTZ worldwide, 65 (2/2004) 4, pp. 13–16

[81] Stegemann J, Seebode J, Baltes J, Baumgarten C, Merker GP (2002) Influence of Throttle Effects at the Needle Seat on the Spray Characteristics of a Multihole Injection Nozzle. ILASS Europe, Zaragoza, pp 31–36

[82] Steiger W (2003) Neue Kraftstoffe für zukünftige Brennverfahren?. 5. Dresdner Motorenkolloquium (in German), 5th–6th June 2003, Dresden, pp 143–150

[83] Stiebels B, Schweizer MJ, Ebus F, Pott E (2003) Die FSI-Technologie von Volkswagen – nicht nur ein Verbrauchskonzeot. In: Spicher U (editor): Direkteinspritzung im Ottomotor IV. Haus der Technik Fachbuch 24, Expert-Verlag, 2003, pp 157–174

[84] Stiesch G (2003) Modeling Engine Spray and Combustion Processes. Springer Verlag, Berlin Heidelberg New York

[85] Stockinger M, Schärpertöns H, Kuhlmann P (1992) Versuche an einem gemischansaugenden Verbrennungsmotor mit Selbstzündung. Motortechnische Zeitschrift (MTZ) 53, pp 80–85

[86] Stranglmaier RH, Li J, Matthews RD (1999) The Effect of In-Cylinder Wall Wetting on the HC Emissions from SI Engines. SAE paper 1999-01-0502

[87] Stranglmaier RH, Roberts, CE (1999) Homogeneous Charge Compression Ignition (HCCI): Benefits, Compromises, and Future Engine Applications. SAE paper 1999-01-3682

[88] Takeda Y, Keiichi N, Keiichi N (1996) Emission Characteristics of Premixed Lean Diesel Combustion with Extremely Early staged Fuel Injection. SAE paper 961163

[89] Tanabe K, Kohketsu S, Mori K, Kawai K (2000) Innovative Injection Rate Control with Next Generation Common Rail Fuel Injection System. FISITA World Automotive Congress, Seoul, paper F2000A055

[90] Tayama K, Tateishi M, Tosa Y, Nagae Y, Ura A, Ishida M, Motomura O (1995) Water Mixing Combustion for Low NO_x Diesel Engine. Trans JSME, no 590B, p 3548

[91] Thring RH (1989) Homogeneous Charge Compression Ignition (HCCI) Engines. SAE paper 892068

[92] Tsurushima T, Shimazaki N, Asaumi Y (2000) Gas Sampling Analysis of Combustion Processes in a Homogeneous Charge Compression Ignition Engine. Int. J. Engine Research, vol 1, no 4, pp 337–352

[93] van Basshuysen R, Schäfer F (2002) Handbuch Verbrennungsmotor. Vieweg-Verlag, Braunschweig, Wiesbaden, ISBN 3-528-03933-7

[94] Velji A, Günthner M, Spicher U (2003) Direkteinspritzung im Ottomotor mit Fremd- und Kompressionszündung – Brennverfahren der Zukunft?. 5. Tagung Direkteinspritzung im Ottomotor (in German), In: Spicher U (ed) Direkteinspritzung im Ottomotor IV, Haus der Technik Fachbuch Band 24, Expert-Verlag, pp 1–25

[95] Warnatz J (1983) Hydrocarbon Oxidation at High Temperatures. Berichte der Bunsen-Gesellschaft für Physikalische Chemie 87, pp 1008–1022

[96] Warnatz J (1992) Resolution of Gas Phase and Surface Combustion Chemistry into Elementary Reactions. 24th International Symposium on Combustion, The Combustion Institute, Pittsburgh, pp 553–579

[97] Warnatz J, Maas U, Dibble RW (1999) Combustion. Second Edition, Springer-Verlag

[98] Westbrook CK (2000) Chemical Kinetics of Hydrocarbon Ignition in Practical Combustion Systems. Proceedings of the Combustion Institute, vol 28, pp 1563–1577

[99] Wirbeleit F, Enderle C, Lehner W, Raab A, Binder K (1997) Stratified Diesel/Water/Diesel Injection (DWD) Combined with EGR, the Most Efficient In-Cylinder NO_x and PM Reduction Technology. SAE paper 972962

[100] Wirth M, Zimmermann D, Friedfeldt R, Caine J, Schamel A, Storch A, Reis-Müller K, Gansert KP, Pilgram G, Ortmann R, Würfel G, Gerhardt J (2003) The Next generation of Gasoline Direct Injection: Improved Fuel Economy and Optimized System Cost. 12th Aachener Kolloquium Fahrzeug- und Motorentechnik, Aachen, Germany

[101] Wolters P, Salber W, Krüger M, Körfer T, Dilthey J (2003) Variable Ventilsteuerung – Schlüsseltechnologie für Homogene Selbstzündung. 5. Dresdner Motorenkolloquium (in German), 5th–6th June 2003, Dresden, pp 161–175

[102] Wolters, P., Salber, W., Dilthey, J.: Radical Activated Combustion, A new Approach for Gasoline Engines, in A New Generation of Engine Combustion Processes for the Future?, P. Duret (Editor) and Editions Technip, Paris, 2001, pp. 153–162

[103] Yanagihara H (2001) Ignition Timing Control at TOYOTA "UNIBUS" Combustion System. In: Duret p (ed) A New Generation of Engine Combustion Processes for the Future?, Paris, pp 35–42

[104] Zhao H, Peng Z, Milliams J, Ladommatos N (2001) Understanding the Effects of Recycled Burnt Gases on the Controlled Autoignition (CAI) Combustion in Four-Stroke Gasoline Engines. SAE paper 2001-01-3607

[105] Zhao, FQ, Lai MC, Harrington DL (1997) A Review of Mixture Preparation and Combustion Control Strategies for Spark-Ignited Direct-Injection Gasoline Engines. SAE paper 970627

[106] Zheng J, Yang W, Miller L, Cernansky NP (2001) Prediction of Pre-ignition Reactivity and Ignition Delay for HCCI Using a Reduced Chemical Kinetic Model. SAE paper 2001-01-1025

[107] Zheng J, Yang W, Miller L, Cernansky NP (2002) A Skeletal Chemical Model for the HCCI Combustion Process. SAE paper 2002-01-0423

7 Conclusions

The internal combustion engine is by far the most important power train for all kind of vehicles today. Up to now there is no alternative to this kind of engine, and it is for sure that it will keep its leading position for at least the next three to five decades. However, it has to be continuously improved, and great efforts have to be made in order to increase efficiency and to fulfill future emission legislation. In this context, the reduction of engine-out raw emissions by applying improved or new mixing formation and combustion concepts will be one of the key measures to keep the internal combustion engine up to date.

A systematic and precise control of mixture formation with modern high-pressure injection systems, including fully variable rate shaping, variable nozzle geometry, pressure-modulated injection etc., will become crucial for realizing future combustion concepts. However, due to of the growing number of free parameters, the prediction of spray and mixture formation is becoming increasingly complex. For this reason, the optimization of the in-cylinder processes using 3D computational fluid dynamics (CFD) is becoming increasingly important. In this book, the state of the art in modeling in-cylinder spray and mixture formation processes in internal combustion engines has been presented and discussed. This includes the description of the mathematical treatment of two-phase in-cylinder flows consisting of a continuous and a dispersed phase, the derivation of the relevant conservation equations, and the detailed description and discussion of all CFD models needed to describe the relevant sub-processes during spray and mixture formation.

The research work that is necessary to develop and to continuously improve these models is of great importance for several reasons. Simulation models, which have been carefully adjusted to a specific range of boundary conditions, can be used to perform extensive parametric studies, which are usually faster and cheaper than experiments. Much more important is the fact that despite the higher uncertainty compared to experiments, numerical simulation can give much more extensive information about the complex in-cylinder processes than experiments could ever provide. Using numerical simulations, it is possible to calculate the temporal behavior of every variable of interest at any place inside the computational domain. This allows getting a detailed knowledge of the relevant processes, and is a prerequisite in order to improve them. Furthermore, the numerical simulation can be used to investigate processes that take place at time and length scales or at places that are not accessible and thus cannot be investigated experimentally. Last but not least, the research work that is necessary to increase our knowledge about the relevant processes and to improve the CFD models, reveals new and unknown mechanisms, and is also the source of new unconventional ideas and improvements.

Summarizing the situation, it must be pointed out that the predictive quality of the models currently used in CFD codes has already reached a very high level, and that the use of CFD simulations for research and development activities, concerning enhanced and new mixture formation and combustion concepts, is not only useful but already necessary today. For this reason, a continuous improvement of existing CFD models as well as the development of models describing new effects, which arise from the development of new injection systems and injection strategies, is absolutely necessary in order to guarantee a detailed and accurate modeling of the relevant sub-processes and to increase the predictive quality of CFD calculations in the future.

Based on the critical analysis of the present state of spray and mixture formation modeling as well as the future demands concerning enhanced and new mixture formation concepts, as discussed in this book, the following aspects represent the most important limitations today. For this reason, further research should meet the following challenges.

- A major source of uncertainty in CFD calculations using the Eulerian-Lagrangian description is still the fact that the results are dependent on grid size, and that in contrast to the pure Eulerian description of continua a grid refinement does not automatically result in more accurate results. As described in Chap. 5, the inter-phase coupling of the liquid fuel drops (Lagrangian description) and the gaseous environment (Eulerian description) is the reason for this grid dependency. The stronger the local gradients in flow quantities, the more important these effects. The Eulerian-Lagrangian description is based on the assumption that the void fraction inside all gas cells is close to one. For this reason, it is not possible to increase grid resolution until the strong gradients in flow quantities can be accurately resolved near the nozzle, even if the required computer power would be available. If larger or smaller grid sizes than the recommended ones, for which the sub-models are calibrated, are used, the prediction of inter-phase mass, energy and momentum exchange deteriorates. Furthermore, statistical convergence in the dense spray near the nozzle is usually not given. However, there are already promising approaches to eliminate the effect of grid dependency, Chap. 5. The first group of approaches aims at introducing appropriate sub-models in the Eulerian-Lagrangian calculation in order to provide a more accurate inclusion of sub-grid scale gradients of flow quantities and to limit numerical diffusion. The second group of approaches aims at overcoming the numerical problems in a more direct way by describing the liquid phase in the dense spray region using the Eulerian instead of the Lagrangian description. Both approaches show promising results and should be enhanced in the near future. The reduction of grid dependency is especially important, because the more the predictive quality of the CFD models increased, the stronger the relative effect of grid dependency.

- Apart from the problem of inter-phase coupling, the numerical grid also influences the prediction of heat transfer from the in-cylinder gas to the wall. The velocity and thermal boundary layers cannot be resolved by the gas phase grid and have to be described by so-called wall functions. Usually the grid cells at the wall are too coarse, and the range of applicability of the wall-functions is exceeded. A possible approach is to refine the mesh near the walls until the desired grid spacing is obtained. However, due to the strongly changing flow quantities, the boundary layer thickness is highly transient and the sizes of all wall cells must be adapted dynamically to the required values. This still causes considerable problems today. For this reason, a new and completely grid-independent approach would be highly desirable.

- As discussed in Chap. 6, future mixture formation and combustion concepts for diesel as well as gasoline engines will utilize almost exclusively high-pressure direct injection. The higher the injection pressure, the more energy for spray and mixture formation must be provided by the injection system itself, and the more important the influence of internal nozzle flow on primary and secondary spray break-up, Chap. 2. Especially the primary break-up, which is responsible for the starting conditions of the liquid droplets that penetrate into the cylinder volume, is significantly influenced by the three-dimensional turbulent and cavitating flow that emerges from the nozzle holes. Primary break-up models, which describe the relevant effects during this first disintegration of the liquid jet in the vicinity of the nozzle, are highly desirable in order to eliminate uncertainties regarding the starting conditions of the spray in CFD calculations. Today, there are already some promising approaches to solve this problem in the case of diesel injection, Chap. 4. However, further research work is necessary to extend them for the use in the case of high-pressure gasoline injection. Furthermore, effects like needle seat throttling, flash-boiling, pressure-modulated injection, variable nozzle geometry, highly transient operation etc. have to be included. The most important problem regarding the modeling of primary break-up however is the validation of new models. Due to the very dense primary spray and the small dimensions, this region is hardly accessible by optical measurement techniques. Thus, it is difficult to understand the relevant processes and to verify CFD models.

- The basic challenge in describing the evaporation process of fuel droplets is the choice of an appropriate reference fuel that represents the relevant behavior of the fuel. Especially in the case of new tailored synthetic fuels, e.g. for HCCI applications, a single reference fuel can often not predict the relevant sub-processes during evaporation, ignition, and combustion with adequate accuracy, and multi-component fuel models become increasingly desirable. A possible approach to describe such a multi-component fuel has been presented in Chap. 4. Compared to the conventional auto-ignition combustion concepts, the ignition delays of HCCI concepts are significantly longer, and in this case detailed ignition chemistry models that can predict the two-stage ignition of realistic multi-component fuels with sufficient accuracy are also needed.

- Another challenge is the improvement of turbulence modeling. Turbulence greatly influences the spray and mixture formation process as well as the subsequent combustion. The turbulence models used today are usually of semi-empirical nature, Chap. 3. Different specific flow configurations require different sets of model constants. Due to the complexity of turbulence it has up to now not been possible to develop a single universal model capable of predicting the turbulent behavior for all kinds of turbulent flow fields. Hence, the present models can only be regarded as approximations and not as universal laws.

- A general task is of course the continuous enhancement of all existing models as well as the development of new ones in order to achieve a more accurate and complete description of the relevant physical and chemical sub-processes during spray and mixture formation. However, it must be kept in mind that more detailed models are more expensive regarding the consumption of computer power and time. Further on, more detailed experimental data are necessary to calibrate and verify these models. For this reason, the enhancement of CFD models is usually directly coupled with the development of advanced measurement techniques. Thus, a general task for future research regarding CFD-modeling will also be the continuous enhancement of experimental investigations.

Altogether, it can be concluded that the numerical simulation of spray and mixture formation processes has already reached such a high level that its use in the various tasks related to the prediction and investigation of in-cylinder processes is already extremely useful today. Due to of the growing number of free parameters of modern mixture formation concepts, the optimization of in-cylinder processes is becoming increasingly complex and can often no longer be achieved without numerical studies. For this reason, the use of 3D computational fluid dynamics will become more and more important in the future.

Index

amplified pressure CR system (see injection systems)
auto-ignition (see ignition)

Bernoulli equation 18, 54, 60, 61, 86, 87
blob-method (see primary break-up modeling)
boundary layer 75, 76, 139, 165
break-up
  length 5, 7, 16, 112, 120
  modeling (see primary/secondary break-up modeling)
  primary 11, 16, 24, 85
  regimes of liquid drops 8
  regimes of liquid jets 5
  secondary 11, 24
  time 112, 115, 121, 127, 129, 177
  zone 16

cavitation 17, 18, 19, 20, 21, 37, 88, 90, 98, 99, 100, 102, 103, 105, 232
  number 21
cavitation-induced break-up (see primary break-up modeling)
coalescence (see droplet coalescence)
collision (see droplet collision)
common rail injection 32, 33
conservation equations 49
  conservation of energy 53, 61, 70, 80
  conservation of mass 49, 55, 80
  conservation of momentum 52, 56, 70, 80
continuity equation (see conservation of mass)

continuous thermodynamics 145
continuum 47
controlled auto-ignition (CAI) 254, 278

diffusion flame 227
direct injection
  diesel engines 32, 226
  gasoline engines 41, 242
direct numerical simulation (DNS) 66
discharge coefficient 88, 110
drag coefficient (see droplet drag)
droplet
  coalescence 169
  collision 11, 169
  collision modeling 172, 220
  collision regimes 169
  drag 136
  drag coefficient 136, 138
  drag modeling 136
droplet deformation and break-up model (DDB) (see secondary break-up modeling)
droplet evaporation 139
  multi-component 144
  single-component 140
dual-mode operation (see HCCI)

eddy viscosity 71
Einstein notation 49
energy equation (see conservation equations)
enhanced TAB model (see secondary break-up modeling)

Euler equation 60
Eulerian description 47, 48
Eulerian-Eulerian description 221
Eulerian-Lagrangian description 83,
    212
evaporation (see droplet evaporation)
exhaust gas recirculation 232, 242, 246,
    263, 268, 272

Favre-averaging 67, 69
flash-boiling 24, 158
full-cone spray (see spray)

gamma-function 146
grid
    dependencies 211
    spray adapted grid 214

HCCI 253
    chemistry 256
    dual-mode operation 255, 266, 268,
        270, 279
    emission behavior 261
    high temperature oxidation 256, 260,
        276
    influence parameters 270
        air excess ratio 275
        compression ratio 270
        engine speed 275
        exhaust gas recirculation 272
        fuel composition 275
        homogeneity 278
        inlet air temperature 274
        variable valve timing 271
    low temperature oxidation 256, 257,
        276
    mixture formation 264
    negative temperature coefficient 256,
        259
high temperature oxidation 260
hollow-cone spray (see spray)
homogeneous charge compression
    ignition (see HCCI)

ignition 197
    auto-ignition 197
    laser-induced 252
    spark-ignition 200
    delay 197, 227, 231, 268
index notation (see Einstein notation)

injection
    direct injection 32, 41, 226, 242
    direct injection spark ignition 41
    main injection 232
    multiple injection 230, 231
    multi-point injection 38, 39
    port fuel injection 38
    post injection 233
    pre injection 231, 232
    pressure modulation 239
    pressure-swirl injector 24, 25, 27, 42,
        109
    pulsed injection 265
    rate 35
    rate shaping 230, 232
    single-point injection 38, 39
injection systems 32
    amplified pressure CR system 238
    common rail injection 32, 33
    piezo injectors 234
    unit injector system 34
    unit pump system 36

Kelvin-Helmholtz break-up (see
    secondary break-up modeling)
knock 243, 246, 249, 250, 253, 268,
    273
k-ε model 73, 106

Lagrangian description 47, 48
Laplace number 182
large eddy simulation 66
law of the wall 78
low temperature oxidation 257

mass conservation (see conservation
    equations)
material derivate 47, 48
main injection (see injection)
maximum entropy formalism (MEF) 92
Miller cycle 230
momentum equation (see conservation
    equations)
Monte-Carlo method 82, 215
multiple injection (see injection)
multi-point injection (see injection)

Navier-Stokes equations 70
negative temperature coefficient 197,
    256, 259

nozzle
  k-factor 38
  sac hole nozzle 37
  types 32, 37, 38, 40, 42
  variable nozzle concept 236
Nusselt number 141, 150, 192

Ohnesorge 5
  diagram 5
  number 6, 126, 182

penetration length (see spray
  penetration)
piezo injectors (see injection systems)
port fuel injection (see injection)
post injection (see injection)
Prandtl number 79, 150
  turbulent Prandtl number 71, 79, 165
Prandtl mixing-length model 72
pre-injection (see injection)
premixed peak 227, 231, 232
pre-spray 25, 114, 133
pressure modulation (see injection)
pressure-swirl atomizer 24, 25, 27, 42,
  109
primary break-up (see break-up)
primary break-up modeling
  blob-method 86, 130, 131, 132
  cavitation-induced break-up 98, 100
  distribution functions 90
  maximum entropy formalsim (MEF)
    92
  sheet atomization model 109, 133,
    135
  turbulence-induced break-up 94, 100
pulsed injection (see injection)

rate shaping (see injection)
Rayleigh-Taylor break-up (see
  secondary break-up modeling)
Reynolds
  number 6
  number (droplet collision) 170
  averaged Navier-Stokes equations 66
  averaging 67
  decomposition 66
  stress tensor 69, 71
  stress models 75
RNG k-ε model 75

sac hole nozzle (see nozzle)
Sauter mean diameter 14, 27
Schmidt number 150
secondary break-up (see break-up)
secondary break-up modeling 114
  droplet deformation and break-up
    model (DDB) 122, 131, 132, 135,137
  enhanced TAB model 119
  Kelvin-Helmholtz model 125, 130,
    131, 132
  phenomenological models 115
  Rayleigh-Taylor model 128, 130, 132
  Taylor-Analogy break-up 116, 132,
    133, 136
shear velocity 77
sheet atomization model (see primary
  break-up modeling)
Shell model 198
Sherwood number 143, 150
single-point injection (see injection)
soot-NO_x trade-off 229, 253
Spalding transfer number 142, 149
spark-ignition (see ignition)
spray
  angle 13, 86, 96, 99, 118
  break-up zone 16
  core length 8, 15
  full-cone 10, 90
  hollow-cone 23, 25, 26, 41, 91, 109
  penetration length 11, 13
  Sauter mean diameter 14
  wall impingement (see wall
    impingement)
  equation 81
  -wall interaction 29
stochastic parcel method 82, 215
stratified charge 243, 244, 246
  air-guided technique 248
  spray-guided technique 249
  wall-guided technique 247
stress tensor 59, 60
  Reynolds stress tensor 69, 71
substantial derivate (see material derivate)
swirl 248

Taylor number 126
Taylor-Analogy break-up (see
  secondary break-up modeling)
thermal energy equation 65

tumble 248
turbulence models 72
  k-ε model 73
  Prandtl mixing-length model 72
  Reynolds stress models 75
  RNG k-ε model 75
turbulence-induced break-up (see
  primary break-up modeling)
turbulent
  dispersion 166
  flow 66
  heat flux 69, 70, 71
  kinetic energy 74, 167
  length scale 95, 97
  Prandtl number 71, 79, 165
  stress 68, 70, 72
  time scale 95, 97
  viscosity 71, 73

unit injector system (see injection
  systems)

unit pump system (see injection
  systems)

valve covered orifice nozzle 37
vena contracta 13, 19, 87, 88

wall
  film 12, 29, 38, 162
  film evaporation 162
  film modeling 191
  function 76, 164, 165
  impingement 138, 180
  impingement modeling 183
  impingement regimes 181
Weber number
  droplet collision 170
  gas phase 7, 8, 9, 114, 117
  liquid phase 5
  wall impingement 138, 182,
    183